Kinematics and Dynamics of Mechanisms

JACQUES GROSJEAN
Formerly Head of Applied Mechanics
University of Bath

McGRAW-HILL BOOK COMPANY

London · New York · St Louis · San Francisco · Auckland · Bogotá · Caracas · Lisbon
Madrid · Mexico · Milan · Montreal · New Delhi · Panama · Paris · San Juan · São Paulo
Singapore · Sydney · Tokyo · Toronto

Published by
McGRAW-HILL Book Company (UK) Limited
SHOPPENHANGERS ROAD · MAIDENHEAD · BERKSHIRE · ENGLAND
TEL 0628-23432; FAX 770224

British Library Cataloguing in Publication Data
Grosjean, J. (Jacques)
Kinematics and dynamics of mechanisms.
1. Mechanisms. Kinematics
I. Title
621.811

ISBN 0-07-707242-1

Library of Congress Cataloging-in-Publication Data
Grosjean, J. (Jacques)
Kinematics and dynamics of mechanisms/Jacques Grosjean.
p. cm.
Includes bibliographical references and index.
ISBN 0-07-707242-1
1. Machinery Kinematics of. I. Title.
TJ175.G69 1991
621.8′11 dc20
90-6551
CIP

12345 CL94321

Typesetting and technical drawings were done by Interprint Limited, Malta.
Printed in England by Clays Ltd, St Ives plc.

To the memory of my parents,
to my children Peter and Caroline,
and to
Rita for her patience and understanding
during the preparation of this text

CONTENTS

FOREWORD

The main difference between *homo sapiens* and other mammals of the ape family is that we humans have developed in the course of our early evolution a quite remarkable ability to conceive tools and mechanisms, however elementary, to achieve imagined objectives.

From time immemorial, and certainly from the days of early civilizations whether Chinese, Egyptian, Roman, Greek, or other, examples are to be found of interesting solutions to problems through the use of mechanisms to perform sometimes simple tasks but at times highly sophisticated objectives particularly when this was associated with wartime pursuits.

An object lesson in the manifestations of mechanisms and their practical possibilities are to be found in the legacies of Leonardo da Vinci's sketches in 15th century Italy. Never before had there been such a display of natural and innate understanding of what could be done with links, cams, screws, rope, etc. It is surely worth recalling that since da Vinci's period one of the most spectacular advances in linkage mechanisms, cams, etc., was in the field of clocks, chronometers and watches. A visit to Strasbourg cathedral and its historic (and monumental) clock will convince the most expert of engineers of the sophistication attained in the use of kinematics of mechanisms to achieve very high time accuracy together with planetary and astronomical information regarding our solar system. In more recent times, the development of the (mechanical) wrist chronometer and stop watch epitomizes the complexity to which the development of this side of engineering science has risen. However, although all this has now been overtaken by electronic systems of startling accuracy, the technology of this bygone age is not wasted. The combination of computer technology with that of advanced mechanical systems and actuators, etc., has opened new horizons which include examples such as robotics for advanced manufacturing techniques in which the science of dynamics plays a crucial part.

Our every-day lives are heavily dependent on the motor car. The days of simple chassis design are over, yet many of us are perhaps unaware of the refinement achieved in the last two or three decades to give us the present 'independent' suspension systems and steering geometry. These systems have given us the dynamic stability, safety and comfort we currently enjoy both at high and low speeds. Further advancement in this engineering science must surely emphasize the need to be fully conversant with the fundamentals covered by Dr Grosjean in this book.

Nowadays, we seem so besotted with the new technologies of the chip or solid state derivatives that it is a welcome step when a substantial book is written to give the budding engineer or mechanical scientist the wherewithal to approach the highly important world of mechanisms that surround him or her at every turn in 20th century life. To engineer mechanisms for new applications correctly or to improve the performance of existing systems demands the utmost understanding of the underlying principles that govern the parameters of motion of each item. Jacques Grosjean has provided us with just the book needed for this. He covers in his studies a wide range of examples such as the linkages found in recent designs of undercarriage mechanisms of aircraft, or those of a complex earth digger/shifter, or the suspension system of a modern car or any of the hundred other types of equipment around us.

Progress in this field is largely evolutionary in nature and the scope for invention and innovation is almost unlimited if we are to continue with the refinements to which we have become accustomed and which have given us an undreamt-of quality of material life.

To the student who succeeds in penetrating the initial, and at times somewhat forbidding, phase of kinematics and dynamics the challenge could be very exciting. Now Jacques Grosjean has provided us with an excellent rationalized treatment of this wide subject. He has also brought to bear modern techniques for finding solutions and this must surely remove the tedious work usually encountered by using old-established graphical or analytical methods. I heartily commend the book both to engineering students as well as to the fully qualified professional engineer.

Dr Waheeb Rizk CBE FEng FIMechE,
Past President
The Institution of Mechanical Engineers

PREFACE

This text is intended for students preparing for a degree in mechanical engineering at a university, a polytechnic, or a technical college, or for those who intend to sit the qualifying examinations of an engineering institution. It is hoped that it will prove useful to those studying on their own and to designers in industry who wish to acquire a knowledge of the subject, or who need such a knowledge during the execution of a design project.

The text is based on lectures and tutorials given at the School of Mechanical Engineering of the University of Bath, at ECP (Ecole Centrale des Arts et Manufactures), Paris, and at ENSEM (Ecole Nationale Supérieure d'Electricité et de Mécanique) in Nancy, France, during the past 10 years.

The four-bar linkage is the simplest of all linkages; it is, in fact, the building brick of many linkages and mechanisms. Its uses are many and varied. It operates satisfactorily in all kinds of environments to transmit power, large or small, generates functions to a high degree of accuracy, and guides objects along predetermined paths. Its manufacture is relatively inexpensive and maintenance is minimal.

A detailed knowledge of its properties and performance is therefore essential if the student and designer are to make good use of it. There are two aspects to consider:

1. Analysis, which is concerned with the displacement, velocity, acceleration, and power transmission of a given linkage. The value of the 'jerk' (rate of change of acceleration) is also required in some cases since it leads to wear, vibration, and noise.
2. Synthesis, which is concerned with the design of a linkage to perform a given function.

This text investigates these two aspects both graphically and analytically for the four-bar linkage and is the result of my industrial experience and a long association with Professor Jacques Hervé of Ecole Centrale to whom I am most grateful.

The slider–crank mechanism and the six-bar linkage are also included since both are derived from the four-bar linkage. Manufacturing tolerances are also investigated because they can adversely affect the performance of precision mechanisms.

There is an elementary treatment of the four-bar spatial linkage in Chapter 7 since such linkages are to be found in many industries. If a mechanism is to be designed properly account must be taken of the static and dynamic forces; inertia effects are particularly important in high-speed mechanisms. This aspect is covered in Chapter 8. Chapters 4 and 6 deal mainly with the kinematics of cams and gears respectively, since these are extensively employed in many industries and devices. Chapter 9 is a short introduction to the dynamics of robotic manipulators using the Gibbs–Appell method of analysis as an alternative to Lagrange; I am grateful to my colleague Dr C. W. Stammers for having brought it to my notice and I have found it most useful in some cases.

It is my opinion and experience that models play an important part in the study and design of mechanisms and a few examples are illustrated in Chapter 5. I am very grateful to my colleague and friend, David Tallin, whose remarkable skill as a tookmaker made these models and many others which are in constant use at the University of Bath.

I am also grateful to my colleagues: Dr D. K. Longmore for his valuable advice and for writing the FORTRAN version of the program on the analysis of the four- and six-bar linkages, and to Dr J. Vogwell for his useful advice concerning cams.

I also acknowledge with thanks the permission granted by the Authorities of the University of Bath to use some of the problems and exercises taken from past examination papers and for the photographs of the models referred to above.

The final manuscript was reviewed in detail by Professor D. T. Pham, Professor of Computer-Controlled Manufacture, of the University of Wales, Cardiff. I very much appreciate the time and effort he devoted to this task. Finally I would like to express my appreciation to all the manufacturers who kindly supplied diagrams and photographs to illustrate real mechanisms which should be helpful to the reader.

Jacques Grosjean
Bath

CHAPTER
ONE
INTRODUCTION

1.1 MECHANISMS

Mechanisms play a major role in practically every branch of engineering. The following are typical examples: excavators, cranes (see Fig. 1.7), engine mechanisms (see Fig. 1.11*a*), quick-return mechanisms (see Fig. 1.12*a*), packing machines, wrapping machines, bottling machines, machine tools, steering gear, car suspensions, agricultural machines, manipulators, robotic devices, artificial limbs, sewing machines, textile machinery, timing mechanisms, undercarriages of, e.g., aircraft, space shuttles, etc., and so on, right down to door hinges, such as those shown in Fig. 1.1, as well as mechanisms to help the disabled. The list is virtually endless.

Some mechanisms are very simple, e.g., the engine mechanism; others are very complicated, e.g., looms, which consist of links, slides, gears, and cams operating singly or in combination, modify motion, transmit power, work at high or low speeds, and operate in all kinds of environments (cold, hot, humid, dusty, oily, etc.).

Many of the mechanisms mentioned are made up of simple links or bars joined together by means of pins, slides, or balls and sockets in the case of spatial mechanisms; the purpose of this text is to analyse and synthesize such mechanisms, which are generally referred to as 'linkages'.

The problem in practice, from a design point of view, is the choice of a mechanism to perform a given task. It may be possible to use a familiar mechanism, to modify and adapt it

Figure 1.1 Examples of door hinges that use simple links and slides.

for the job in hand. In some cases, however, we require to find or even to discover a mechanism to do the job if none of the well-known ones are suitable. In either case it will be necessary at some stage in the design process to analyse the mechanism that has been chosen in order to calculate the velocities and accelerations (the 'jerk' in some cases) of its various elements. A knowledge of the velocities or velocity ratios (outputs/inputs) is required for calculating power transmission, and knowledge of accelerations in order to work out inertia forces and torques.

All the elements that make up a mechanism will be assumed to be rigid bodies in the sense that their deformations due to the application of forces, torques, and other effects such as temperature will be small enough not to affect their kinematic performance. However, the more precise the mechanism, the more these deformations may have to be taken into account.

Graphical or mathematical methods may be used to calculate displacements, velocities, and accelerations. Graphical techniques are quick and visual for one position of a mechanism, but become tedious if many positions are required to obtain a complete picture of the performance of a given mechanism during one cycle of operation; their accuracy is also limited.

Mathematical methods express the displacements, velocities, and accelerations (and jerk if required) by means of equations which can be easily and speedily handled by computers to a high degree of accuracy for all possible positions of the mechanism. Programmable calculators are also extremely effective. Visual presentation of the results is restored when using a computer with graphical display (see Fig. 5.4); this would very likely be the case in those industries concerned with the design and manufacture of mechanisms. In some cases an engineer may have to analyse a mechanism as part of a design project, which will necessitate reliance on whatever calculator or computer with no visual display is available. In such a situation simple checks on calculations can be carried out graphically using the methods discussed in this text.

We will concern ourselves essentially with plane mechanisms; spatial mechanisms, although less utilized in practice, are also important but they require special methods of analysis, e.g., vector calculus and Chapter 7 of this text considers the spatial four-bar linkage as an example. The reader should realize that a plane mechanism becomes a spatial one if allowed to rotate about an axis inclined at some angle to the input axis, e.g., the crane in Fig. 1.7 in which the movements about two axes A_0 and B_0 and the vertical axis occur simultaneously. Such mechanisms are then easier to analyse as well as to manufacture. The choice between a plane and a spatial mechanism depends very much on the particular requirement.

1.2 MOTION TRANSFORMATION

The examples in the following table illustrate the way in which mechanisms transform motion.

Mechanism	Motion transformation
(a) Simple lever Load $= Q$, Δy, B, Output, $OB = d_2$, $OA = d_1$, β, θ, Input, Δx, $Q = \Delta x P/\Delta y$, O, P, A, α	*Rectilinear → Rectilinear* $\Delta y = k\Delta x$, displacement relationship. k is a constant which depends on the shape of the crank. If a force applied at A moves through a small displacement Δx and the load Q at B moves through a small displacement Δy, then $Q\Delta y = P\Delta x$. But $Qd_2 \sin\beta = Pd_1 \sin\alpha$, hence $k = d_2 \sin\beta / d_1 \sin\alpha$.

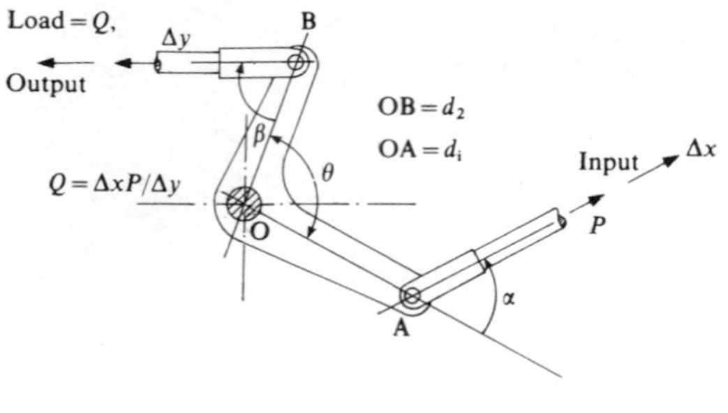

Mechanism	**Motion transformation**

(b) Contacting cylinders; gears

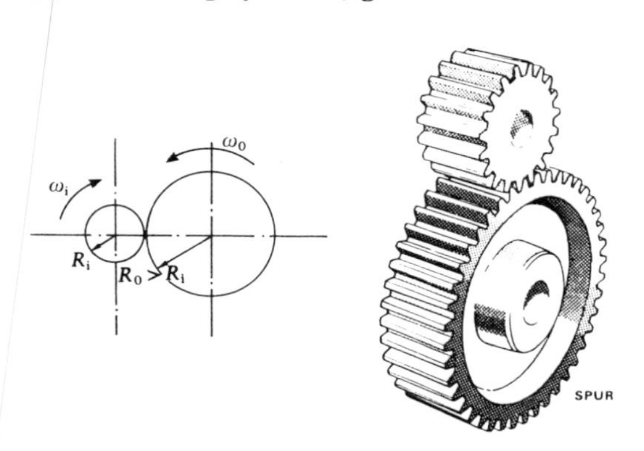

Rotary → *Rotary + a change in direction*
Velocity ratio:
$\omega_O/\omega_i = R_i/R_O$ for cylinders
or $\omega_O/\omega_i = T_i/T_O$ for gears
where T_i = number of teeth on input gear and T_O = number of teeth on output gear.

Note: To avoid a change in direction another gear, known as an idler, is introduced between the input and output gears.

(c) Cam

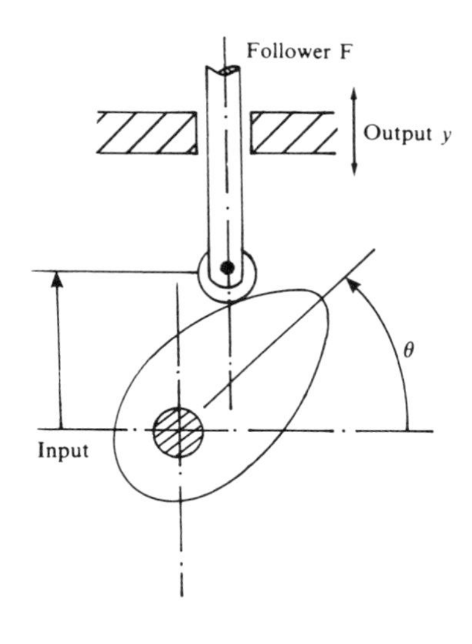

Rotary → *Oscillatory with or without dwells*
$y = G(\theta)$
The motion of the follower depends on the cam profile. By a suitable choice of cam profile the follower can be made to dwell during part of the cycle. Figure 4.3 in Chapter 4 shows a variety of cam follower configurations.

(d) Slider–crank

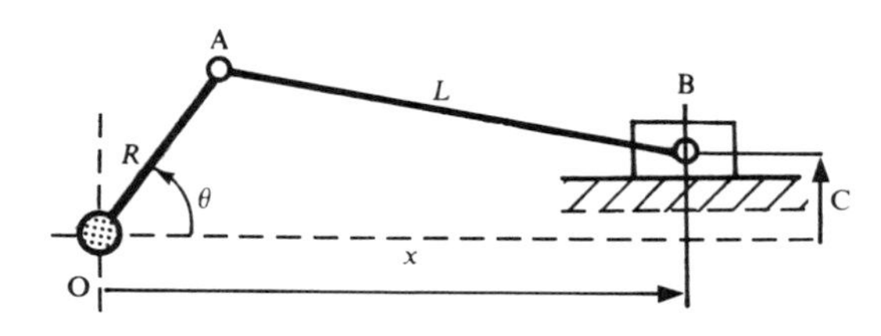

Rotary → *Oscillatory* if crank OA is the driver

$x = f(\theta)$
Oscillatory → *Rotary* if slider B is the driver

$\theta = g(x)$

for given R, L, C.

Mechanism	Motion transformation
(e) Hooke's joint	*Rotary → Rotary* $\omega_0 = \omega_i f(\phi)$ where ϕ = angle between the two shafts, ω_i = input velocity, ω_0 = output velocity
(f) Four-bar linkage	(i) *Rotary → Rotary* (ii) *Rotary → Oscillatory* (iii) *Oscillatory → Rotary* *+a change in direction if required* $\psi = G(\phi)$. See Eqs (5.11) and (5.12). Such a mechanism can coordinate input and output positions, velocities, or accelerations or generate mathematical functions, e.g., $\psi = 240 + 0.095\phi^{1.5}$ degrees, to within 0.05° accuracy
(g) Six-bar linkage	As for the four-bar linkage, (f) above. The illustration shows a stitching mechanism as used in the shoe industry. As the crank A_0A rotates the needle fixed to the output crank D_0D oscillates.

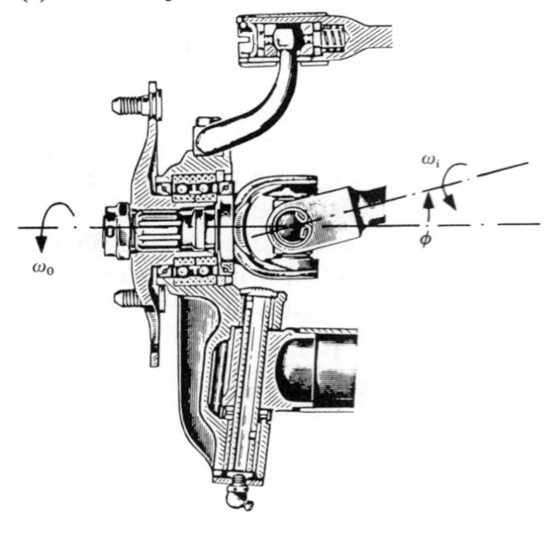

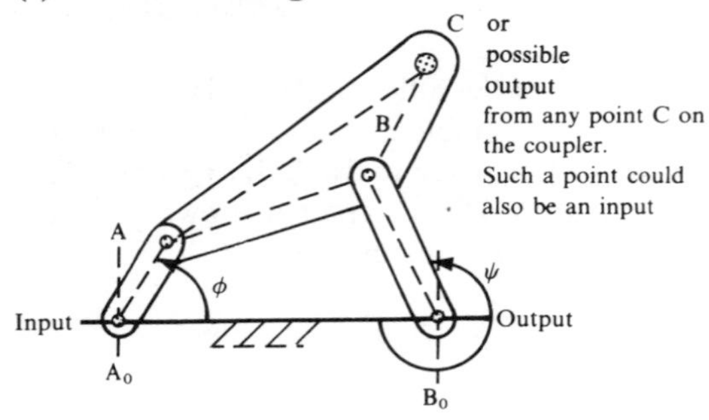

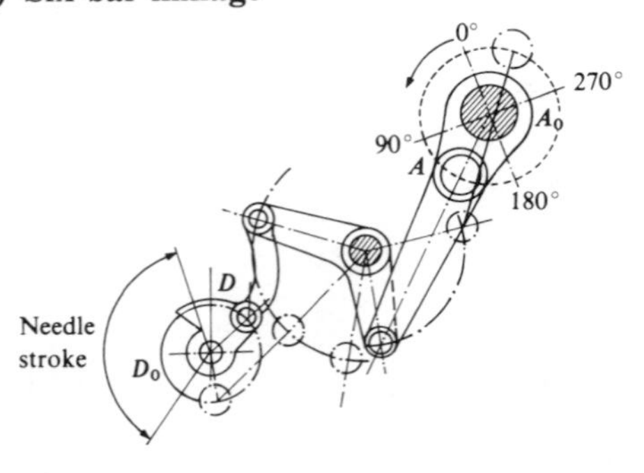

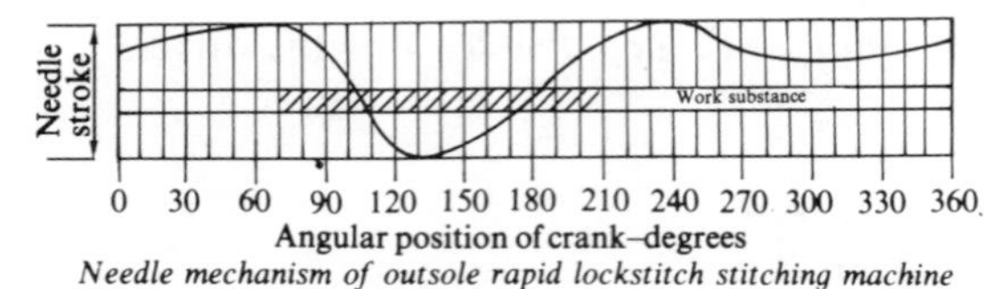

Needle mechanism of outsole rapid lockstitch stitching machine

(h) Six-bar linkage

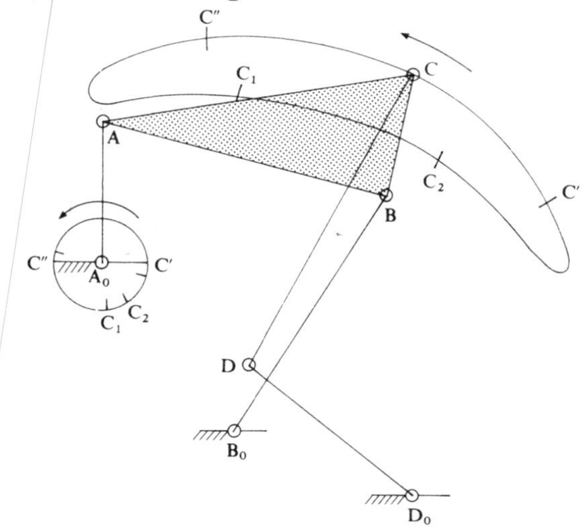

Rotary → Oscillatory
with dwells

The coupler point C on coupler ABC describes a closed loop as the crank A_0A rotates. Portions $C'C''$ and C_1C_2 are nearly arcs of circles so that as C travels along these arcs the output link DD_0 is stationary. Thus the mechanism provides two dwells of different duration during one revolution of the crank, and behaves like a cam.

(j) Slotted wheel or Geneva mechanism

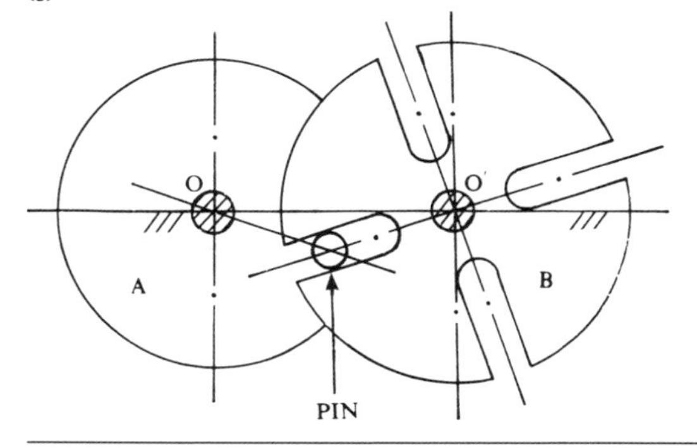

Rotary → Intermittent Rotary

As the driver A rotates the pin engages in a slot and drives wheel B. When the pin is out of the slot wheel B is stationary. Wheels with as many as 18 slots are used in practice.

To generalize the fact that a mechanism transforms motion let ϕ, $\dot{\phi}$, $\ddot{\phi}$, and $\dddot{\phi}$ represent the input position, velocity, acceleration, and jerk (rate of change of acceleration), and ψ, $\dot{\psi}$, $\ddot{\psi}$, and $\dddot{\psi}$, the output position, velocity, acceleration, and jerk shown below:

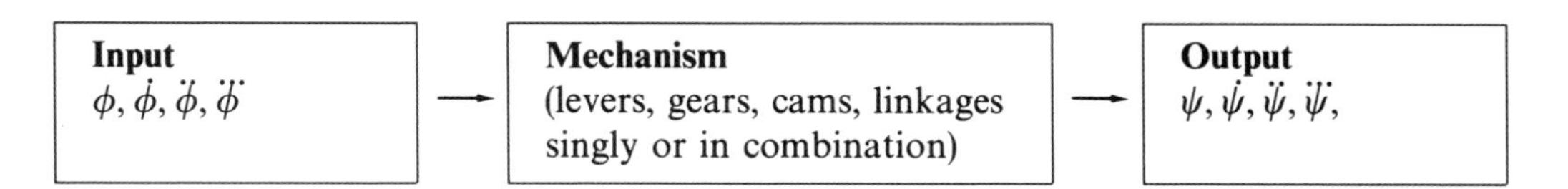

In general:

Displacement,	$\psi = G(\phi)$
Velocity,	$\dot{\psi} = G'\dot{\phi}$
Acceleration,	$\ddot{\psi} = G''\dot{\phi}^2 + G'\ddot{\phi}$
Jerk,	$\dddot{\psi} = G'''\dot{\phi}^3 + 3G''\ddot{\phi} + G'\dddot{\phi}$

Notes:

$$\dot{\psi}=\frac{d\psi}{d\phi}\cdot\frac{d\phi}{dt}=\frac{d\psi}{d\phi}\dot{\phi}$$

$$G'=\frac{f(\phi,\psi)}{g(\phi,\psi)} \quad \text{in general (e.g., see Sec. 5.4)}$$

where $G'=\dfrac{dG}{d\phi}$, $\quad G''=\dfrac{d^2G}{d\phi^2}$, etc.

If $\dot{\phi}$ is constant, which is often the case in many practical situations, e.g., where a motor is driving the input at constant velocity, then $\ddot{\phi}=\dddot{\phi}=0$, hence

Output velocity,	$\dot{\psi}=G'\dot{\phi}$
Output acceleration,	$\ddot{\psi}=G''\dot{\phi}^2$
Output jerk,	$\dddot{\psi}=G'''\dot{\phi}^3$

Furthermore, if G is independent of ϕ, e.g., as with gears or simple levers, then $\psi=k\phi$, where k is a constant, and hence $\dot{\psi}=k\dot{\phi}$, $\ddot{\psi}=k\ddot{\phi}$, $\dddot{\psi}=k\dddot{\phi}$.

1.3 LINKAGES

As mentioned in Sec. 1.1, we are primarily concerned in this text with mechanisms whose members are 'rigid' links or bars and with those which include a sliding member.

Figure 1.2(*a*) shows two rigid links, 1 and 2, each with two holes so that they can be connected to each other or to other links by means of pins as in Fig. 1.2(*c*). A link with *two* holes, e.g., link 1 with holes A and B, is referred to as a *binary* link. A link with *three* holes is called a *ternary* link, as in Fig. 1.2(*d*).

By putting a pin through B and C we form an *open kinematic chain*, as in Fig. 1.2(b). Relative motion between the two links is possible. We say that the two links are paired and the joint formed is a *pin joint*, also known as a *revolute* and indicated by the letter 'R'. Such a connection is also referred to as a lower pair. From now on in this text links will be represented by straight lines, and small circles will represent revolute joints.

If we now take two more links 3 and 4 and join them to our open chain in such a way that we close the chain we obtain the kinematic chain 1234 shown in (*c*). We can go on adding links to make up complex chains; we may also include ternary links as in (*d*) and (*e*). The chain shown in (*d*) is due to Watt and that in (*e*) is due to Stephenson.

In (*f*) and (*g*) double joints have been introduced; there is one double joint linking 2, 3, and 6 in (*f*), and two double joints are shown in (*g*). In order to obtain a linkage capable of transmitting motion we can fix any one link, i.e., if link 4 of Fig. 1.2(*c*) is fixed or grounded then we obtain Fig. 1.3(*a*) and if link A_0A is an input then an output can be taken from link BB_0 or a shaft at B_0 fixed to BB_0.

If we fix an alternative link such as AB the absolute motion of the links will change but their relative motion remains the same. Fixing other links of a kinematic chain is known as *inversion.*

The linkages shown in Fig. 1.3 are all formed from the kinematic chains of Fig. 1.2 by fixing a particular link; the readers can verify for themselves.

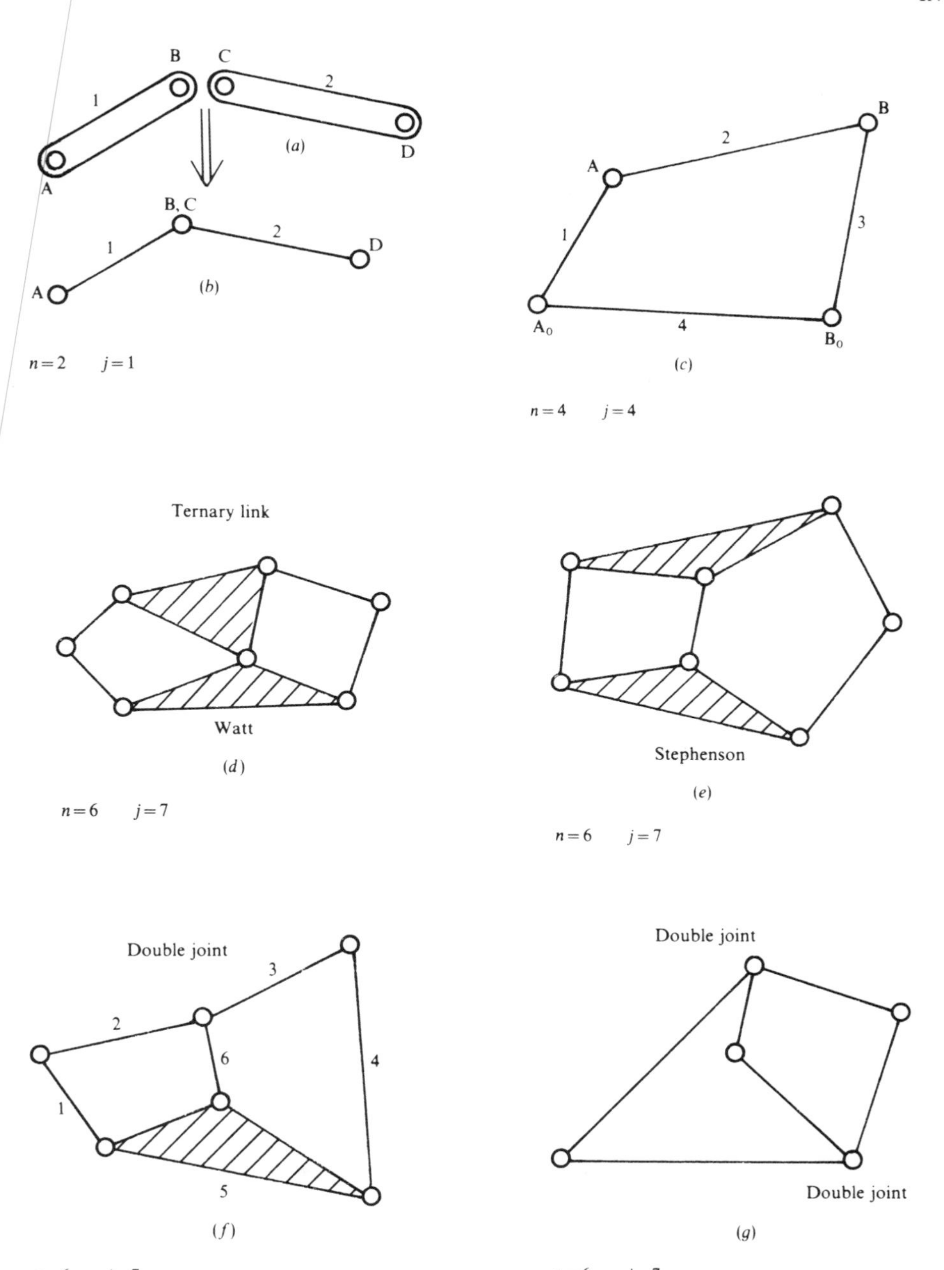

Figure 1.2 Kinematic chains.

If we replace a link by a slider we obtain mechanisms such as those shown in Fig. 1.4; for example (*a*) is the slider–crank mechanism, which is the familiar engine mechanism; A_0A is the crank while AB is the connecting rod. The link BB_0 of Fig. 1.2(*c*) has been replaced by the slider; or length BB_0 is infinite. The slider at B and the fixed frame B_0 constitute a *prismatic pair*, denoted by the letter P; it is a *lower pair*.

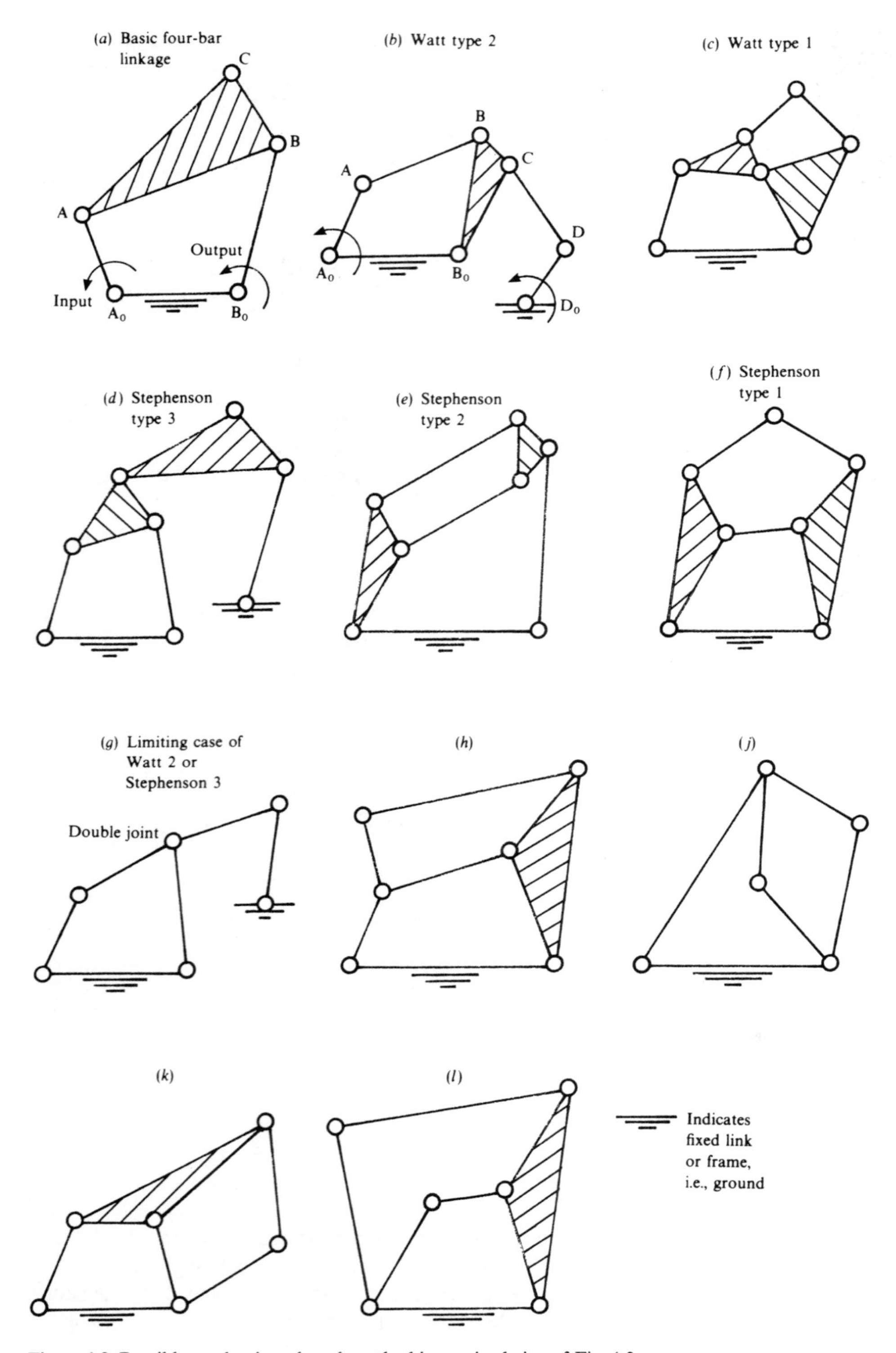

Figure 1.3 Possible mechanisms based on the kinematic chains of Fig. 1.2.

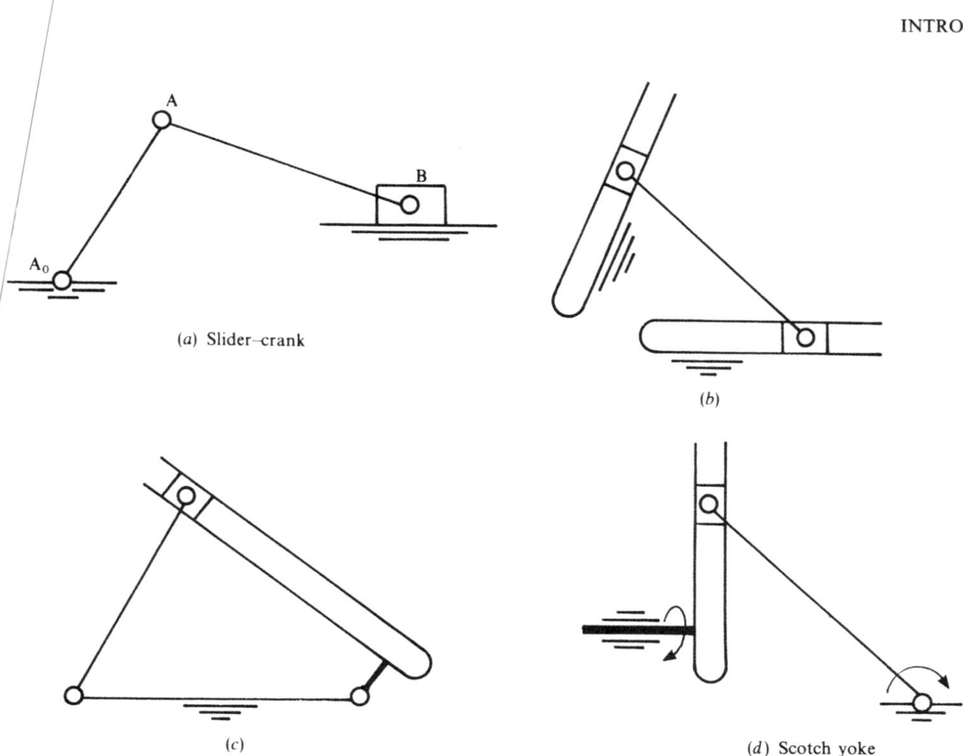

Figure 1.4 Mechanisms formed by replacing a link by a slider.

Figure 1.5 shows a practical application of one of the linkages shown in Fig. 1.3. It is suggested that readers discover for themselves which of the linkages has been used in this case by considering the linkage $A_0ABB_0B'FE$, leaving out H_0HA' which is a means of introducing power to elevate or lower the platform P; notice also that the platform remains horizontal throughout its motion.

If no relative motion is possible between the links then the linkage is a structure: thus if we add a link between A and D in Fig. 1.2(*b*) we obtain a triangular chain or frame. Such a triangle forms the basis of many practical structures.

In a linkage, therefore, both relative and absolute motion are necessary for the transmission

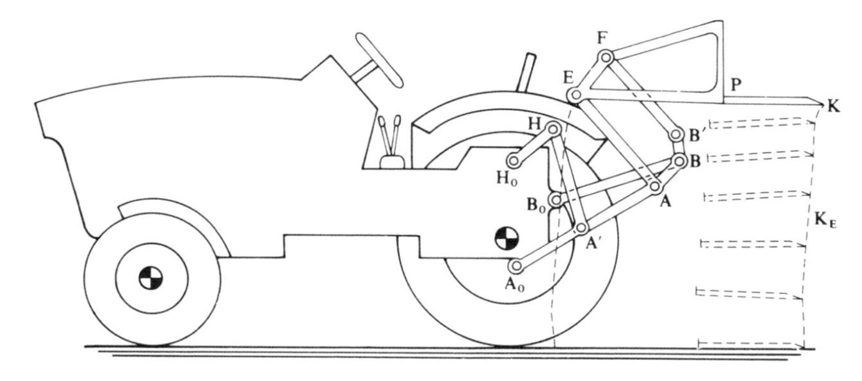

Figure 1.5 Practical application of one of the mechanisms shown in Fig. 1.3.

of power. Apart from transmitting power, a linkage is also a device that transforms the motion of an input, as we saw in Sec. 1.2; e.g., in the case of the slider–crank mechanism the rotation of the crank is transformed into the translation of the slider and vice versa. Consequently a linkage has *mobility* or freedom of movement.

A linkage is said to have one degree of freedom if a single coordinate is sufficient to define its position, two degrees of freedom if two independent coordinates are necessary to define its position, and so on. For example, the unconstrained link AB in Fig. 1.2(*a*) has three degrees of freedom in the plane: two of translation and one of rotation. In this text we consider linkages that have one degree of freedom only, a requirement for many linkages.

In Fig. 1.2(*a*), links AB and CD each have three degrees of freedom (DOF); by connecting B and C with a pin we remove two DOF, hence the kinematic chain in (*b*) has four DOF. Suppose now we fix AB, i.e., AB is grounded: it follows that link CD with C pivoting at B has one DOF since it can only rotate about C. Thus fixing a link removes three DOF.

Let us now consider the four-bar chain in Fig. 1.2(*c*): with four links and four joints the number of DOF of this kinematic chain will be $4\times3-2\times4=12-8=4$ DOF. By fixing one link, say AB, we remove three DOF, hence the four-bar linkage with 1 link fixed has one DOF.

We can express the above arguments using the following notation:

let f = number of DOF
n = number of links including the fixed link or frame
j = number of joints of the revolute type, i.e., type R

Then

$$f=3(n-1)-2j$$

For a mechanism to have one DOF, i.e., $f=1$,

then

$$1=3n-3-2j$$

hence

$$j=\tfrac{3}{2}n-2$$

is the number of joints required. It follows from this equation that there must be an even number of links including those fixed.

For example, consider the mechanism shown in Fig. 1.3(f):

Number of links $=6$ Number of joints $=7$

Hence

$$\text{DOF}=3(6-1)-2\times7=15-14=1$$

Readers should satisfy themselves that the mechanisms shown in Fig. 1.3 satisfy the above relationship.

Of all the linkages the four-bar one is extremely important; for this reason we propose to examine its properties in some detail.

1.4 THE FOUR-BAR LINKAGE

The four-bar linkage or four-bar chain is fundamental in the design of mechanisms. It is widely used and takes many different shapes and sizes. It is simple, fairly inexpensive, and easy to maintain; it can work at high or low speeds and transmit small or large powers and operate in all kinds of environments.

The four-bar linkage consists of four bars or links 1, 2, 3, and 4, pinned as shown in Fig. 1.6(*a*), thus forming a closed chain. The lengths of the links are: link $1=a$; link $2=b$; link $3=c$; link $4=d$.

In operation any one of the links may be fixed, resulting in different output–input relationships depending on the link lengths. This is illustrated in Figs 1.6(*b*), (*c*), (*d*), and (*e*) and demonstrates the four *inversions* of the four-bar linkage.

(*b*) and (*c*) are referred to as crank and rocker mechanisms; link 1 is the crank and link 3 the rocker. In these cases a rotary motion of the crank leads to an oscillatory motion of the rocker.

(*d*) is a *double rocker*; links 2 and 4 oscillate while link 1 rotates.

(*e*) is known as the *drag link* mechanism; links 2 and 4 rotate.

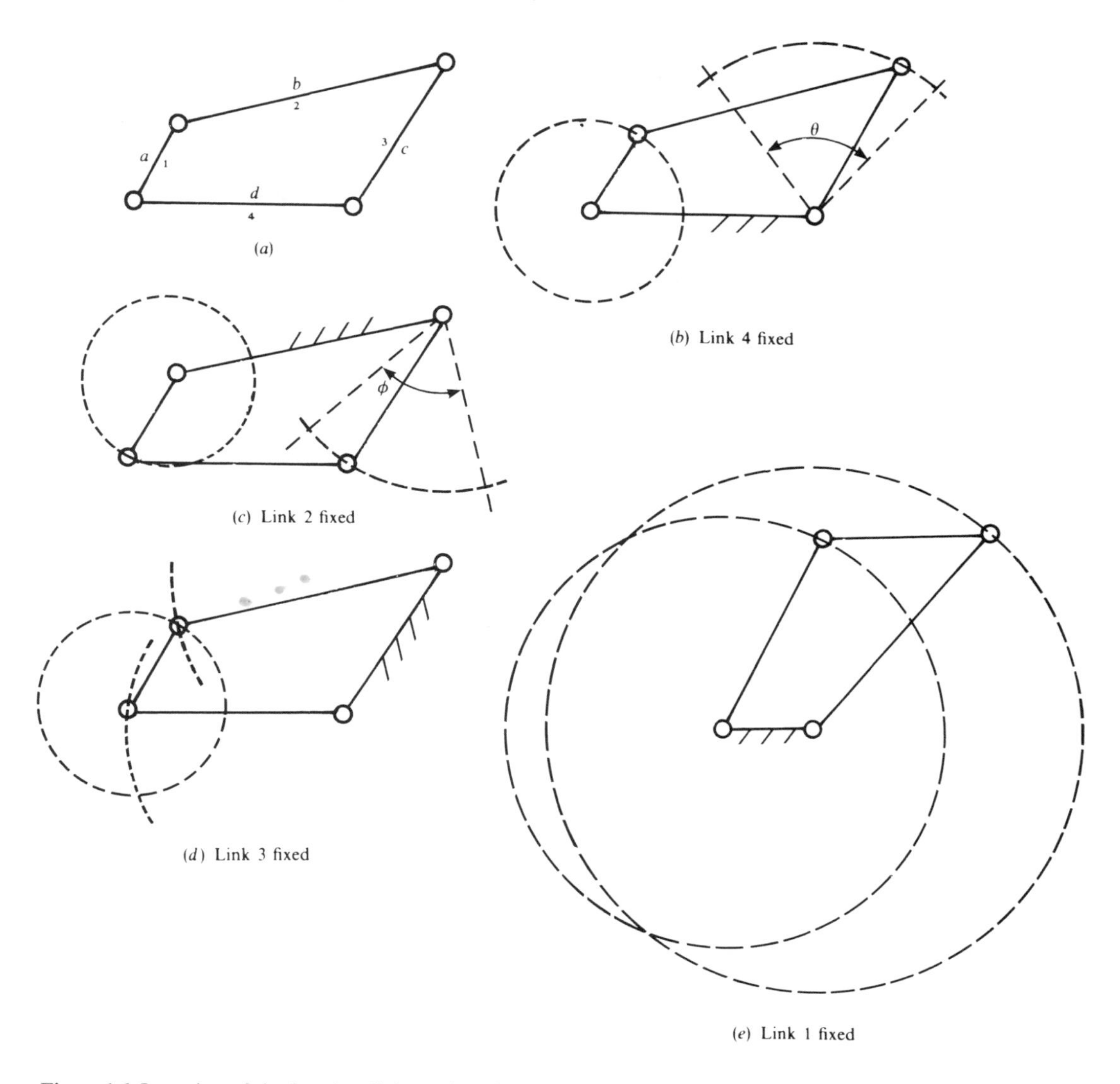

Figure 1.6 Inversion of the four-bar linkage that obeys Grashof's criterion.

(*a*)

(*b*)

Figure 1.7(*a*) Four-bar linkage applied to the design of a crane: lemniscate configuration. 25 tonne × 12–40 metres (*Courtesy of Figee-Haarlem, Holland.*) **1.7**(*b*) Lemniscate type floating grabcrane. (*Courtesy of Figee-Haarlem, Holland.*)

In cases (*b*) and (*c*) the angles of swing of link 3 are not necessarily equal, i.e., $\phi \neq \theta$, in general.

A designer who wants a four-bar linkage to be driven by an electric motor, for example, as is often the case in practice, will have to ensure that one crank, the input, can rotate through a complete revolution, otherwise the linkage will not be of much use in such a situation. There is a criterion due to Grashof (1883) which will ensure that such a requirement is possible. The criterion is as follows: For continuous relative rotation between two links of a four-bar linkage the length of the shortest link plus that of the longest link must be less than or equal to the sum of the other two links. Thus referring to Fig. 1.6 we must have

$$a+b \leqslant c+d$$

The link joining the crank to the rocker, also known as the follower, is called the coupler, e.g., *b* in Fig. 1.6(*a*).

In some applications the output of a four-bar linkage may be taken at the coupler instead of at the follower. An example of this is shown in Fig. 1.7 where the jib of a floating crane is the coupler of a four-bar chain in which the link A_0B_0 is fixed; the point C on the coupler of a four-bar chain carries the pulley to enable the load to be raised or lowered. The lengths of the links are such that C follows a straight horizontal path CC_1 over a length L of 28 m as the input link AA_0 is rotated through an angle ϕ of 89°. It is also an example of a double rocker mechanism.

Extensive use is made of the properties of coupler curves in every kind of engineering situation, from the crane just mentioned above, through automatic assembly manipulators, and on down to robotic devices for helping the handicapped.

An atlas containing thousands of curves generated by points on the coupler or the coupler extended of a four-bar linkage was produced by Hrones and Nelson[1] for a very wide range of link lengths; it is extremely valuable in the design of any linkage that is required for guiding an object along a particular path. An example taken from this atlas is shown in Fig. 1.8; point C, for instance, on the coupler ABC will trace out the path shown during one complete rotation of the input link A_0A. All the curves in the atlas are produced by linkages that obey Grashof's criterion.

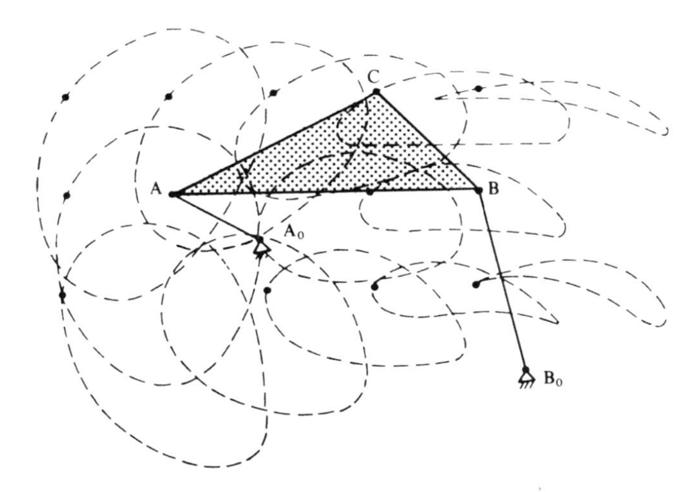

Figure 1.8 An example of curves generated by the coupler of a four-bar linkage.

1.5 FURTHER EXAMPLES OF THE USE OF LINKAGES

1.5.1 In the Building Industry

Figure 1.9(*a*) shows a JCB 415 wheeled loader which is hydraulically operated. The bucket can be replaced by a platform as can be seen in the diagram. It will be noticed that the bucket or the platform is the follower of a four-bar linkage; the input link or crank is activated by the hydraulic ram. Figure 1.9(*b*) is a JCB Reach Master 12 mechanical digger showing the large

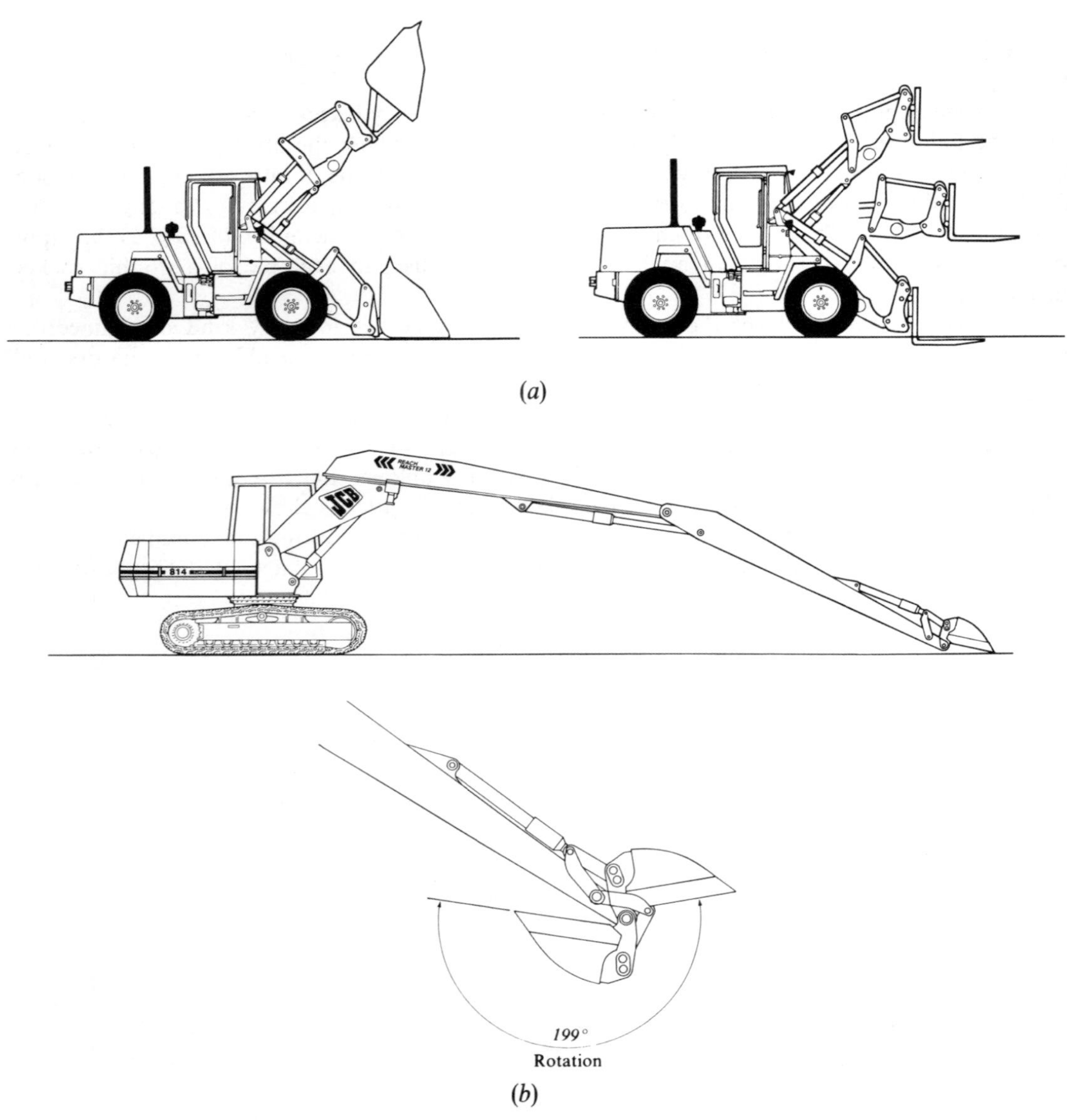

Figure 1.9(*a*) Hydraulically operated linkages of a wheeled loader. (*b*) Mechanical digger showing the open linkage for the jib and the four-bar linkage for operating the bucket. (*Courtesy of JCB, Rochester.*)

movement of the bucket (199°) obtained by means of the four-bar linkage. The boom itself is an open linkage activated by means of the hydraulic rams shown. Both machines are made up of *revolute* (R) and *prismatic* (P) pairs, i.e., the four-bar linkage and the hydraulic rams respectively.

1.5.2 In the Aircraft Industry

Figure 1.10 shows the main landing gear fitted to the A310 Aircraft (known as the Airbus). Also shown is the gear uplock in the locked and unlocked positions which is made up of simple links and operated by an electric actuator.

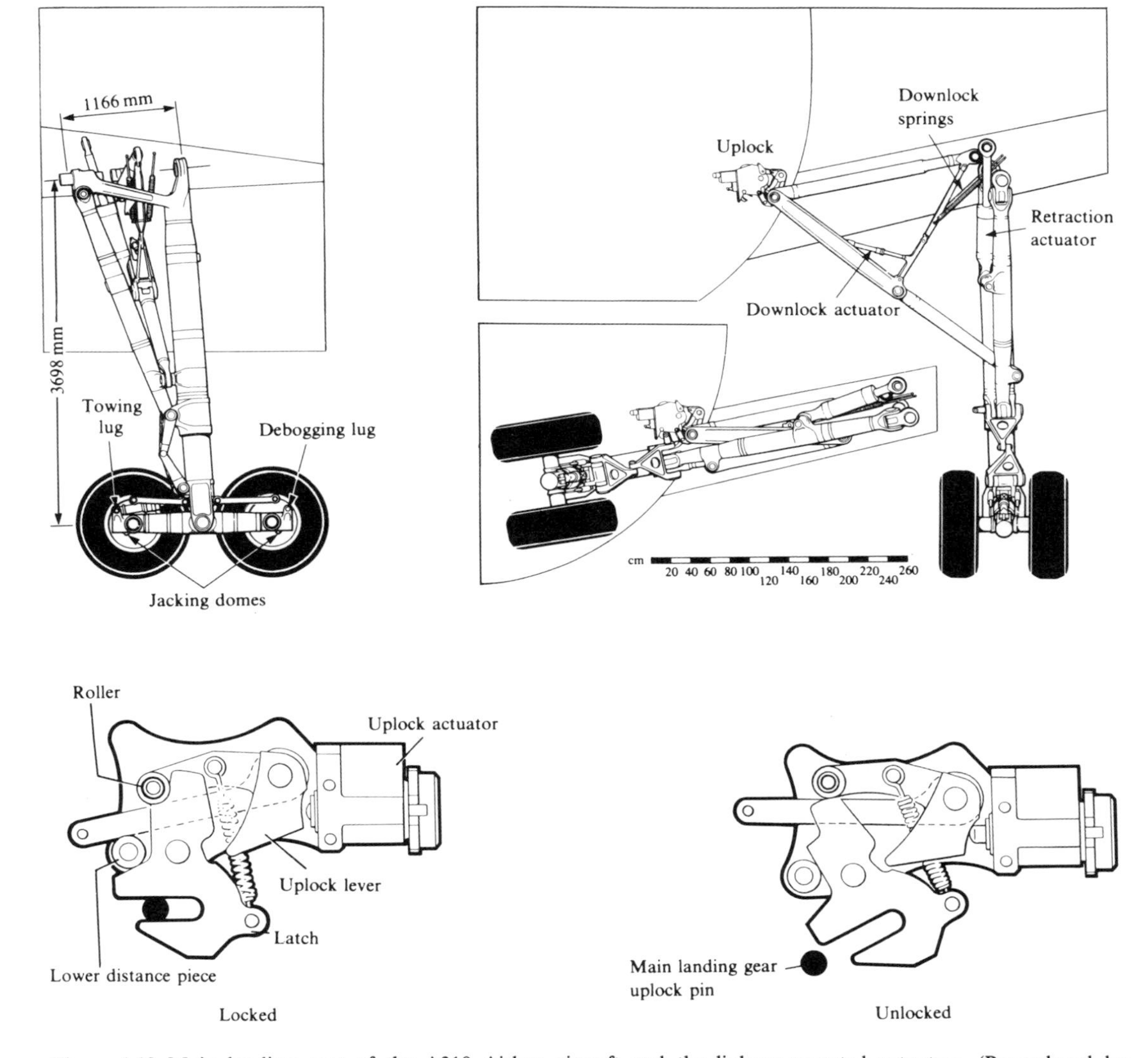

Figure 1.10 Main landing gear of the A310 Airbus aircraft and the linkage-operated actuators. (Reproduced by permission of the Council of the Institution of Mechanical Engineers from D. W. Young, 'Aircraft landing gears', **200**, 2, 1986.)

1.5.3 In the Automobile Industry

Figure 1.11(*a*) is a drawing of the Ford turbo-diesel engine used for passenger cars and commecial vehicles. It shows the slider-crank mechanism: the piston is the slider and AB the connecting rod and the crakshaft is the crank. The overhead-cam configuration can also be seen, the valve being the follower (see Chapter 4).

Figure 1.11(*b*) shows the double-wishbone suspension of the Renault 18; it is a four-bar linkage with the hub as the coupler.

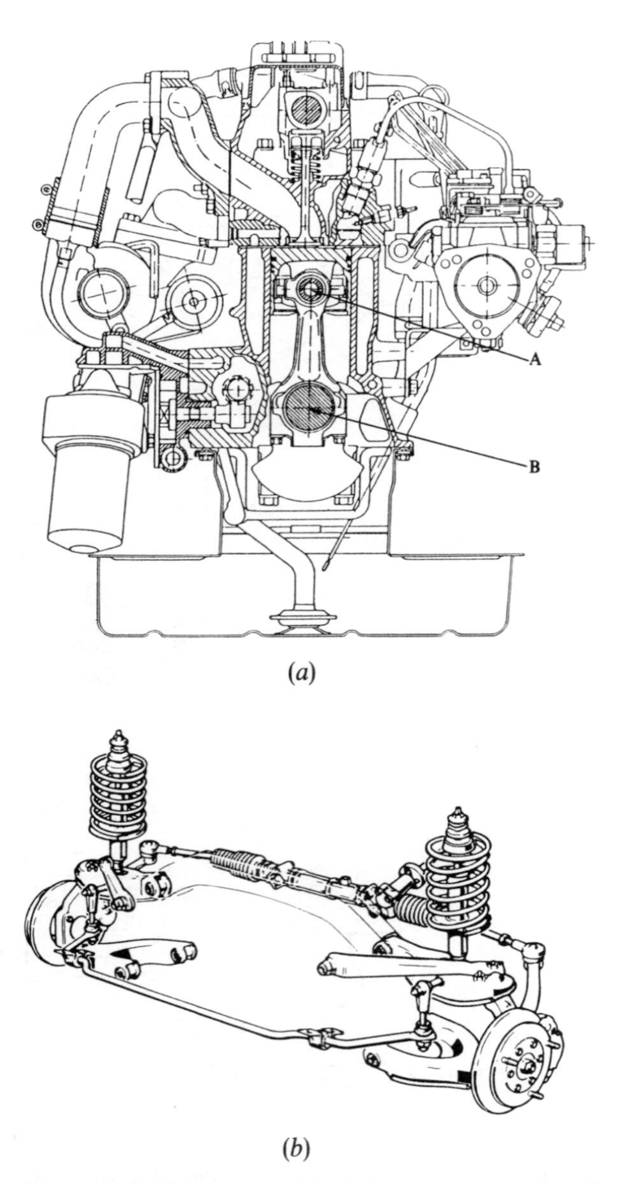

Figure 1.11(*a*) Slider–crank mechanism as used in the engine of a motor vehicle. (*Courtesy of the Ford Motor.*); (*b*) Four-bar linkage used in the suspension of the Renault 18 motor car (*Courtesy of Régie Renault, France.*)

1.54 In the Machine Tool Industry

Figure 1.12(*a*) shows the assembly drawing of one version of the quick-return mechanism used for removing material by means of a cutting tool. The oscillating link is held at A as shown in the diagram (Fig. 1.12*b*), and drives the ram to which the tool is attached. The bottom of the

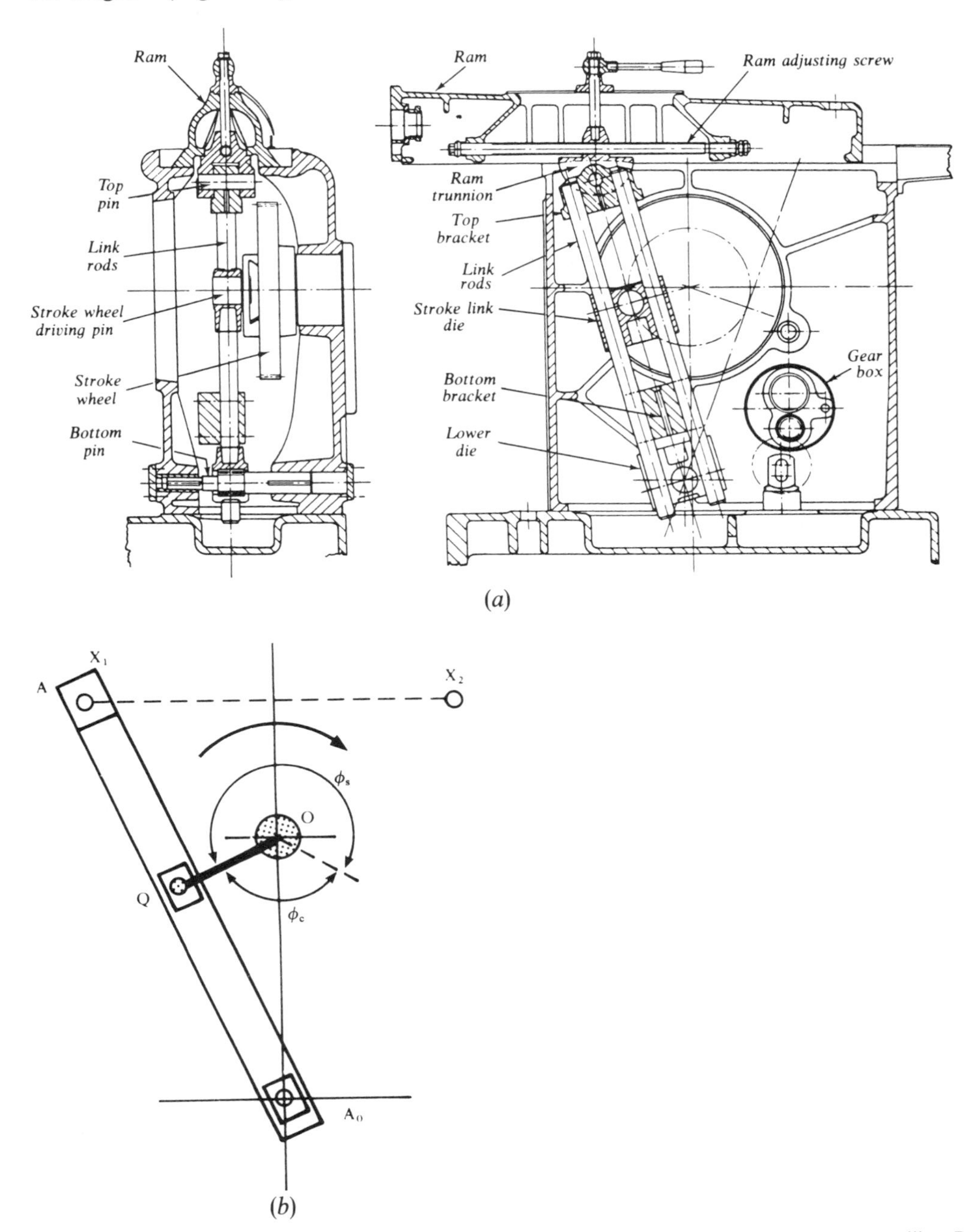

Fig. 1.12 (*a*) Shaping machine. (From G. D. Redford, *Mechanical Engineering Design*, 2nd edn, Macmillan, Basingstoke, 1973.) (*b*) One version of the quick-return mechanism as used in a shaper.

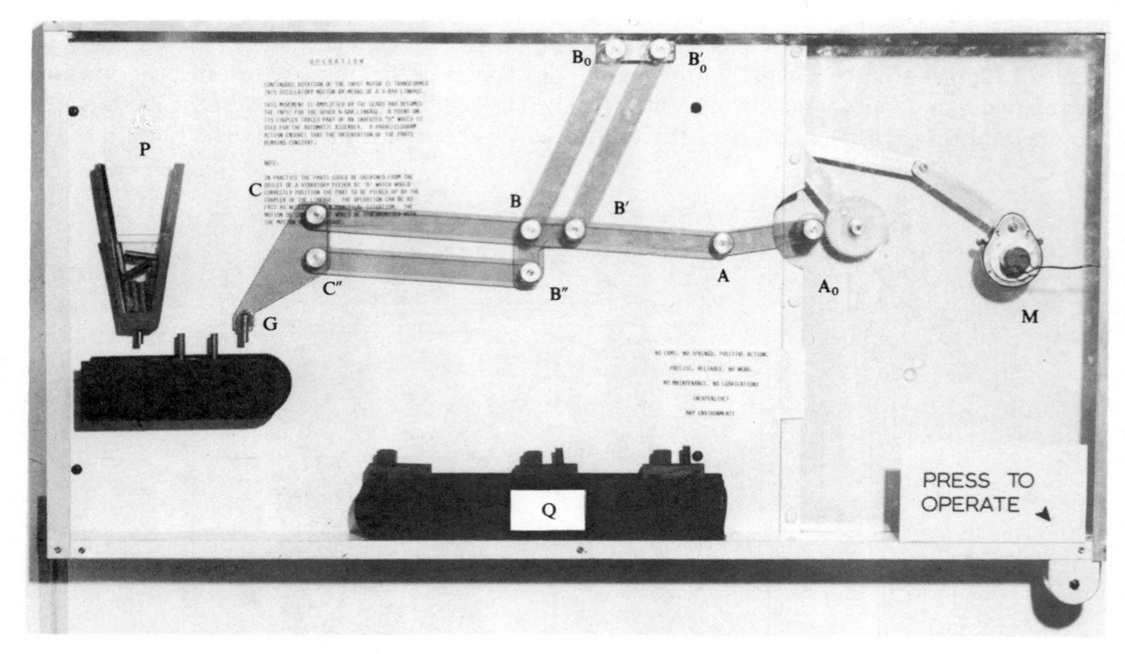

Figure 1.13 Two four-bar linkages linked by means of a set of gears in the design of a mechanism for the automatic assembly of parts.

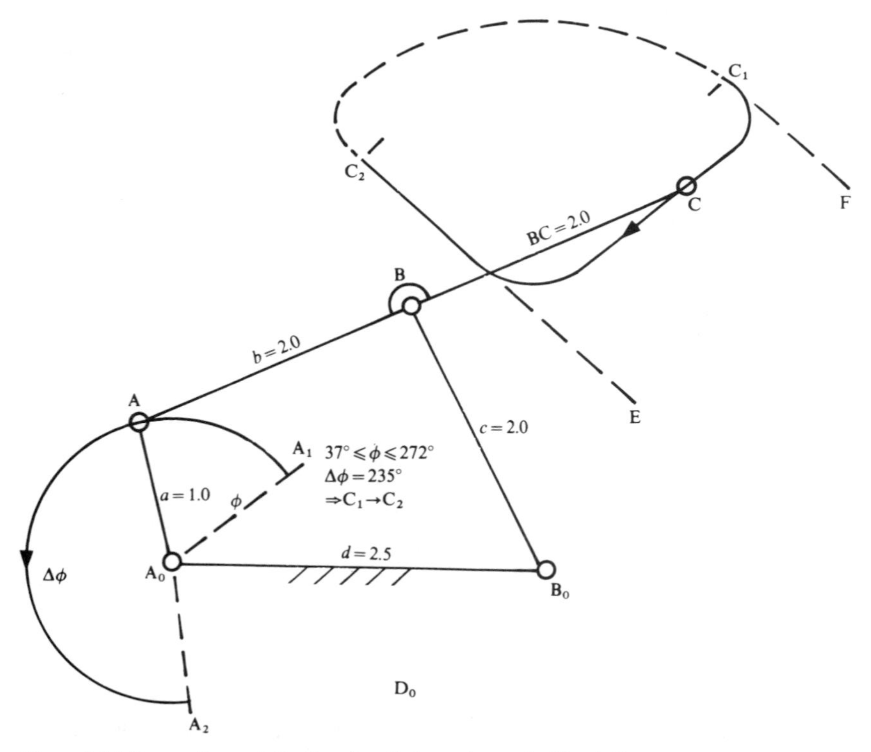

Figure 1.14 Proportions of the four-bar linkage shown in Fig. 1.13.

link slides in a trunnion mounted on a shaft at A_0. As the driver OQ rotates, the link A_0A oscillates between X_1 and X_2, and the motion is transferred to the ram. The motion is such that during the cutting stroke, X_1 to X_2, the velocity of the ram is slow, but the return stroke, X_2 to X_1, is faster since $\phi_s > \phi_c$ and the speed of the driving motor is constant. This shaper operates at 150 cycles per minute with a stroke $X_1X_2 = 152.4$ mm and a cutting speed of 711 mm per second.

1.5.5 In Automatic Assembly

Another example which uses linkages and gears is shown in Fig. 1.13, which is a display model of a mechanism for the automatic assembly of parts to be fitted to castings at Q, only one being shown. The parts are taken from a feeder at P and accurately positioned into the castings. The path followed by each part is that portion C_1C_2 of the trajectory traced by the coupler point C of the linkage shown in Fig. 1.14, which is one of the symmetrical coupler curves found in

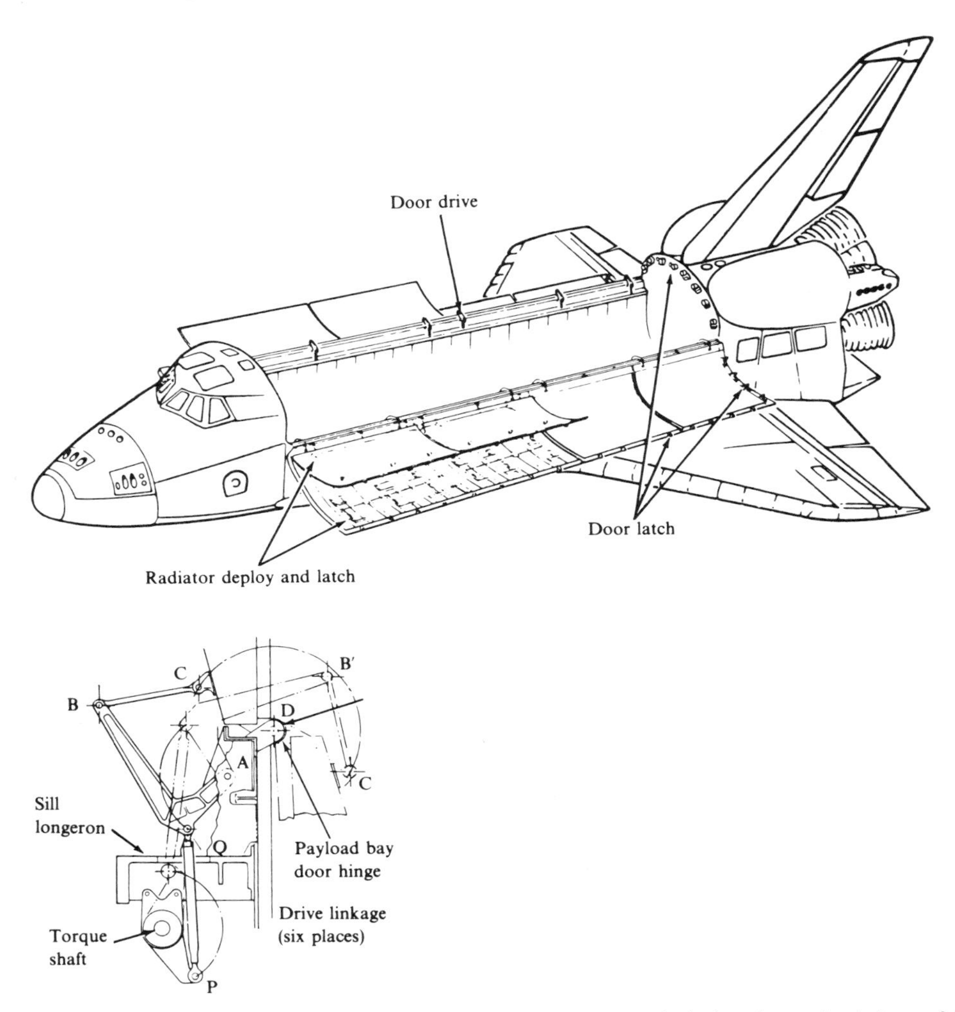

Figure 1.15 Two four-bar linkages in series used for opening and closing the payload door of the space shuttle. (Reproduced by permission of the Council of the Institution of Mechanical Engineers from G. D. Griffin, *Mechanical Engineering Aspects of the Space Shuttle Orbiter*, 1984.)

the book by Hrones and Nelson[1] (see Sect. 1.4). An examination of the path shows that the tangents C_1F and C_2E at C_1 and C_2 respectively are parallel and this property is made use of to ensure that the parts will follow a vertical path with the linkage placed in the inverted position as shown in Fig. 1.12. In this particular application the driving crank A_0A rotates through an angle of 235° from A_1 to A_2 while the point C moves from C_1 to C_2.

In order to obtain this motion the linkage is driven by another four-bar linkage via a set of precision gears while the driving motor M rotates continuously at a uniform speed. In the actual industrial case the motion to the right of the castings is synchronized with the motion of the linkage A_0ABCB_0. Referring again to Fig. 1.13, we see that the parallelogram $B_0'B'BB''C''$ ensures that the gripper *G*, and the parts will maintain the desired orientation.

1.5.6 In Linkages in Outer Space

Figure 1.15 of the space shuttle *Orbiter* illustrates further applications of linkages. The figure shows one of the four-bar linkages ABCD operating the payload bay door, which is 18.29 m long, and hinged at 13 points. AD is the fixed link, i.e., the shuttle's structure; AB is the input link in the shape of a bellcrank, and is actuated by means of a link PQ; BC is the coupler; and CD is the follower, or output link, which is the bay door itself. In the open position B moves to B′ and C to C′ corresponding to an angular movement of 180°, while the input link rotates through an angle of 103°. The torque shaft for each bay door is actuated by two motors by means of an epicyclic gear system.

Other bellcranks and linkages operated by actuators are used in the various latch systems required to secure the payload bay doors and the umbilical doors underneath *Orbiter*, and are shown in Fig. 1.16.

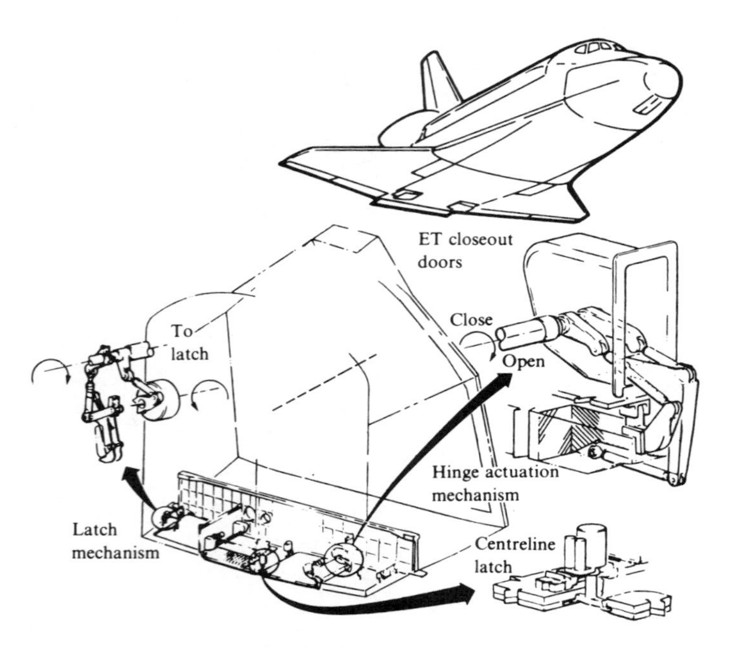

Figure 1.16 Latching mechanisms to secure the payload door of the space shuttle. (Reproduced by permission of the Institution of Mechanical Engineers from G. D. Griffin, *Mechanical Engineering Aspects of the Space Shuttle Orbiter*, 1984.)

CHAPTER

TWO

GRAPHICAL ANALYSIS: VELOCITIES

2.1 VELOCITIES OF POINTS IN LINKAGES

Consider the motion of the four-bar linkage shown in Fig. 2.1, in which A_0B_0 is the fixed link, A_0A the driver, B_0B the follower or output link, and AB the coupler. The displacement of link A_0A leads to corresponding displacements of the other two links, and thus every point of every link of the linkage or mechanism is compelled to follow a definite path. For instance, point A of link A_0A follows a circular path of radius a, the length of the link, with centre A_0; similarly point B of link B_0B of length c, centre B_0. Thus both links are in pure rotation, whereas the displacement of link AB has both translation and rotation since any point on the link will follow a path such as those shown in Fig. 1.8. If we consider the velocity of point B on link AB, it must have the velocity of the point A plus its velocity relative to A. The velocity of A is tangential to the circle of radius a and therefore perpendicular to A_0A. Similarly, if we observe the motion of B relative to A, then B can only move in a circle of radius $AB = b$, centre A. Hence the velocity of B relative to A is perpendicular to the link AB. Thus the velocity of B can be expressed vectorially as follows:

$$\mathbf{V}_B = \mathbf{V}_A + \mathbf{V}_{BA} \tag{2.1}$$

Equation (2.1) is a vector equation, which means that since a vector has both magnitude and

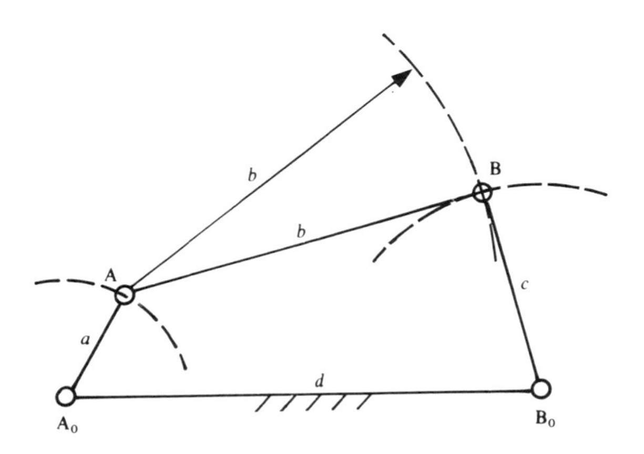

Figure 2.1 Four-bar linkage.

direction, addition of two vectors must be in accordance with the parallelogram law or triangle law.

The acceleration of points such as A and B has both a tangential component and a radial or centripetal component, the latter being always directed towards the centre of rotation or of curvature.

Velocities and accelerations of points in a linkage can be obtained using various graphical methods. In the case of velocity the most utilized are as follows:

1. the instantaneous-centre method;
2. the relative-velocity method;
3. the rotated-vectors method.

Graphical methods for accelerations are considered in Chapter 3.

2.2 INSTANTANEOUS-CENTRE METHOD

This method is based on the fact that at any particular instant in time the motion of any rigid member or link is equivalent to a rotation of the member as a whole about a point in the plane of the link in the case of plane motion.

Consider the link AB in Fig. 2.2, and suppose that in a short time δt it moves to position A′B′ such that the displacements AA′ and BB′ are infinitesimal. The lines that bisect AA′ and BB′ meet at a point I as shown. As the time interval is reduced further, these lines become normal to the paths of A and B respectively, and it therefore follows that at the instant when the link is in the position AB it is turning about the point I which is the point of intersection of the lines AI and BI drawn normal to the velocities of A and B. This point is called the *instantaneous centre*, *instant centre*, or *virtual centre*. It is a point that is instantaneously at rest. Hence the velocity of any point on the link AB must be proportional to its distance from that point and perpendicular to the line joining it to the instant centre.

Thus in Fig. 2.3 the velocities of points A, B, and C rigidly attached to AB are at right angles to AI, BI, and CI and their magnitudes are such that they satisfy the following relationship:

$$\frac{\mathbf{V}_A}{AI}=\frac{\mathbf{V}_B}{BI}=\frac{\mathbf{V}_C}{CL}=\omega \tag{2.2}$$

where ω is the angular velocity of the link at that instant.

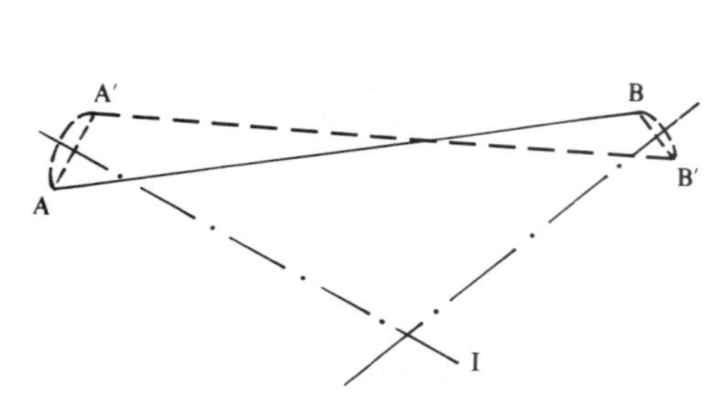

Figure 2.2 Position of the instantaneous centre of a moving link.

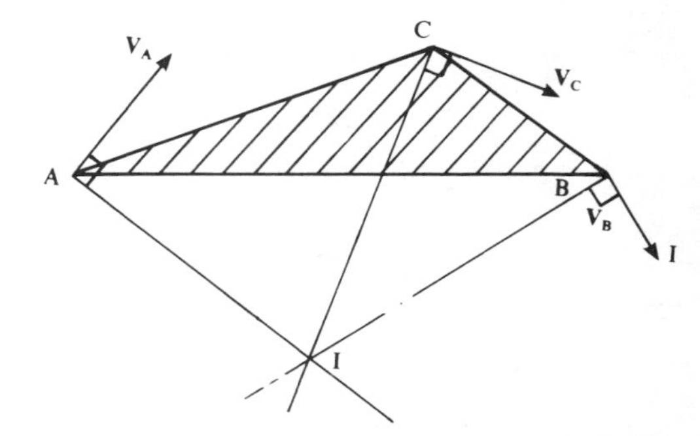

Figure 2.3 Velocity of a point on a rigid link from a knowledge of the position of the instantaneous centre.

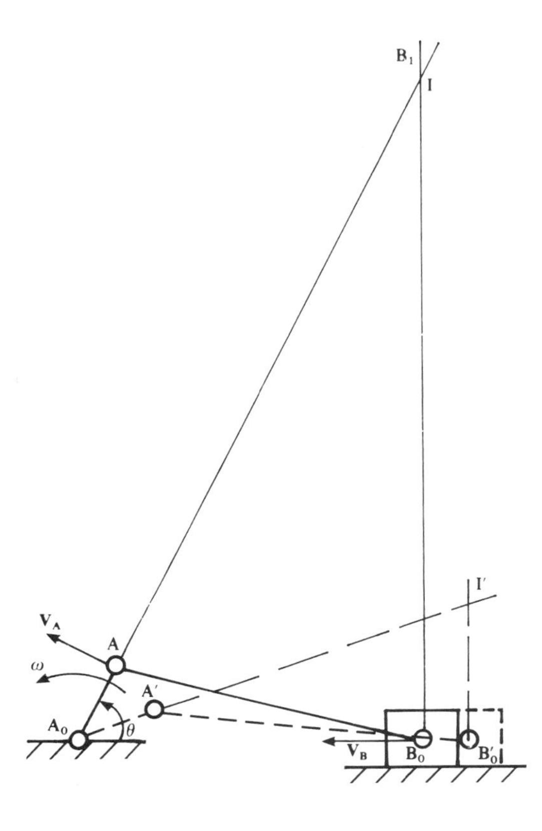

Figure 2.4 Position of the instantaneous centre of the slider–crank mechanism.

Since the link AB may move in any manner whatsoever, the position of the instant centre I will change for every position of the link.

As an application consider the slider–crank or engine mechanism shown in Fig. 2.4. We will use the instant-centre method to calculate the velocity of the slider B for a particular position θ of the crank A_0A which rotates at ω rad/s. The velocity V_A of the point A is at right angles to the crank A_0A so that the instant centre I of the connecting rod AB_0 lies on A_0A produced. The velocity V_B at the slider B is along the line A_0B_0, so that the instant centre of the connecting rod AB also lies on a line B_0B_1 drawn through B_0 at right angles to A_0B_0. The position of the instant centre is thus given by I, the point of intersection of A_0A produced and B_0B_1. Hence for that particular position of the mechanism the connecting rod AB is rotating about I.

If ω is the angular velocity of the connecting rod then Eq. (2.2) gives

$$\omega=\frac{V_A}{AI}=\frac{V_B}{B_0I}$$

hence

$$V_B=V_A\frac{B_0I}{AI} \tag{2.3}$$

When the crank was in the position A_0A' the slider was at B_0' and by a reasoning similar to the above we find that for this position the instant centre was at I′, illustrating quite clearly that the instant centre is not a fixed point. The locus of the point I is referred to as a *centrode.*

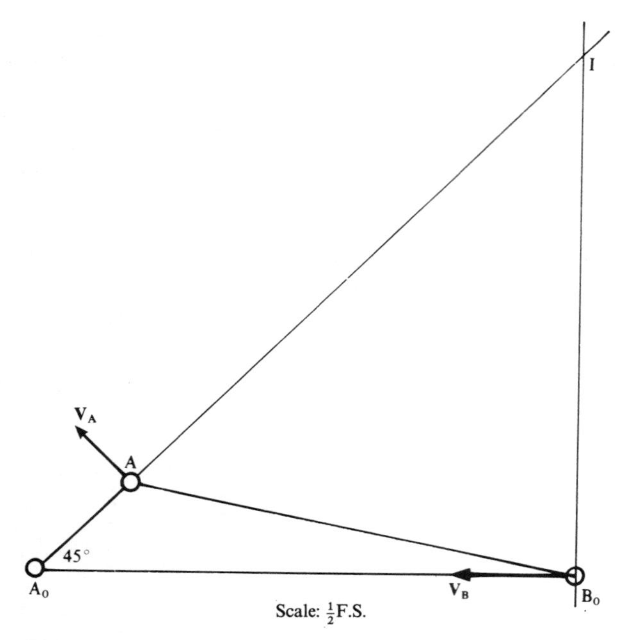

Figure 2.5

Example 2.1 The crank of a particular engine mechanism has a length of 35 mm and the connecting rod a length of 124 mm. Calculate the velocity of the piston when the engine rotates at 2500 rev/min and the angle between the crank and the line of stroke is 45°.

SOLUTION The mechanism is drawn to scale as in Fig. 2.5. The velocity V_A of the crankpin A is

$$V_A = \frac{2\pi \cdot 2500}{60} \times \frac{35}{1000}$$
$$= 9.16 \text{ m/s}$$

Scaling off the lengths AI and BI, we find

$$AI = 85 \text{ mm} \qquad B_0I = 72 \text{ mm}$$

Hence applying Eq. (2.3) we get the velocity of the piston,

$$V_B = 9.16 \times \frac{72}{85} = \underline{7.76 \text{ m/s}}$$

in the direction shown, i.e., to the left towards A_0.

Let us now consider the velocities of the four-bar linkage shown in Fig. 2.6. Let ω_1 be the angular velocity of a shaft A_0 to which the link A_0A is fixed and ω_2 the angular velocity of shaft B_0 to which the link B_0B is fixed and AB a link joining A and B. Furthermore let A_0 be the input shaft and B_0 the output shaft. We require the relationship between input and output.

A_0B_0 is the fixed link and A_0A the input crank makes an angle ϕ at a particular instant.

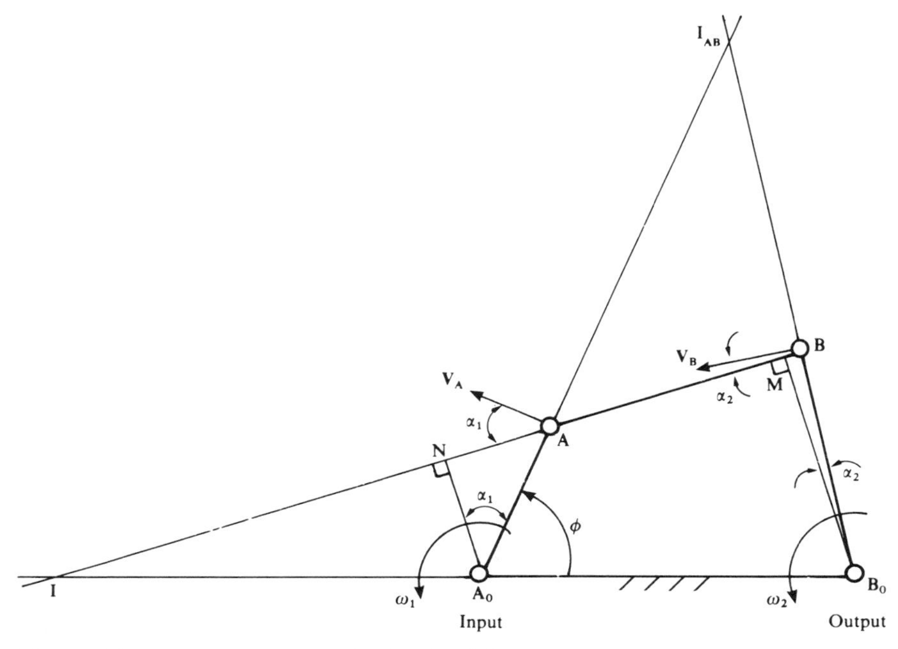

Figure 2.6 Instantaneous centres of the four-bar linkage.

Consider the points A on link A_0A and B on link B_0B. The velocity of A is

$$\mathbf{V}_A = \omega_1 \overline{A_0A}$$

in magnitude and in a direction perpendicular to A_0A, the sense being as shown by the vector $\mathbf{V}_A$.

Similarly, for the point B on link B_0B,

$$\mathbf{V}_B = \omega_2 \overline{B_0B}$$

The components of the velocities of A and B along AB are

$$\mathbf{V}'_A = \omega_1 \overline{A_0A} \cos \alpha_1 = \omega_1 \overline{A_0N}$$

and

$$\mathbf{V}'_B = \omega_2 \overline{B_0B} \cos \alpha_2 = \omega_2 \overline{B_0M}$$

Two lines drawn along AB and A_0B_0 meet at the point I; points N and M are such that A_0N and B_0M are perpendicular to IAB.

Since the link AB is rigid it follows that

$$V'_A = V'_B$$

hence

$$\omega_1 \overline{A_0N} = \omega_2 \overline{B_0M}$$

or

$$\frac{\omega_2}{\omega_1} = \frac{A_0N}{B_0M}$$

Also, because the triangles IA_0N and IB_0M are similar we have

$$\frac{A_0N}{B_0M}=\frac{IA_0}{IB_0}$$

The required velocity ratio is therefore given by

$$\frac{\omega_2}{\omega_1}=\frac{IA_0}{IB_0} \tag{2.4}$$

I is in fact the instantaneous centre for the links A_0A and B_0B, and I_{AB} is the instantaneous centre for link AB.

The angular velocity of link AB, the coupler, is such that

$$\omega_{AB}=V_A/AI_{AB}=V_B/BI_{AB} \tag{2.5}$$

As ϕ takes on various values (0 to 2π in many cases), I and I_{AB} will change position.

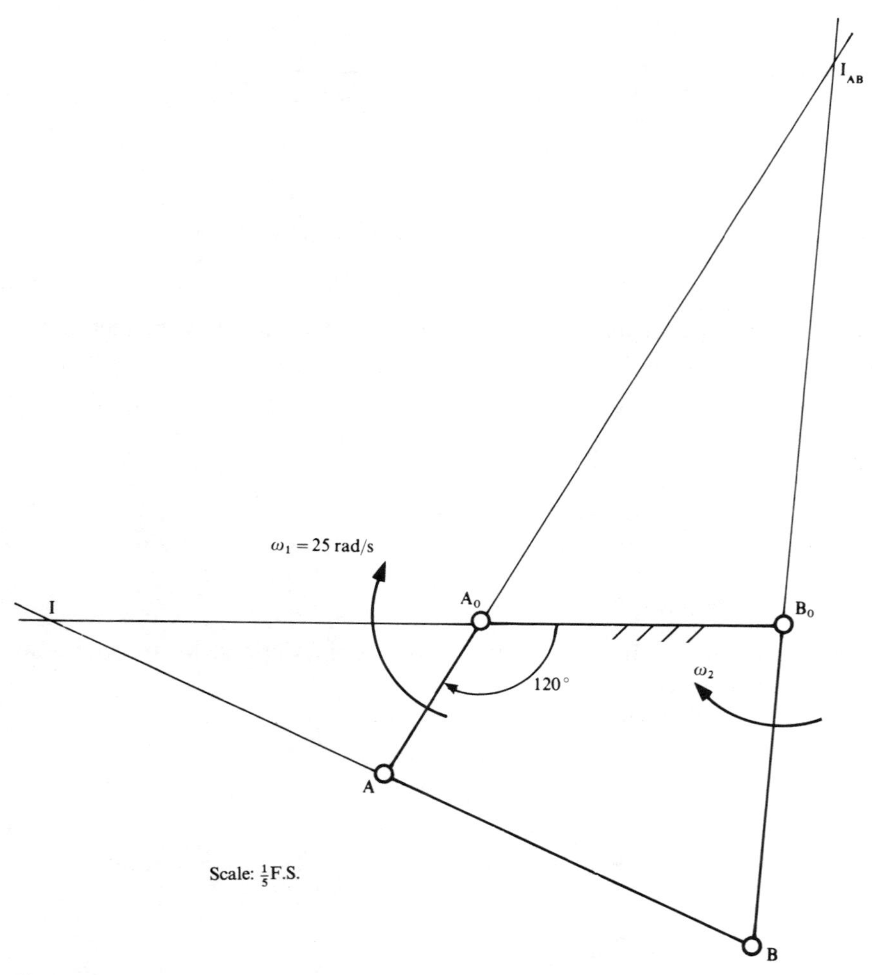

Figure 2.7

Example 2.2 The four-bar linkage shown in Fig. 2.6 has the following link lengths: $A_0A = 125$ mm, $AB = 275$ mm, $B_0B = 225$ mm and $A_0B_0 = 200$ mm. For the position when $\phi = -120°$ calculate the angular velocity of the output and of the coupler if the input angular velocity is 25 rad/s clockwise.

SOLUTION First we have to draw the linkage to some suitable scale in the required position as shown in Fig. 2.7, noting that with $\phi = -120°$ the link A_0A points downwards. As we do not know where the instantaneous centres will be located it may take one or two sketches to ensure that this will lie within the space of the paper available. In our case a scale of $\frac{1}{5}$ full size is found to be suitable.

Note: You should always indicate the scale on the drawing.
The point I is located at the intersection of AB and A_0B_0 prolonged, and I_{AB} at the intersection of A_0A and B_0B prolonged. ω_2 will have the same direction as ω_1, and

$$\omega_2 = \omega_1(IA_0/IB_0) = 25 \times 59.5/99 = 15.03 \text{ rad/s} \qquad \text{(from 2.4)}$$

$$\omega_{AB} = V_A/AI_{AB} = \omega_1\, A_0A/AI_{AB} = 25 \times 125/117 \times 5 = 5.34 \text{ rad/s, clockwise} \qquad \text{(from 2.5)}$$

As a check,

$$\omega_2 = V_B/BB_0 = \omega_{AB} BI_{AB}/B_0B = 5.34 \times 126 \times 5/225 = 14.95 \text{ rad/s}$$

which is near enough in view of the small-scale diagram.

2.3 VELOCITY DIAGRAMS

2.3.1 Relative Velocity

Consider a body moving in a plane and two points A and B within it as shown in Fig. 2.8(*a*). Let A have an absolute velocity $\mathbf{V}_A$ and B an absolute velocity $\mathbf{V}_B$ in the directions shown relative

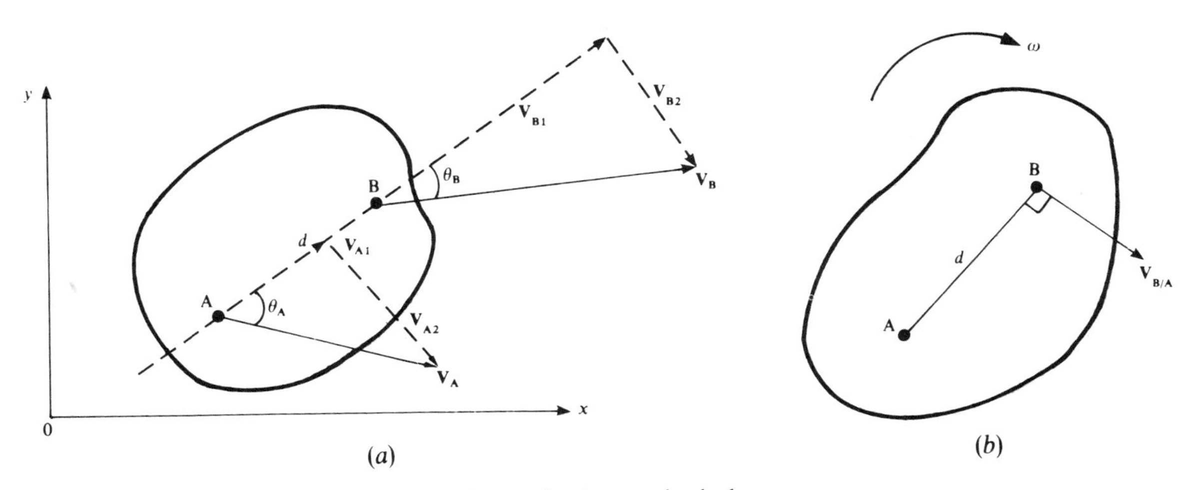

Figure 2.8 Absolute and relative velocities of two points in a moving body.

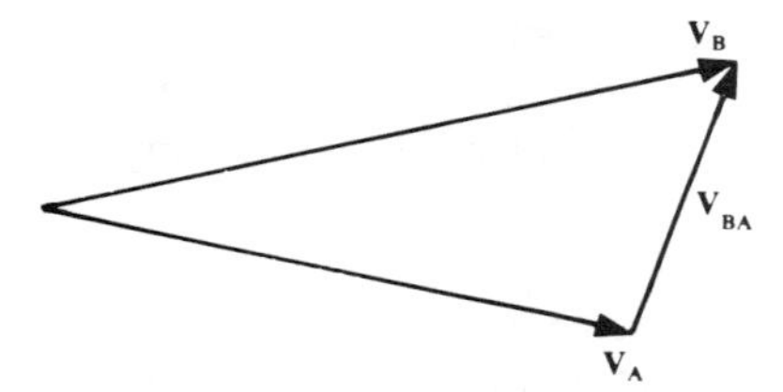

Figure 2.9 Vector diagram for velocities.

to a Newtonian frame oxy. Resolving these velocities along AB and perpendicular to AB gives

$$\mathbf{V}_{A1} = \mathbf{V}_A \cos \theta_A$$

$$\mathbf{V}_{B1} = \mathbf{V}_B \cos \theta_B$$

$$\mathbf{V}_{A2} = \mathbf{V}_A \sin \theta_A$$

$$\mathbf{V}_{B2} = \mathbf{V}_B \sin \theta_B$$

Since the body is rigid it follows that

$$\mathbf{V}_{A1} = \mathbf{V}_{B1}$$

and the velocity of B relative to A, written $\mathbf{V}_{B/A}$ or more simply $\mathbf{V}_{BA}$, is given by

$$\mathbf{V}_{BA} = \mathbf{V}_{B2} - \mathbf{V}_{A2}$$

Its direction is perpendicular to AB; see Fig. 2.8(*b*). Hence the motion of B as seen by A is one of rotation about A; it follows that the angular velocity ω of the body is given by $\omega = \mathbf{V}_{BA}/d$, where $d = \overline{AB}$. The absolute velocity of the point B is then given by

$$\mathbf{V}_B = \mathbf{V}_A + \mathbf{V}_{BA} \tag{2.6}$$

These must be added vectorially as shown in Figure 2.9.

Equation (2.6) states that the absolute velocity of B is equal to the absolute velocity of A together with the velocity of B relative to A.

2.3.2 Velocity Diagrams

The construction of velocity diagrams is based on Eq. (2.6). To illustrate the method let us consider the simple linkage OAB (Fig. 2.10), where the link OA rotates as shown. The velocity V_A of A is known in magnitude and direction, while the velocity V_B of B is known in direction only since it is constrained to move in the slot SS′. We now wish to calculate the magnitude of the velocity of B as well as the angular velocity ω_{AB} of the link AB.

To do so we proceed as follows:

1. From an arbitrary point o (known as a pole) we draw the line oa to represent the magnitude, to some convenient scale, and the direction of the absolute velocity of A. Thus oa is parallel to the velocity vector $\mathbf{V}_A$.
2. From o we draw a line ob′ parallel to the slot SS′ which defines the direction of the velocity of B.
3. We know that the velocity of B relative to A is perpendicular to AB, hence through a we draw a line at right angles to AB to cross the line ob′ at b. Then ob $= V_B$, the absolute velocity of B, to the scale of the diagram, and its direction is from o to b.
4. The line ab represents the velocity of B relative to A so that the angular velocity of AB is

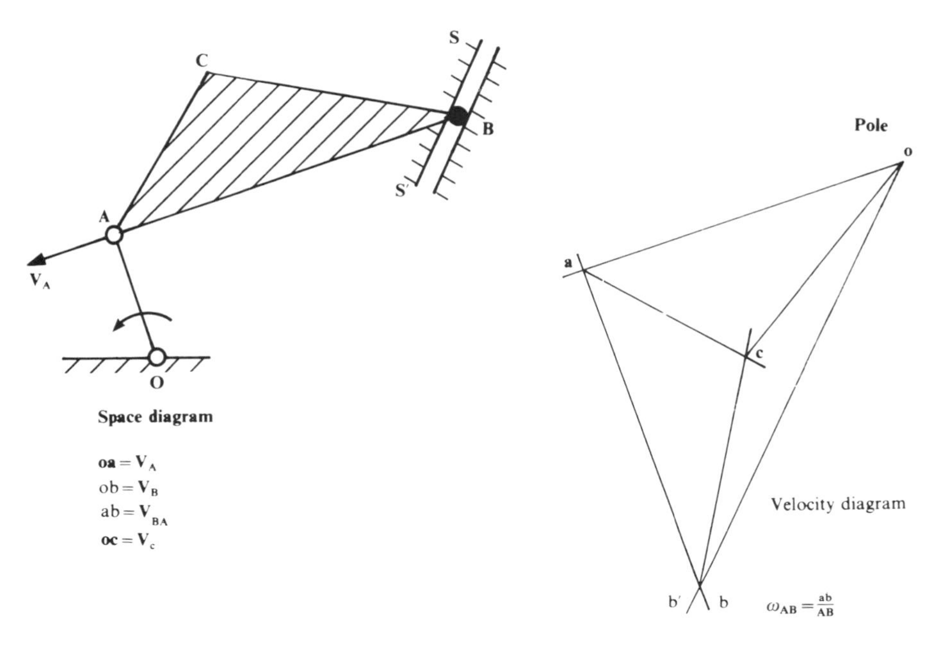

Figure 2.10 Velocity diagram: velocity of diagram; velocity of points in a typical linkage.

given by

$$\omega_{AB} = \frac{ab}{AB}$$

If C is any point on the link AB its velocity is obtained by drawing a line through a perpendicular to AC and a line through b perpendicular to BC. The velocity $\mathbf{V}_C$ of C is given by oc to the scale of the diagram. By virtue of the construction we see that

$$\triangle ABC \text{ is similar to } \triangle abc$$

since each side of one is perpendicular to a side of the other. The triangle abc is the *velocity image* of the triangle ABC. Hence any point on the space diagram will appear on the velocity diagram as a geometrically similar image.

Example 2.3 Let us calculate the velocity of the piston B of Example 2.1.

SOLUTION The engine mechanism is first drawn to a convenient scale as shown in Fig. 2.11. From a point o, oa is drawn to a suitable scale such that

$$oa = V_A = 9.16 \text{ m/s}$$

Since B is constrained to move along the line of stroke a line parallel to B_0A_0 is drawn through o. From a a line is drawn perpendicular to AB and the point b is therefore located. Hence $ob = V_B = \underline{7.8 \text{ m/s}}$ by scaling.

Example 2.4 Calculate the angular velocity of the output link and of the coupler in Example 2.2.

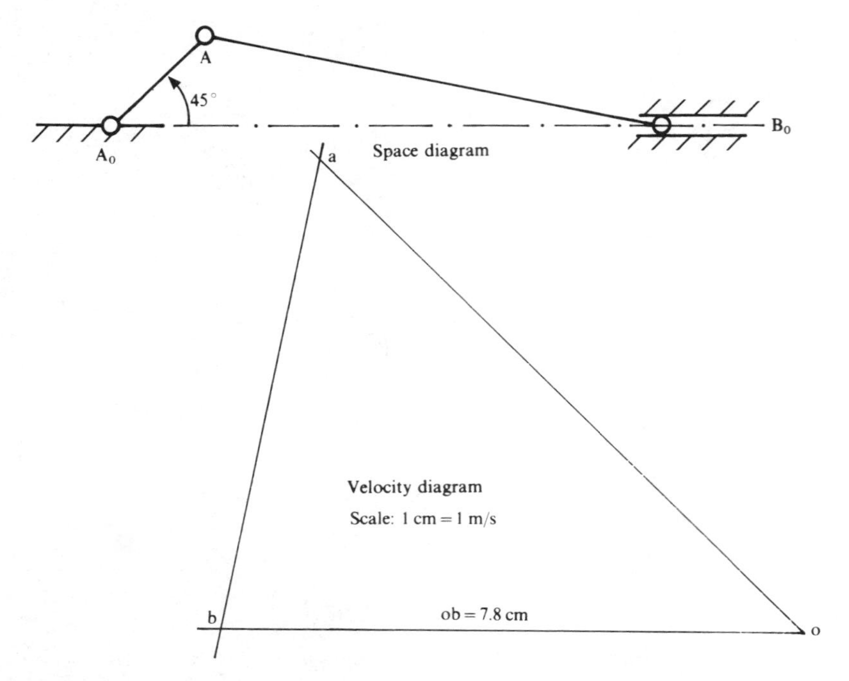

Figure 2.11

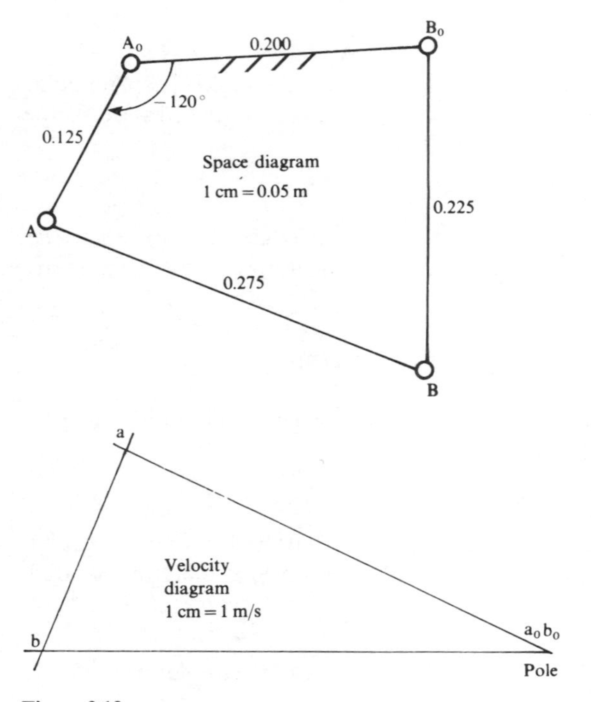

Figure 2.12

SOLUTION Draw the linkage A_0ABB_0 to some convenient scale (Fig. 2.12).

$$V_A = 25 \times 0.125 = 3.13 \text{ m/s}.$$

From the pole a_0, b_0 (since A_0 and B_0 are fixed points) draw $a_0a = 3.13$ m/s to a suitable scale. Then draw b_0b perpendicular to B_0B, and a line ab perpendicular to AB to locate b. By scaling we find $b_0b = 3.4$ m/s and $ab = 1.5$ m/s.
Hence the angular velocity of the output link B_0B is given by

$$\omega_{B_0B} = \frac{b_0b}{B_0B} = \frac{3.4}{0.225} = \underline{15.11 \text{ rad/s}} \qquad \text{clockwise}$$

and the angular velocity of the coupler AB is given by

$$\omega_{AB} = \frac{ab}{AB} = \frac{1.5}{0.275} = \underline{5.45 \text{ rad/s}} \qquad \text{clockwise}$$

2.3.3 Velocity Diagrams of Mechanisms with Slides

When a mechanism has slides on rotating members we have to take into account the motion of the slide relative to the member. The construction of the velocity diagram then follows the same procedure as in Sec. 2.3.2.

Consider, for example, the Geneva mechanism shown as in (*j*) on page 5 and reproduced in Figure 2.13. The pin P on the driver D slides in one of the slots of the follower F. At a particular instant corresponding to the angle ϕ of the driver arm OP there is a point Q on the follower F which at that instant coincides with P. Therefore Q relative to P has a velocity V_{QP} in the direction of the slot. To construct the velocity diagram we simply have to solve Eq. (2.6) graphically. In this case the equation is

$$\mathbf{V}_Q = \mathbf{V}_P + \mathbf{V}_{QP}$$

From poles O, O′, which are coincident since O and O′ are fixed points, we draw a line $op = V_p$ in a direction perpendicular to OP. We also know that Q is moving at right angles to O′F; we therefore draw a line from O′ perpendicular to O′F. Since Q relative to P is sliding in the direction O′F, then from p we draw a line parallel to O′F to locate the point q on the velocity diagram as shown. If Ω is the angular velocity of the follower F, then

$$\Omega = \frac{o'q}{O'Q}$$

and the velocity of sliding, important from a wear and lubrication point of view, is given by

$$V_{QP} = qP \text{ to the scale of the diagram.}$$

Example 2.5 The simple offset circular cam of 45 mm radius and a 10 mm eccentricity rotates at 1400 rev/min. When the line OC makes an angle of 60° as shown in Fig. 2.14, calculate the velocity of the follower F and the rubbing velocity between the follower and the cam.

SOLUTION Let P be the contact point on the cam surface and Q the coincident point on the follower. Before we can draw the velocity diagram for the position shown in Fig. 2.14 we need the velocity V_p of P. If ω is the angular velocity of the cam then

$$V_p = \omega\overline{OP} = \omega y$$

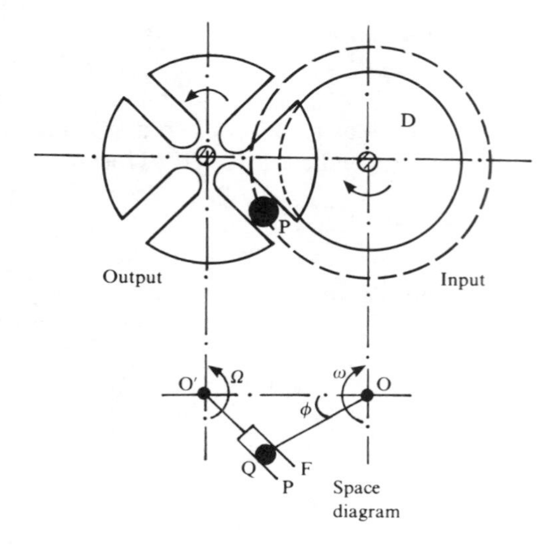

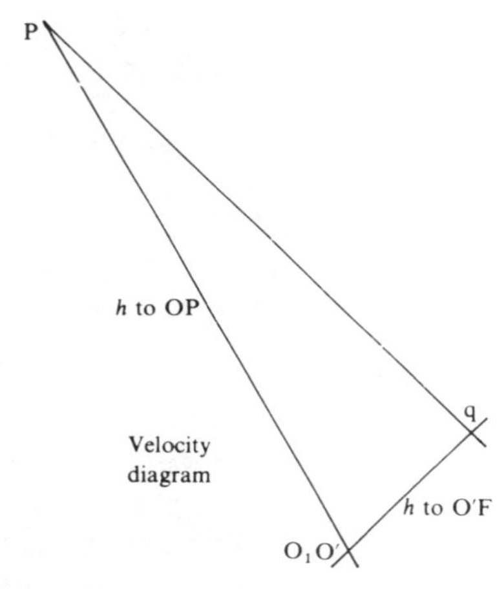

Figure 2.13 Geneva mechanism: velocity diagram.

From the triangle OCP we have

$$R^2 = e^2 + y^2 - 2ey\cos\theta$$

By the cosine rule, rearranging we have

$$y^2 - 2e\cos\theta\ y - (R^2 - e^2) = 0$$

and solving for y yields

$$y = e\cos\theta \pm \sqrt{e^2\cos^2\theta + (R^2 - e^2)}$$

Substituting numerical values we find

$$y = 10\cos 60 + \sqrt{100\cos^2 60 + (45^2 - 10^2)}$$

Hence

$$y = 49.16\ \text{mm}$$

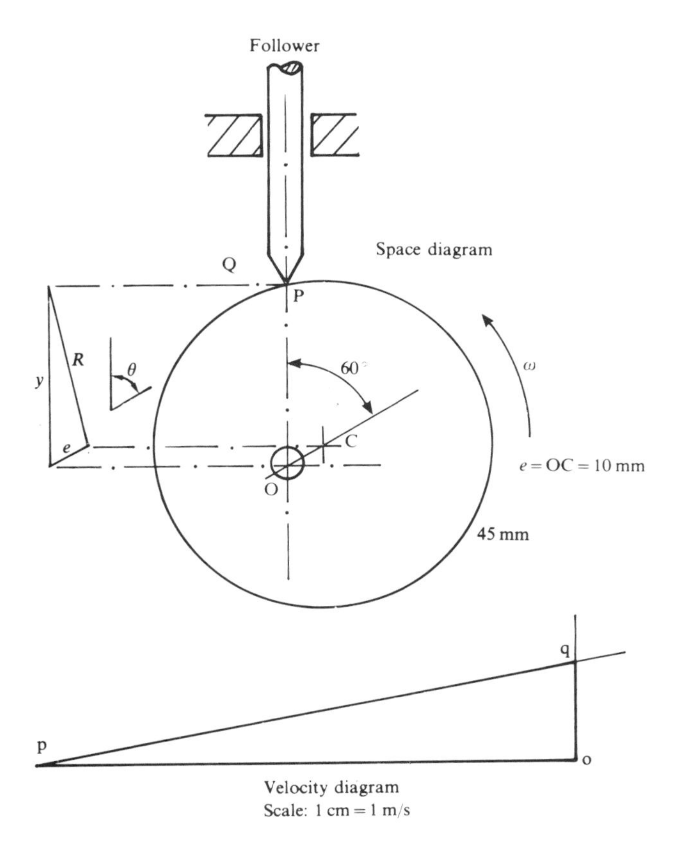

Figure 2.14

Also

$$\omega = \frac{2\pi \cdot 1400}{60} = 146.6 \text{ rad/s}$$

therefore

$$V_{\text{p}} = 146.6 \times 49.16/1000 = 7.21 \text{ m/s}$$

From the pole o drawn op $= V_{\text{p}}$ perpendicular to OP to a suitable scale. Since the follower is constrained to move along the line OP draw a line through o parallel to OP. The point Q on the follower has a velocity relative to P which must be along the tangent to the cam profile at P. Hence through p draw a line parallel to this tangent, i.e., perpendicular to PC, and locate the point q as shown.

By scaling off the velocity diagram we find

$$V_{\text{Q}} = \text{oq} = \underline{1.35 \text{ m/s}}$$

and the rubbing velocity

$$V_{\text{QP}} = \text{pq} = \underline{7.3 \text{ m/s}}$$

2.4 ROTATED VECTORS (ORTHOGONAL VECTORS)

The method is based on rotating the velocity vectors of points of a mechanism through an angle of 90°, the direction of rotation being arbitrary. It does not require a separate diagram such as the velocity diagram, and the figure obtained by joining the heads of the arrows representing the rotated vectors of the points is similar to the figure formed by joining those points in the actual mechanism.

Figure 2.15 shows a body moving in the plane and three points A, B, and C within it whose velocities are $\mathbf{V}_A$, $\mathbf{V}_B$, and $\mathbf{V}_C$ respectively.

If we now rotate each vector through 90° as shown then the lines representing the directions of these rotated vectors all pass through the instantaneous centre I. These vectors are denoted by $\overline{\mathbf{V}_A}$, $\overline{\mathbf{V}_B}$, and $\overline{\mathbf{V}_C}$.
Hence

$$\overline{\mathbf{V}_A}\,\overline{\mathbf{V}_B} \text{ is parallel to AB}$$
$$\overline{\mathbf{V}_A}\,\overline{\mathbf{V}_C} \text{ is parallel to AC}$$
$$\overline{\mathbf{V}_C}\,\overline{\mathbf{V}_B} \text{ is parallel to CB}$$

In the triangle $\overline{\mathbf{V}_A}\,\overline{\mathbf{V}_B}\,\overline{\mathbf{V}_C}$ the angles are equal and the lengths proportional to those of the triangle ABC. The triangles are homothetic.

In a particular case, suppose that the velocity $\mathbf{V}_A$ of a point A is known in magnitude and direction, and that the velocity of another point B is known in direction only; then to obtain its magnitude we proceed in the following way.

In Figure 2.16 let A and B be the two points, $\mathbf{V}_A$ be the velocity of A, and xx represent the direction of the velocity $\mathbf{V}_B$ of B. Rotate the vector $\mathbf{V}_A$ through 90° so that it lies along the line

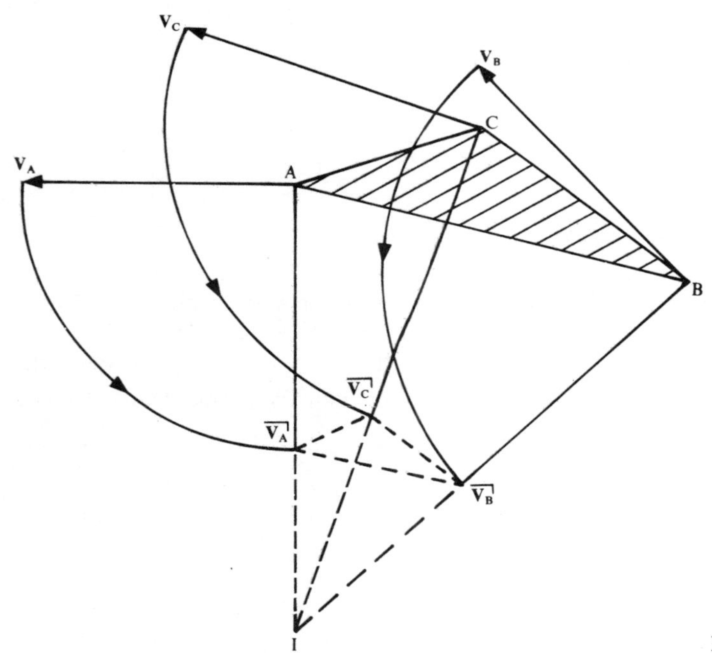

Figure 2.15 Orthogonal or rotated vectors.

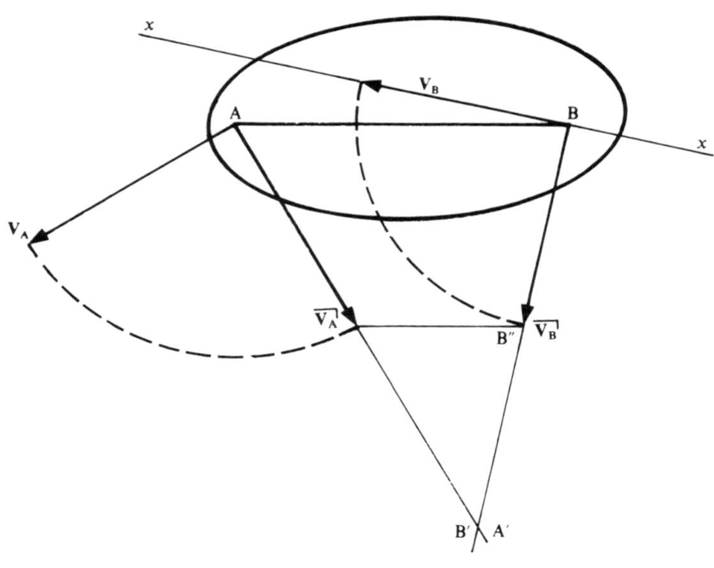

Figure 2.16 Determination of the velocity of a point in a body by means of the rotated-vectors method.

AA′; then draw a line BB′ at 90° to xx. From the tip of the rotated vector $\overline{\mathbf{V}_A}$ draw a line parallel to AB to cut the line BB′ at B″; then $BB'' = \overline{\mathbf{V}_B}$ is the rotated vector for the velocity V_B of B.

Example 2.6 For the six-bar linkage (Stephenson type 1) shown in Fig. 2.17 calculate the angular velocity of the output link DD_0 in magnitude and direction. The input link A_0A rotates at 25 rad/s in the clockwise direction. Dimensions: $A_0B_0 = 40$ mm; $A_0A = 50$ mm; $AB = 75$ mm; $AC = 150$ mm; $CB = 80$ mm; $B_0B = 60$ mm; $CD = 120$ mm; $DD_0 = 80$ mm; $B_0E = 85$ mm; $ED_0 = 15$ mm; $\phi = 70°$.

Solution The linkage is drawn to a suitable scale so that it will occupy as much as possible of the space available on a given size of paper.
Velocity of A = angular velocity of link $\omega \times \overline{AA_0}$
Hence

$$V_A = 25 \times 0.05 = 1.25 \text{ m/s}$$

This vector is drawn to a convenient scale as shown and then rotated through 90° in the direction indicated (arbitrary choice) so that it lies on AA_0, then $AA' = \overline{\mathbf{V}_A}$, the rotated vector $\mathbf{V}_A$. From A′ draw one line parallel to AB to cut B_0B at B′, and another line parallel to AD. A line issued from B′ parallel to BC is drawn to meet the line parallel to AC at C′. From this construction we have $BB' = \overline{\mathbf{V}_B}$ and $CC' = \overline{\mathbf{V}_C}$ and all these lines extended would meet at I_{AB}, the instant centre for the link ABC. From C′ draw a line parallel to CD to cut DD_0 at D′, then $DD' = \overline{\mathbf{V}_D}$. The lines CC′ and DD′ extended would meet at I_{CD}, this instant centre for the link CD. Locating these centres was, however, not required for the construction of the rotated vectors. $\overline{\mathbf{V}_D}$ is the rotated vector for the velocity $\mathbf{V}_D$ of the point D, and by rotating $\overline{\mathbf{V}_D}$ through 90° in an anticlockwise direction, i.e., in a direction opposite to the initial rotation of $\mathbf{V}_A$, we find the velocity $\mathbf{V}_D$ of D, showing that the link DD_0 rotates in a

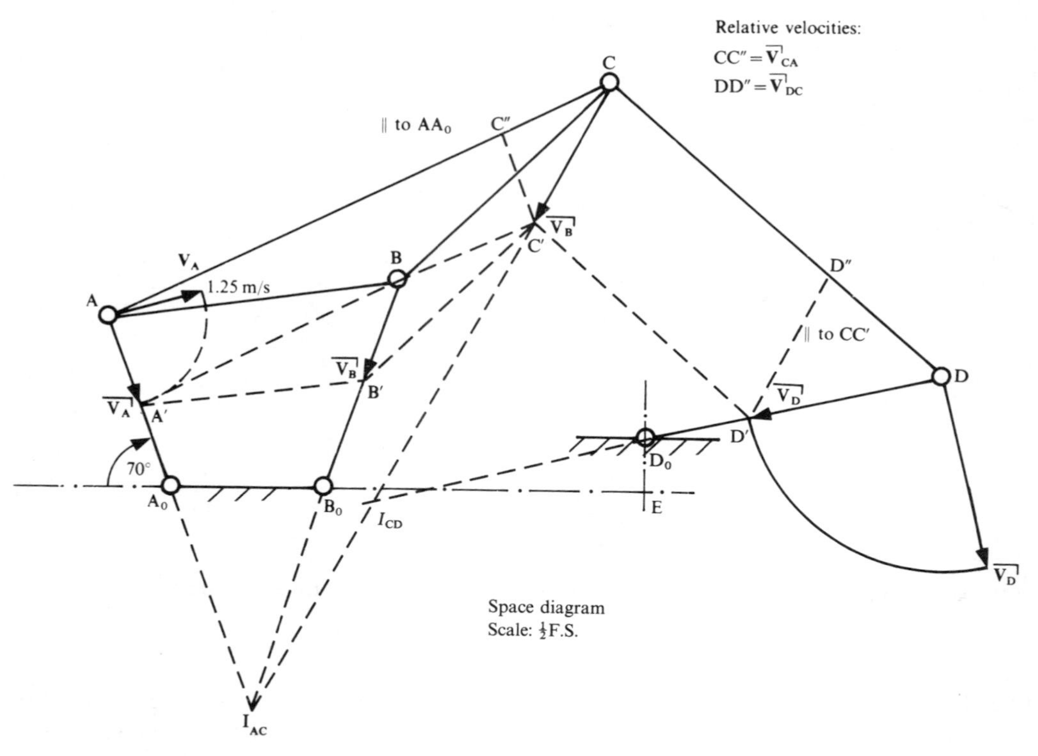

Figure 2.17

clockwise direction. To obtain the magnitude of the velocity of D we simply measure the length DD′ to the scale of the diagram and find

$$DD' = V_D = 2.6 \text{ m/s}$$

(2.6 from 4–6 bar computer programm; see 5.11.5.)

The angular velocity ω_0 of the output link DD_0 is given by

$$\omega_0 = V_D/DD_0 = 2.6/0.08 = \underline{32 \text{ rad/s clockwise}}$$

(32 rad/s from 4–6 bar computer program)

If we require the velocities of the points B and C, all we have to do is to measure the lengths $BB' = \mathbf{V}'_B$ and $CC' = \mathbf{V}'_C$ on the diagram and get

$$\mathbf{V}_B = 1.4 \text{ m/s} \qquad \text{and} \qquad \mathbf{V}_C = 2.2 \text{ m/s}$$

(1.35 m/s and 2.17 m/s from the 4–6 bar computer program)

Also shown on the diagram are the rotated vectors for the relative velocities:

$$\mathbf{V}_{CA} \qquad \text{and} \qquad \mathbf{V}_{DC}$$

Readers will realize that the advantage of this method lies in the fact that a separate velocity diagram is not necessary to calculate the velocities of points in a mechanism, and that, except in special cases, locating the instant centres is not needed.

2.5 LOCATION OF ALL INSTANTANEOUS CENTRES: ARONHOLD–KENNEDY'S THEOREM

We saw in Sec. 2.2 that any link in a mechanism can be considered as rotating about an instantaneous centre at a particular instant in time, e.g., the slider-crank mechanism in Fig. 2.4 and the four-bar linkage in Fig. 2.6. In the case of the latter we found two instant centres which enabled us to obtain the angular velocities of the coupler and the output link. These are not, however, all the instant centres. The instant centre I_{AB} relates the motion of the coupler AB to the fixed frame A_0B_0, while I relates the relative motion between A_0A and B_0B, but link A_0A is also moving relative to A_0B_0 and BB_0 is moving relative to A_0B_0. Hence there are instant centres at A_0 and B_0. Similarly, AB is moving relative to A_0A and relative to BB_0, therefore A and B are also instant centres. Thus we see that in the case of the four-bar linkage there are six instant centres, four of which are 'primary' ones, i.e., at A_0, A, B, and B_0. It is usual to adopt a convention for all instant centres as shown in Fig. 2.18 in the case of the four-bar linkage. The links are denoted by the numbers 1, 2, 3 and 4 and the instant centres by 12, 23, etc, the lower figure first. An examination of the figure shows that three instant centres relating the relative motion of three links lie on a straight line. This follows from Aronhold–Kennedy's theorem:

Theorem When three links, or bodies, are in relative motion, the three instant centres of relative motion lie on the same straight line. (The three links may or may not be connected, e.g., links 1, 3 and 4.)

SOLUTION Consider the three bodies represented by links 1, 3 and 4 (although link 4 is shown grounded it need not be so) in Fig. 2.19. The links may have any shape whatsoever.

The locations of the instant centres 14 and 34 are known. We now need to prove that the instant centre 13 lies on the line joining 14 and 34. Let us assume that P representing

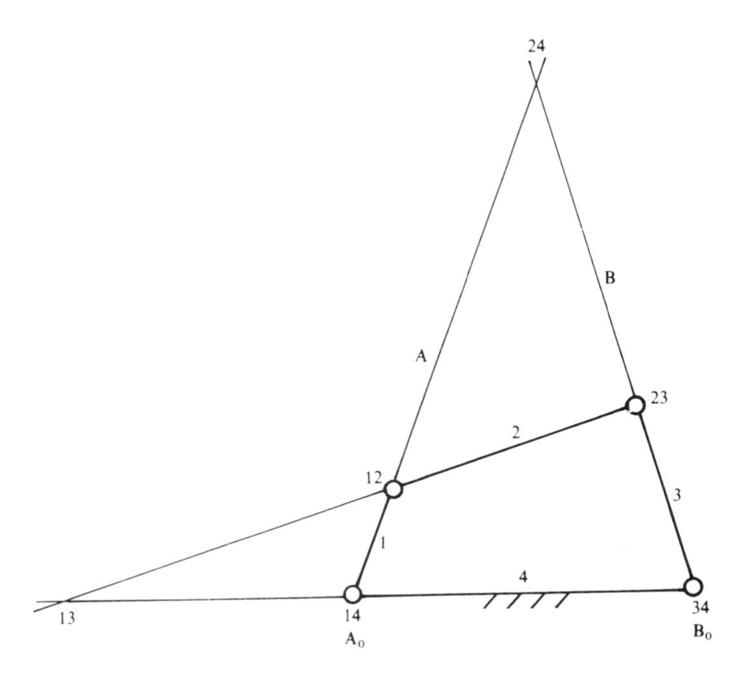

Figure 2.18 The six instantaneous centres of the four-bar linkage.

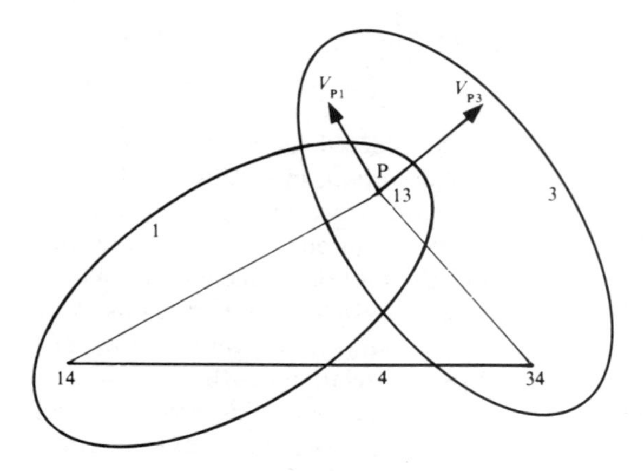

Figure 2.19 Illustrating the Aronhold–Kennedy theorem of the three centres.

the instant centre 13 is as shown. If P is considered as a point on link 1 its velocity would be V_{P1} in the direction indicated and perpendicular to the line 14–P, but the velocity of P as a point on link 3 would have the direction of V_{P3} at right angles to the line 34–P. This is a contradiction to the fact that the instant centre for two links relative to a third must

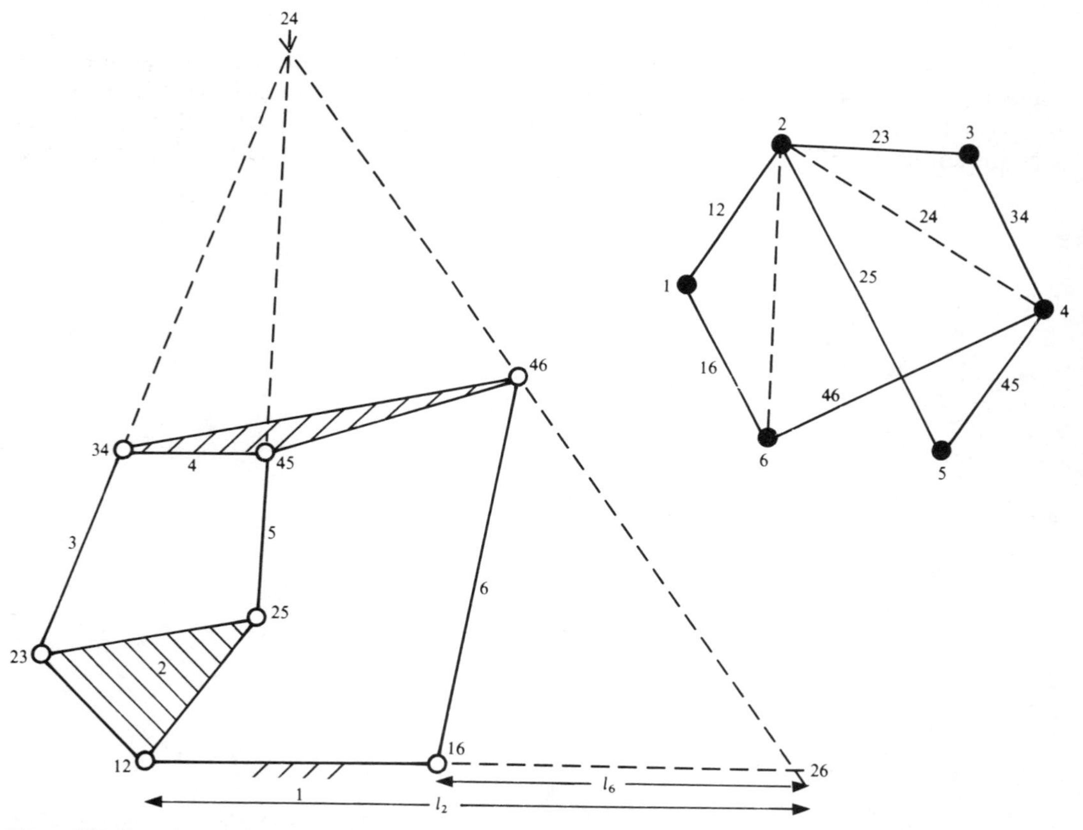

Figure 2.20 Location of all instant centres for a typical linkage.

be the same whether it is considered as a point on the first link or as one on the second. Therefore the velocity of P cannot be perpendicular to both 14–P and 34–P unless it lies on the line 14–34. Hence 14–13–34 lie on the same straight line.

This theorem is particularly useful in enabling us to find instant centres when the linkage is more complex.

For example, consider the linkage shown in Fig. 2.20, in which link 2 is the input and rotating at ω_2 and link 6 the output whose velocity ω_6 is to be found by locating the instant centres. In this case we require the instant centre 26 which relates the relative motion of links 2 and 6. Before proceeding we should note that in any linkage the number of instant centres N corresponds to the number of combinations of n things taken two at a time, i.e.,

$$N = {}^nC_2 = \frac{n(n-1)}{2}$$

Hence in our case since $n=6$, it follows that

$$N = \frac{6 \times 5}{2} = 15$$

Although there are 15 instant centres, we only require two apart from the seven 'primary' ones.

To locate the instant centres it is easier to use a polygon as shown in Fig. 2.20. The dots represent the links and the lines the primary instant centres, and the way we use the polygon is as follows.

As we need to locate the instant centre 24, then from the polygon we see that 24 will lie on the lines joining 23–34 and 25–45. Next 26 must lie on the lines joining 12–16 and 24–46. The two instant centres we require are shown in Fig. 2.20. Since 26 is the instant centre between links 2 and 6 it follows that by definition of an instant centre we must have

$$(12-26)\omega_2 - (16-26)\omega_6 = 0$$

hence

$$\omega_6 = \frac{(12-26)}{(16-26)}\omega_2$$

or

$$\omega_6 = \frac{l_2}{l_6}\omega_2$$

where $l_2 = (12-26)$ and $l_6 = (16-26)$.

EXERCISES

2.1 Figure 2.21 shows the slider–crank mechanism with the slider B driving. Using the data shown, calculate graphically the angular velocity of the crank OA. Check your answers analytically.

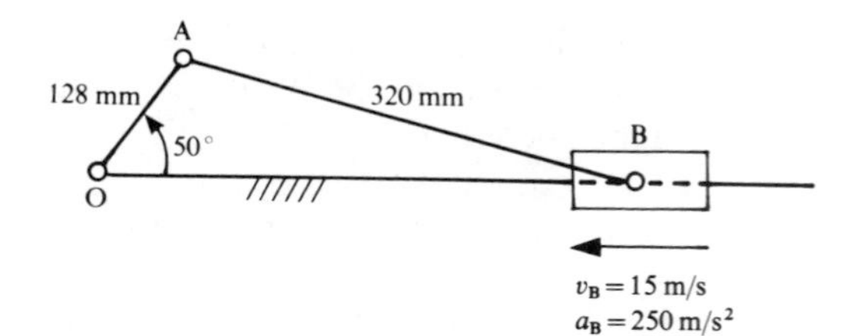

Figure 2.21

2.2 Figure 2.22 shows a crane the dimensions of whose links are: DD′ = 8.6 m; D′C = 7.2 m; BC = 20 m; AB = 8.6 m; AE = 30.0 m; AD = 26 m. Using the instantaneous-centre method, calculate the horizontal velocity of the point E on the jib EAB in the two extreme positions of the crane. The input link BC has an angular velocity of 0.08 rad/s.

Check your answers by drawing the velocity diagrams corresponding to these two positions.

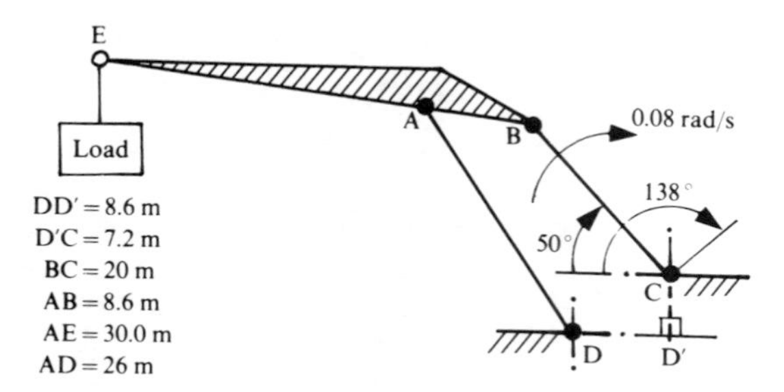

Figure 2.22

2.3 Figure 2.23 shows the offset nose wheel of an airliner. A, D, and E are pivots fixed to the frame of the aircraft. C is a pivot fixed to the undercarriage leg and F is a pivot fixed to the nose wheel door. For the position where the nose wheel makes an angle of 30° with the vertical and the link AB rotates at 0.25 rad/s in the direction shown, calculate graphically the velocity of G, the angular velocity of the leg DG, and the angular velocity of the door EF. The dimensions are AB = 432 mm; BC = 229 mm; CF = 330 mm; EF = 229 mm; DH = 254 mm; CH = 63.5 mm (perpendicular to DH); DG = 990 mm; A is 584 mm to the left of D and 229 mm above D; E is 127 mm to the right of D and 63.5 mm below D.

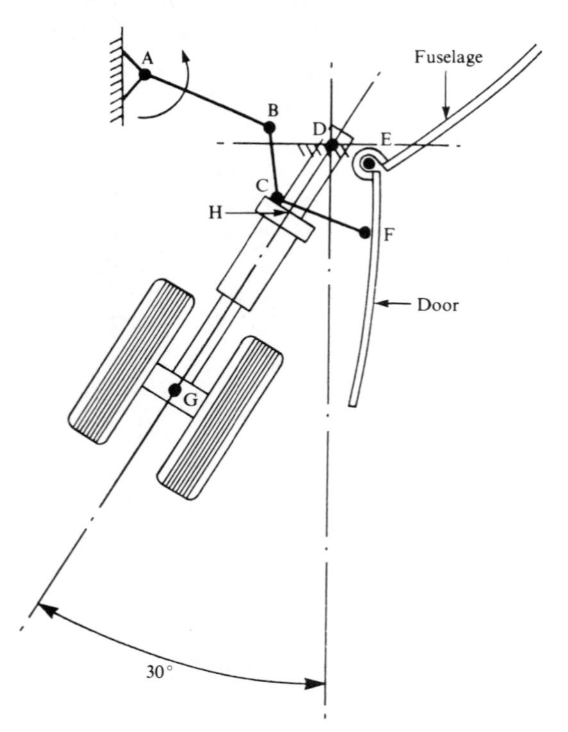

Figure 2.23

2.4 Link AB, the input, of the mechanism shown in Fig. 2.24 rotates at 20 rad/s. There is a resisting torque of 40 Nm at the output shaft E fixed to link DE. Dimensions are: AB = 500 mm; BC = 350 mm; CD = 200 mm; DE = 250 mm; DF = 350 mm.

Calculate using the instantaneous-centres polygon method the velocity of the slider F and the angular velocity of the output link in magnitude and direction. Calculate also, by drawing, the torque required at the input. AB and DE make angles of 60° and 15° respectively with the horizontal.

Check your answers using the orthogonal (rotated-vectors) construction.

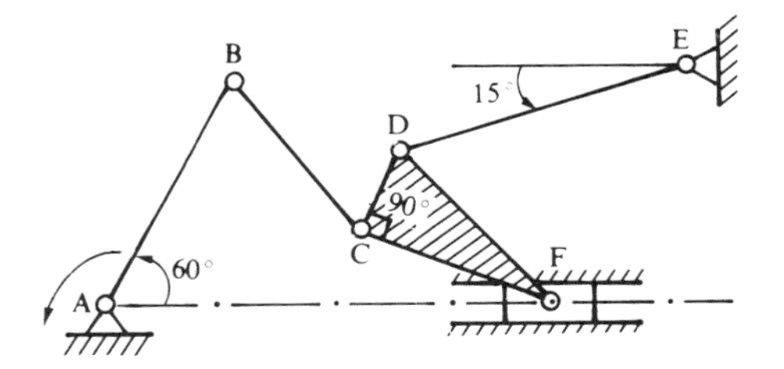

Figure 2.24

2.5 Figure 2.25 shows a mechanism (Stephenson type 1) with the following dimensions: $A_0B_0 = B_0B = AB = DE = 205$ mm; CD = 100 mm; AC = CB = 110 mm; $A_0A = 50$ mm. It produces an oscillatory output with a dwell when the input rotates at a constant speed. Calculate the velocities of B, C, and D in magnitude and direction when the crank makes an angle of 225° with A_0B_0 using

(a) rotated vectors;
(b) the instantaneous centres.

Calculate also the angular velocity of the output link ED. E is 50 mm to the right of B_0 and 150 mm above the datum line.

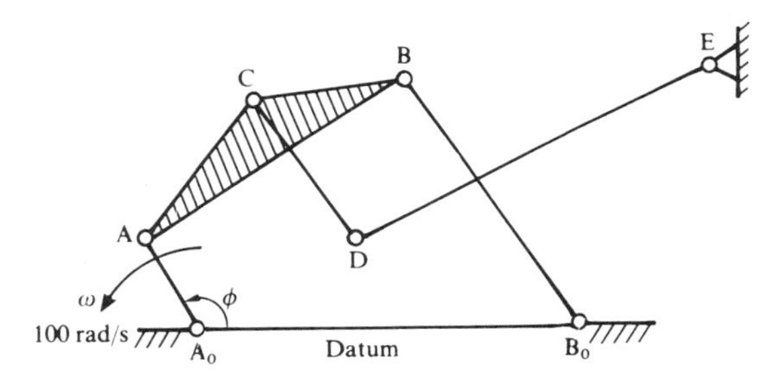

Figure 2.25

2.6 The dimensions of a stone-crusher mechanism are shown in Fig. 2.26. When the crank OA is in the position shown and rotating at the constant angular velocity of 75 rev/min, calculate the velocity of the point X.

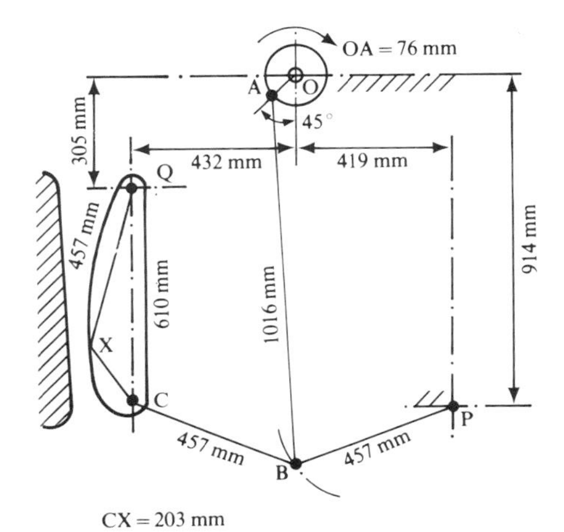

Figure 2.26

2.7 Figure 2.27 shows a type of quick-return mechanism for a slotting machine. The toothed sector gear drives the rack R which carries the tool (not shown). When the crank OP rotates at 65 rev/min, calculate the velocity of the rack R.

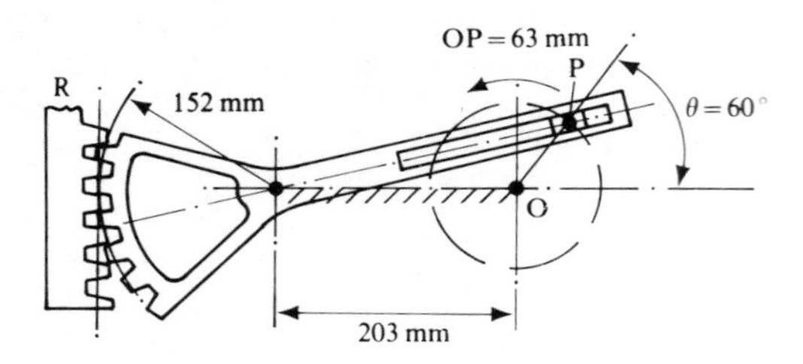

Figure 2.27

2.8 For the linkages shown in Fig. 2.28 calculate ω_6 by locating the instantaneous centres. All dimensions shown are in mm.

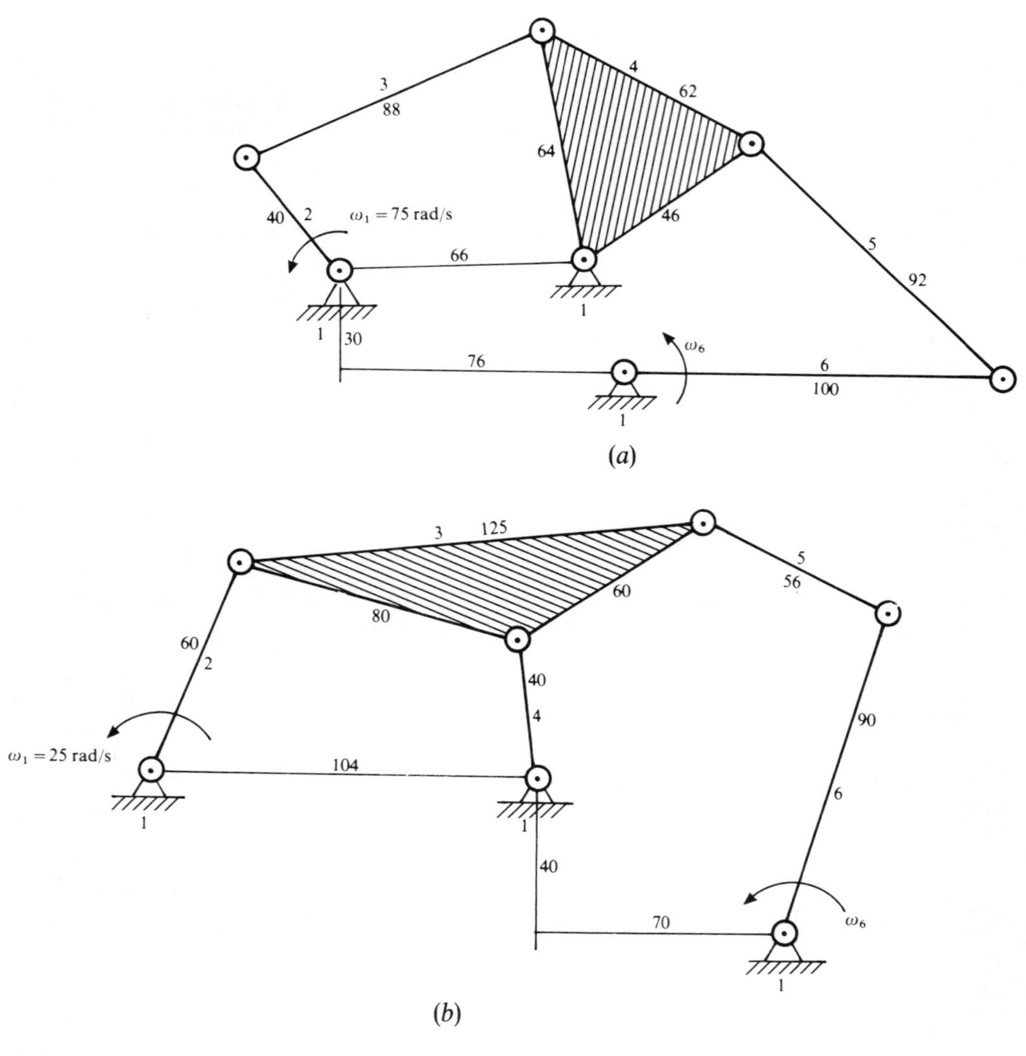

Figure 2.28

CHAPTER

THREE

GRAPHICAL ANALYSIS: ACCELERATIONS

3.1 ACCELERATION OF A POINT ON A ROTATING LINK: RELATIVE ACCELERATION

As a reminder consider the link A_0A rotating about a fixed axis through A_0 as shown in Fig. 3.1. Let ω and α be its angular velocity and angular acceleration respectively. Then we know from kinematics that the point A has two components of acceleration:

1. a tangential or transverse component $a_t = \alpha \overline{A_0A}$ perpendicular to A_0A and in the direction of α;
2. a radial or normal component $a_n = \omega^2 \overline{A_0A}$ directed towards A_0.

Thus in analysing a linkage there will be two such components of acceleration to be considered for each link.

Suppose now that the point A_0 has an acceleration $\mathbf{a}$; then the acceleration of the point A will be given by:

$$\mathbf{a}_A = \mathbf{a} + \text{acceleration of A relative to } A_0$$

i.e.

$$\mathbf{a}_A = \mathbf{a} + \mathbf{a}_{AA_0}$$

or

$$\mathbf{a}_A = \mathbf{a} + \mathbf{a}_{tAA_0} + \mathbf{a}_{nAA_0} \tag{3.1}$$

Equation (3.1) is a vector equation, since each term has magnitude as well as direction.

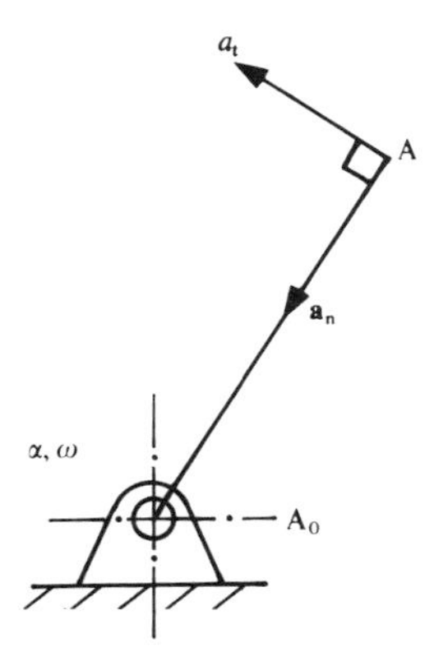

Figure 3.1 Rotating link: acceleration components of a point on the link.

3.2 ACCELERATION OF AN ELEMENT SLIDING ON A ROTATING LINK

Figure 3.2 shows a block sliding relative to the link A_0A which is in rotation. Let ω and $\dot{\omega}$ be the angular velocity and angular acceleration of the link A_0A, and θ its angular position relative to a datum line A_0C. Let **R** be the position vector of the block and **r** its distance from A_0 along the link. Then **R** can be expressed using the complex-number notation, so that

$$\mathbf{R} = \mathbf{r}e^{j\theta}$$

Differentiating once yields the velocity

$$\dot{\mathbf{R}} = (\dot{\mathbf{r}} + j\omega\mathbf{r})e^{j\theta} \tag{3.2}$$

where $\dot{\mathbf{r}}$ is the velocity of B relative to the link, i.e., the velocity of sliding.

Differentiating once more yields the acceleration

$$\ddot{\mathbf{R}} = [\underbrace{(\ddot{\mathbf{r}} - \mathbf{r}\omega^2)}_{\substack{\text{Radial}\\\text{component}\\\text{of the}\\\text{acceleration}}} + j\underbrace{(\mathbf{r}\dot{\omega} + 2\dot{\mathbf{r}}\omega)}_{\substack{\text{Transverse component}\\\text{or the acceleration}}}]e^{j\theta} \tag{3.3}$$

The term $2\dot{\mathbf{r}}\omega$ is the Coriolis component of acceleration, denoted by a_c. The direction of the Coriolis acceleration a_c is perpendicular to the relative velocity vector $\dot{\mathbf{r}}$ and in the direction of the rotation shown in Fig. 3.3.

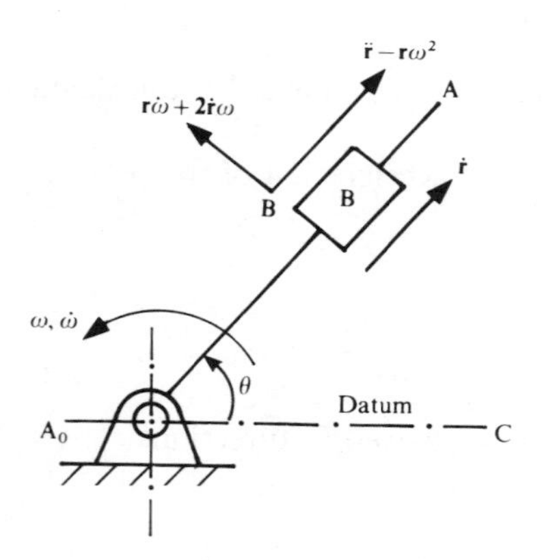

Figure 3.2 Link sliding on a rotating link: Coriolis acceleration.

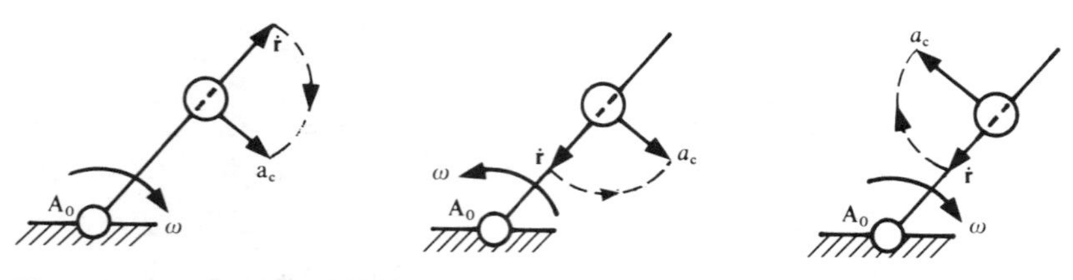

Figure 3.3 Locating the direction of the Coriolis acceleration.

3.3 ACCELERATION DIAGRAMS

3.3.1 Four-bar Linkage as a Typical Case

The linkage is shown in a particular position ϕ of the input in Fig. 3.4. Let ω_1 and α_1 be the angular velocity and angular acceleration respectively of link $A_0A = a$. ω_2 is the angular velocity of the coupler link $AB = b$, and ω_3 the angular velocity of the output link $B_0B = c$.

ω_2 and ω_3 are obtained from a velocity diagram or from an instant-centres diagram or from the rotated-vector method.

The point A has two known components of acceleration:

$$a_{At} = \alpha_1 a \text{ in the tangential direction}$$

and

$$a_{An} = \omega_1^2 a \text{ in the radial direction and towards } A_0$$

Similarly, the point B considered as a point on the link AB has accelerations given by

$$\left.\begin{aligned} a_{BAn} &= \omega_2^2 b \text{ in the radial direction and towards A} \\ a_{BAt} &= \alpha_2 b \text{ in the transverse direction} \end{aligned}\right\} \text{relative to A}$$

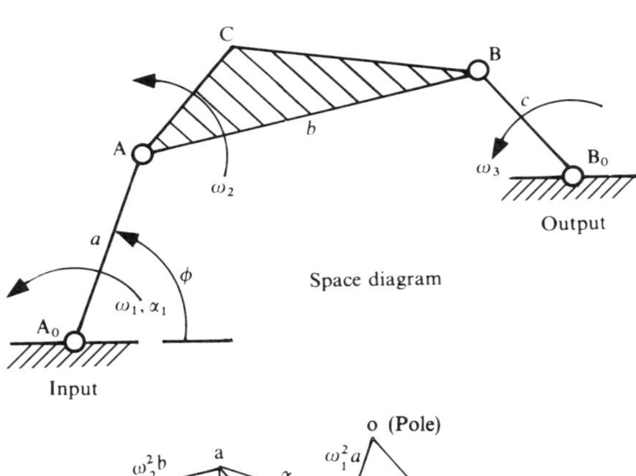

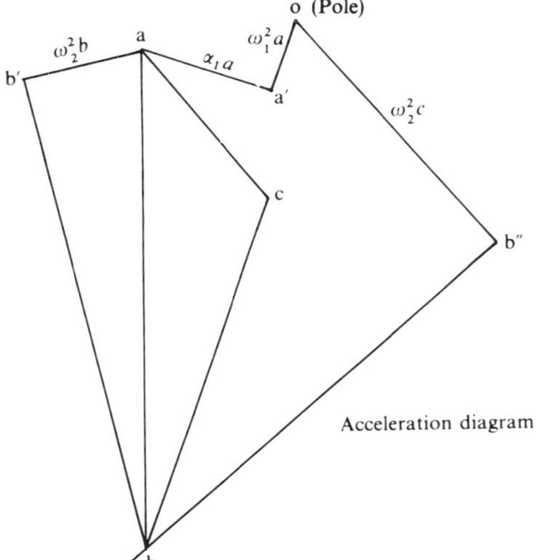

Figure 3.4 Construction of the acceleration diagram for a typical linkage.

α_2, the angular acceleration of the coupler AB, is as yet unknown. The accelerations of the point B considered as a point on the output link BB_0 are

$a_{Bn} = \omega_3^2 c$ in the radial direction and towards B_0

$a_{Bt} = \alpha_3 c$ in the transverse direction

where α_3, the angular acceleration of the output link B_0B, is also unknown at this stage.

The vector equation, Eq. (3.1), for the point B becomes

$$\mathbf{a}_B = \mathbf{a}_A + \mathbf{a}_{AB}$$

i.e.,

$$\mathbf{a}_{Bn} + \mathbf{a}_{Bt} = \mathbf{a}_{An} + \mathbf{a}_{At} + \mathbf{a}_{BAn} + \mathbf{a}_{BAt} \tag{3.4}$$

We will now solve this equation graphically by adopting the following procedure:

1. Starting from an arbitrary point o, known as a pole, draw the vector $oa' = a_{An} = \omega_1^2 a$ in the direction shown, i.e., from o to a′ to correspond to the fact that the radial direction is from A towards A_o;
2. From a′ draw the vector $a'a = a_{At} = \alpha_1 a$ to locate the point a; then oa = total acceleration of the point A on the linkage.
3. From a draw $ab' = a_{BAn} = \omega_2^2 b$ in a direction parallel to the line B towards A since the radial acceleration of B relative to A is towards A. This locates the point b′.
4. From b′ draw a line perpendicular to ab′, i.e., at right angles to AB; the point b must lie somewhere on this line.
5. From the pole o draw the vector $ob'' = a_{Bn} = \omega_3^2 c$ in a direction parallel to BB_0 and towards B_0 to locate the point b″.
6. From this point draw a line perpendicular to ob″; then b will lie somewhere on this line.
7. Finally, where the lines b′b and b″b meet the point b is located, and hence

$$\left.\begin{aligned} b'b &= a_{BAt} = \alpha_2 b \\ b''b &= a_{Bt} = \alpha_3 c \end{aligned}\right\} \quad \text{the two unknown components of the acceleration of B}$$

The angular acceleration of the output link B_0B is

$$\alpha_3 = \frac{bb''}{BB_0}$$

and the angular acceleration of the coupler AB is

$$\alpha_2 = \frac{bb'}{AB}$$

bb″ and bb′ are measured on the diagram and their values will depend on the scale used for the accelerations.

If C is any point on the coupler then the triangle abc on the acceleration diagram is the *acceleration image* of the triangle ABC on the space diagram.

Example 3.1. Figure 3.5 shows one of the four-bar linkages used in a shoe-stitching machine. For the position shown calculate the angular acceleration of the output link BB_0 when the input link A_0A rotates at the constant speed of 500 rev/min.

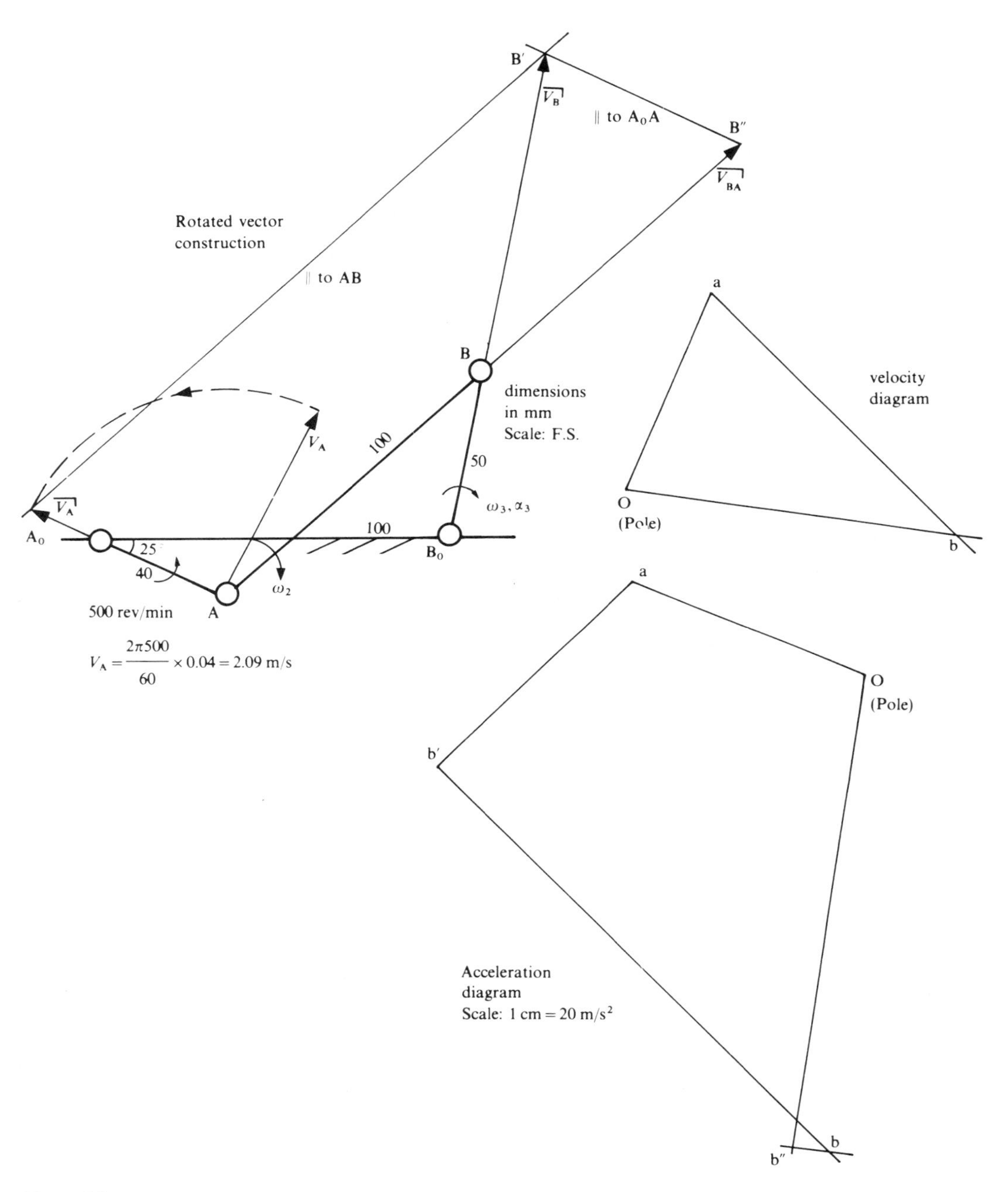

Figure 3.5

SOLUTION Before we can draw the acceleration diagram we need the angular velocity ω_2 of the coupler link AB and the angular velocity ω_3 of the output link B_0B.

To obtain these values we can use any one of the three methods discussed in Chapter 2. First we choose a scale for the space diagram. Next we calculate the velocity of the point A of the input link.

$$V_A = \omega_1 \overline{A_0A} = \frac{2\pi \cdot 500}{60} \times 0.04 = 2.09 \text{ m/s in the direction indicted}$$

The construction (Fig. 3.5) shows two of the methods for calculating the angular velocities of the coupler and the output link.

1. *Velocities.* (*a*) From the rotated vectors construction, we find

$$V_B = 3.28 \text{ m/s} \qquad V_{BA} = 3.45\ m/s$$
$$\omega_3 = BB'/B_0B = 3.28/0.05 = 65.6 \text{ rad/s}$$
$$\omega_2 = 3.45/0.1 = 34.5 \text{ rad/s}$$

(*b*) From the velocity diagram, we find

$$V_B = 3.28 \text{ m/s} \qquad V_{BA} = 3.48 \text{ m/s}$$

$$\omega_3 = 3.28/0.05 = 65.6 \text{ rad/s} \circlearrowright \quad (-66;\ \text{computer value})$$
$$\omega_2 = 3.48/0.1 = 34.8 \text{ rad/s} \circlearrowright \quad (-35;\ \text{computer value})$$

There is no significant difference between the two methods. Accuracy would be improved with a larger diagram.

Note: In this case the instant centre would be a long way off!

2. *Accelerations.* The known accelerations are:

$$a_{An} = \omega_1^2\ \overline{A_0A} = V_A^2/\overline{A_0A} = 2.09^2/0.04 = 109.2 \text{ m/s}^2 \nwarrow$$
$$a_{At} = 0,\ \text{since } \alpha_1 = 0$$
$$a_{Bn} = \omega_3^2\ \overline{B_0B} = 65.6^2 \times 0.05 = 215 \text{ m/s}^2 \swarrow$$
$$a_{BAn} = \omega_2^2\ \overline{AB} = 34.5^2 \times 0.1 = 119 \text{ m/s}^2 \swarrow$$

The acceleration is then drawn as shown after choosing a suitable scale.

$$oa = a_n = 109.2 \text{ m/s}^2$$
$$ab' = a_{BAn} = 119 \text{ m/s}^2$$
$$ob'' = a_{Bn} = 215 \text{ m/s}^2$$

Hence the output angular acceleration α_3 is given by

$$\alpha_3 = \frac{bb''}{B_0B} = \frac{13}{0.05} = \frac{260 \text{ rad/s}^2}{(247 \text{ rad/s}^2,\ \text{computer value})} \qquad \text{anticlockwise}$$

The angular acceleration α_2 of the coupler is given by

$$\alpha_2 = \frac{bb'}{AB} = \frac{234}{0.1} = \frac{2340 \text{ rad/s}^2}{(-2363 \text{ rad/s}^2,\ \text{computer value})} \qquad \text{anticlockwise}$$

3.3.2 Mechanisms with Sliders: Cam-Follower as a Typical Case

Mechanisms with sliders are quite common in practice, e.g., Geneva mechanism, quick-return mechanism, oscillating slider–crank mechanism, cams with sliding followers, etc. When analysing these mechanisms graphically we have to take into account the motion of the slider relative to the member on which it slides giving rise to a relative velocity component as well as a Coriolis acceleration. We have dealt with the relative velocity aspect in Sec. 2.3.3; in what follows we consider the effect of the Coriolis acceleration when drawing an acceleration diagram.

In these cases the vector equation, Eq. (3.1), for a point Q which is sliding relative to another point P becomes

$$\mathbf{a}_Q = \mathbf{a}_P + \mathbf{a}_{QP} + \mathbf{a}_c$$

where $\mathbf{a}_c$ is the Coriolis acceleration.

Each term of this equation, except $\mathbf{a}_c$, has two components as before, a tangential or transverse one and a radial or normal one; hence we have

$$\mathbf{a}_{Qn} + \mathbf{a}_{Qt} = \mathbf{a}_{Pn} + \mathbf{a}_{Pt} + \mathbf{a}_{QPn} + \mathbf{a}_{QPt} + \mathbf{a}_c \tag{3.5}$$

The normal components and the Coriolis component are always known in magnitude and direction and the tangential components in direction only. We are therefore left with two unknown numerical values which can be determined graphically by adopting a procedure similar to the one in Sec. 3.3.1.

Consider a cam having any shaped contour and a point follower as shown in Fig. 3.6. Let P be a point on the cam, Q a point on the follower coincident with P and R the radius of curvature, centre C, of the cam profile in that position. The follower slides in guides on a line

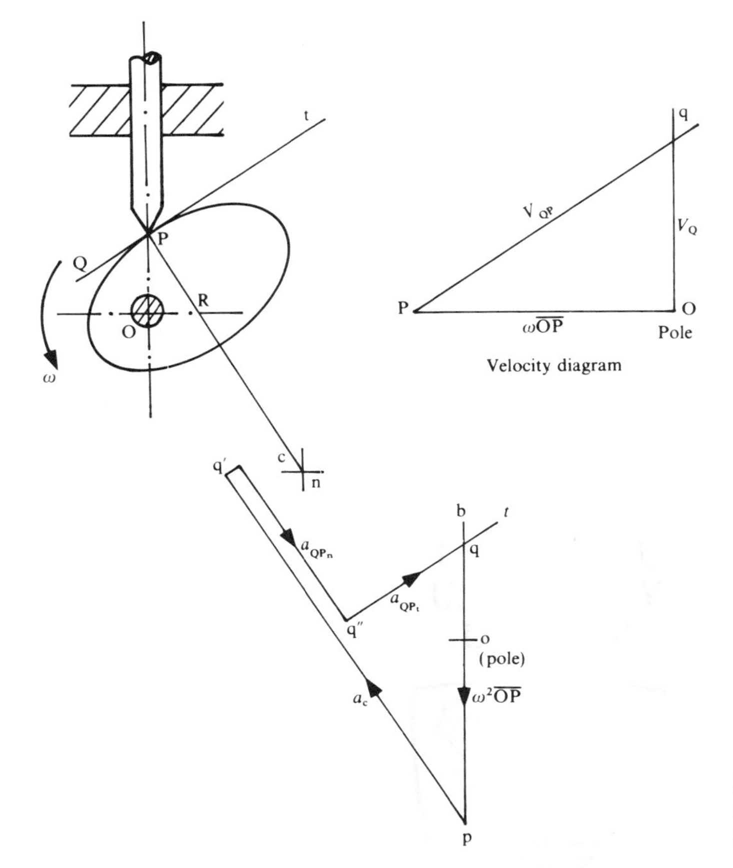

Figure 3.6 Acceleration diagram of a linkage (a cam in this case) with a sliding component.

that passes through the cam axis at O, and the cam rotates at a constant angular velocity ω in an anticlockwise direction.

The first step in our graphical solution for a given position of the cam is to draw the velocity diagram as discussed in Sec. 2.3.3 and shown in Fig. 3.6, where V_{QP} is the velocity of Q relative to P, i.e., the velocity of the follower relative to the cam in the direction indicated.

Before drawing the acceleration diagram we need to calculate the known acceleration components; they are:

$a_{Pn}=$ acceleration of P on the cam

$=\omega^2\overline{OP}$ directed from P to O

$a_{QPn}=$ acceleration of Q relative to P

$=V_{QP}^2/\overline{PC}=V_{QP}^2/R$ directed from Q to C, the centre of curvature of the cam profile at P.

$a_c=$ Coriolis acceleration

$=2\times$ relative velocity $\times$ angular velocity of the cam

$=2V_{QP}\omega$ in a direction perpendicular to the relative velocity and in the same sense as the angular velocity, as in Fig. 3.2

There are two unknown accelerations, namely a_{QPt}, the acceleration of Q relative to P in the tangential direction, and a_Q, the acceleration of Q, i.e., the value we are seeking.

We are now in a position to draw the acceleration diagram as follows:

1. Select a fixed point o, the pole, and draw op$=a_{Pn}$ in a direction parallel to P towards O on the space diagram.
2. From p draw pq$'=a_{c'}$.
3. From q$'$ draw qq$''=a_{QPn}$.
4. From q$''$ draw a line q$''$t perpendicular to PC to represent the unknown tangential acceleration of Q relative to P.
5. From o a line ob parallel to OQ is drawn to represent the absolute acceleration of the follower.
6. Where the lines drawn in (4) and (5) meet locate the point q and, therefore, the acceleration a_Q of Q is given by oq to the scale of the diagram.

If the cam had been accelerating at an angular acceleration α the only effect on the diagram would have been to add a term equal to $\alpha\cdot\overline{OP}$ from p in a direction perpendicular to OP, and in a sense defined by the sense of α.

Example 3.2 Calculate the acceleration of the follower of Example 2.5.

SOLUTION Referring to the solution for the velocity, we have:

$$y=49.16\text{ mm} \qquad V_p=7.21\text{ m/s} \qquad V_{QP}=7.3\text{ m/s} \qquad \omega=146.6\text{ rad/s}$$

Known accelerations are:

$$a_{Pn}=146.6^2\times 0.049\,16=1056\text{ m/s}^2\downarrow$$
$$a_c=2\times 7.21\times 146.6=2114\text{ m/s}^2\uparrow$$
$$a_{QPn}=7.3^2/0.045=1184\text{ m/s}^2\searrow$$

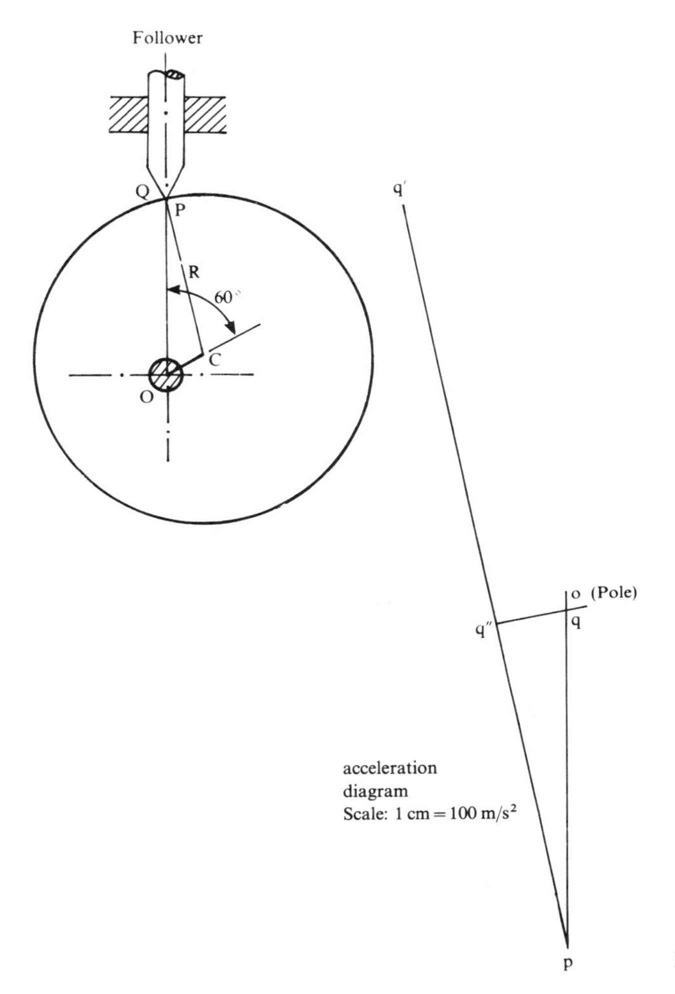

Figure 3.7

Following the above procedure and, hence, from the construction drawn to a scale of 1 cm = 100 m/s^2 (Fig. 3.7), we find $a_p = oq' = 110\,m/s^2$ downwards, i.e., the follower is slowing down.

3.3.3 Equivalent Mechanisms

The calculation of velocities and accelerations of certain mechanisms with contact surfaces, such as cams, can be simplified by replacing these surfaces by four-bar linkages that are kinematically equivalent to them at a particular instant.

Consider the two curved elements E_1 and E_2 shown in Fig. 3.8(*a*), used to transmit motion from shaft A to shaft D and making contact at the point P. Let B and C be the centres of curvature of E_1 and E_2 respectively for that particular position. The mechanism can then be replaced by a four-bar linkage. The original element E_1 is replaced by the link AB connecting the pivot A to the centre of curvature B; similarly, element E_2 is replaced by the link DC connecting D to the centre of curvature C. The coupler of the four-bar linkage connects the two centres of curvature B and C as shown in Fig. 3.8(*b*).

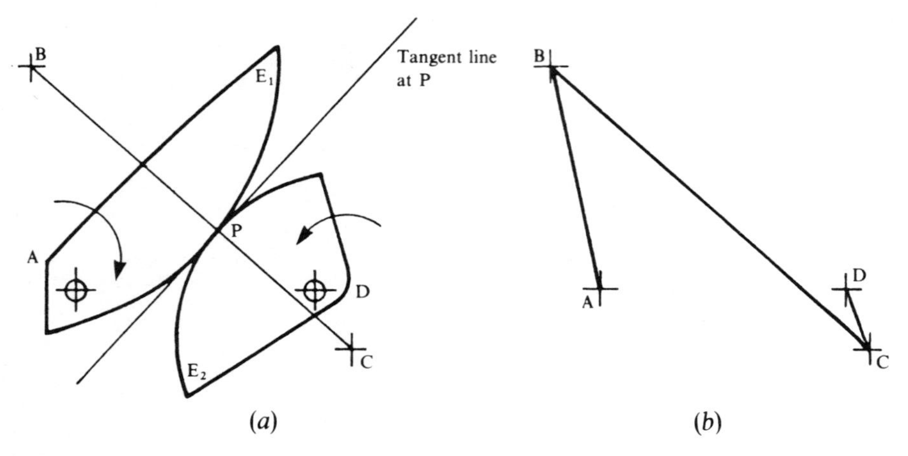

Figure 3.8 Replacing a mechanism by a four-bar linkage.

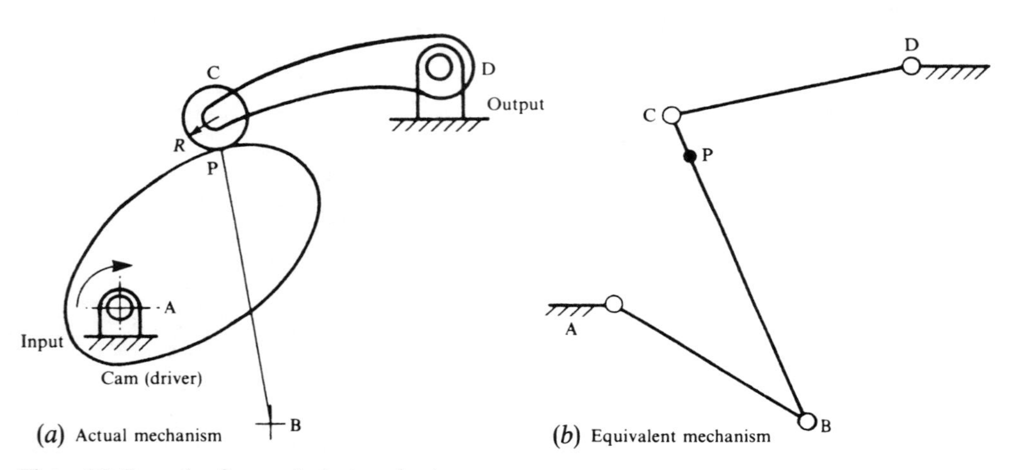

Figure 3.9 Example of an equivalent mechanism.

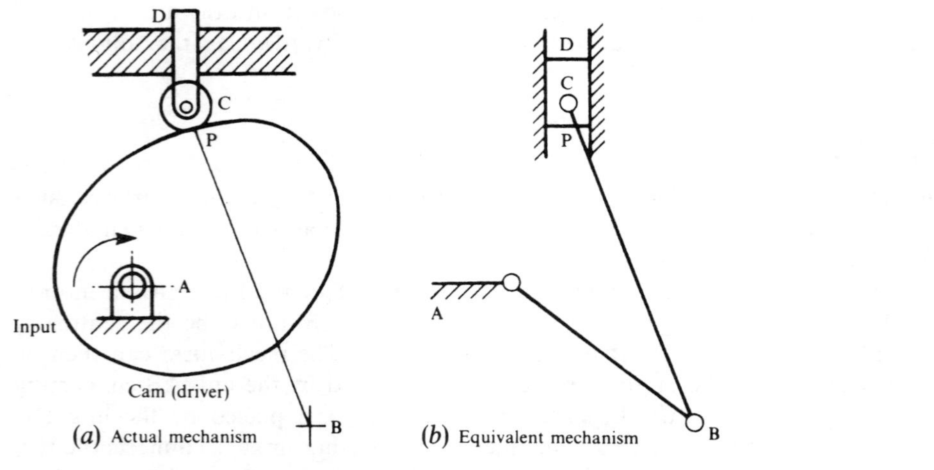

Figure 3.10 Another example of an equivalent mechanism.

It must be remembered that a little later, when the original mechanism has moved to another position, the lengths of the equivalent mechanism will have changed.

Figures 3.9 and 3.10 illustrate the way two different types of cam can be replaced by equivalent mechanisms that are easier to anlayse for a particular position. B is the centre of curvature of the cam at the point of contact P. In Fig. 3.10 the equivalent mechanism is a slider–crank mechanism.

3.4 CARTER–HALL CONSTRUCTION FOR VELOCITY AND ACCELERATION IN THE FOUR-BAR LINKAGE

3.4.1 Characteristic Derivatives

If ϕ is the input, e.g., an angle, of a mechanism and ψ is its output then it is often convenient to express the time derivatives as a function of the 'characteristic derivatives' by differentiating the output with respect to ϕ. There are three such characteristic derivatives that are of practical significance, they are:

$$\left.\begin{aligned} &\text{Characteristic velocity, } n_1=\frac{d\psi}{d\phi}=\psi' \\ &\text{Characteristic acceleration, } n_2=\frac{d^2\psi}{d\phi^2}=\psi'' \\ &\text{Characteristic 'jerk', } n_3=\frac{d^3\psi}{d\phi^3}=\psi''' \end{aligned}\right\} \tag{3.6}$$

The actual velocity, acceleration and jerk are then given by the following formulae:

Velocity,
$$\frac{d\psi}{dt}=\dot{\psi}=\frac{d\psi}{d\phi}\cdot\frac{d\phi}{dt}=n_1\dot{\phi} \tag{3.7}$$

In this case we note that

$$\frac{d\psi}{d\phi}=\frac{d\psi}{dt}\bigg/\frac{d\phi}{dt}=\frac{\omega_0}{\omega_1}=\frac{\text{output angular velocity}}{\text{input angular velocity}}$$

Acceleration,
$$\frac{d^2\psi}{dt^2}=\ddot{\psi}=\frac{d}{dt}(n_1\dot{\phi})=n_1\ddot{\phi}+n_2\dot{\phi}^2 \tag{3.8}$$

Jerk,
$$\frac{d^3\psi}{dt^3}=\dddot{\psi}=\frac{d}{dt}(n_1\ddot{\phi}+n_2\dot{\phi}^2)=n_1\dddot{\phi}+3n_2\dot{\phi}\ddot{\phi}+n_3\dot{\phi}^3 \tag{3.9}$$

If the input velocity $\dot{\phi}$ is constant, these become

$$\dot{\psi}=n_1\dot{\phi}, \qquad \ddot{\psi}=n_2\dot{\phi}^2 \qquad \dddot{\psi}=n_3\dot{\phi}^3 \tag{3.10}$$

The jerk is important in good-quality mechanisms; its value should remain finite. Very high values of the jerk are related to wear, vibration, noise, and bearing failure in some cases; fragile products being handled could also be seriously damaged as a result of high jerks. In the case of the high-speed cams, for example, it is important to avoid abrupt changes in profiles since these would result in n_2 being discontinuous and hence a jerk having an infinite value. (*Note:* Jerk is sometimes referred to as 'pulse'.) In the case of linkages (or equivalent mechanisms; see

Sec. 3.3.3) the characteristic derivatives n_1, n_2, and n_3 can be calculated for a particular position of the linkage by means of a simple graphical construction due to Freudenstein and Carter–Hall.

This construction can also be used to synthesize a four-bar linkage quickly and with reasonable accuracy.

3.4.2 Carter–Hall Construction

Given a particular four-bar linkage, it is possible to calculate quickly the characteristic derivatives of the function $\psi = f(\phi)$, or $\Delta\psi = f(\Delta\phi)$ in the case of finite displacements, generated by the linkage using the simple graphical method below.

The linkage A_0ABB_0 is shown in Fig. 3.11 where $A_0A = a$, $AB = b$, $BB_0 = c$, $A_0B_0 = d$, the fixed link, ϕ is the input angle and ψ the output angle. The lines AB and A_0B_0 meet at the point I_{13} which is the instant centre for the relative rotation between links 1 and 3. The lines A_0A and B_0B meet at I_{24}, which is also an instant centre, that of the rotation of link 2 relative to link 4.

The line joining I_{13} to I_{24} is called the *Pascal line* or *collineation axis*. The angle $\angle AI_{13}I_{24} = \lambda$, between the lines AI_{13} and $I_{13}I_{24}$ is taken positive as shown.

From I_{13} a line $I_{13}H$ is drawn to meet at H the line $I_{24}H$ drawn at 90° to the collineation axis; a circle is then drawn through the three points I_{13}, I_{24} and H of diameter $D_c = I_{13}H$, the diameter of the Carter–Hall circle. If we observe the sign convention shown in the figure the characteristic derivatives are obtained by measuring the following lengths:

$$A_0B_0 = d \quad \text{(always positive)}$$

$$I_{13}H = D_c$$

$$I_{13}A = R$$

and the angle λ.

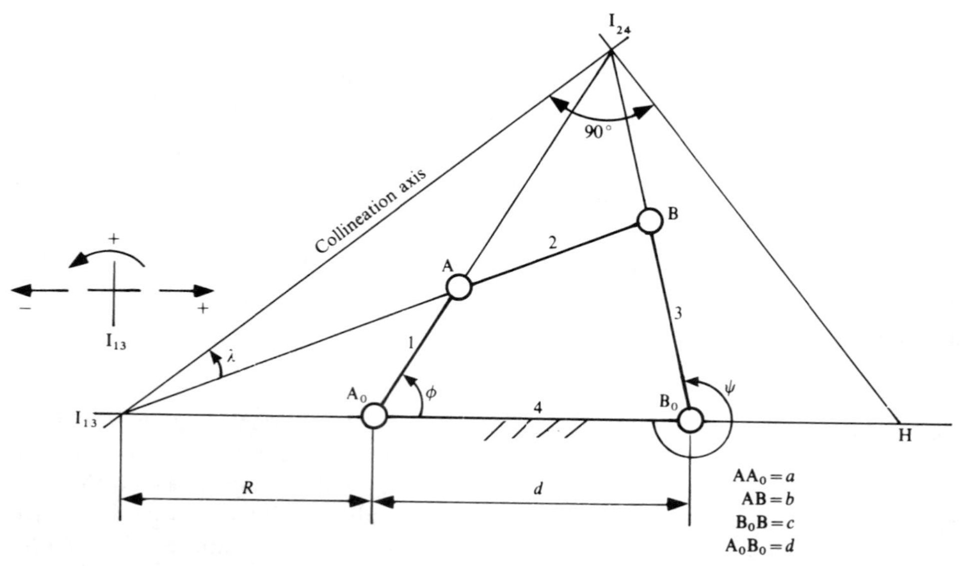

Figure 3.11 Collineation axis in the four-bar linkage.

From this construction it can be shown that:

$$n_1 = \frac{R}{R+d} \tag{3.11}$$

$$n_2 = n_1(1-n_1)/\tan \lambda \tag{3.12}$$

$$n_3 = [3n_1^2 d/D_c - n_1(1-n_1)(3n_1 + (1-2n_1)S^2)/S^2 \tag{3.13}$$

where $S = \sin \lambda$.

We now propose to prove Eqs (3.11) and (3.12) and ask the reader to accept Eq. (3.13), the proof of which is extremely long and complicated; it was derived by Nieto in Spain in 1976.

3.4.3 Proof that the First Derivative $n_1 = R/(R+d)$

Freudenstein's equation (see Sec. 5.2) relating the output ψ to the input ϕ, i.e., $\psi = f(\phi)$, is

$$K_1 \cos \phi + K_2 \cos \psi - K_3 = \cos(\phi - \psi) \tag{3.14}$$

or

$$K_1 \cos \phi - K_2 \cos \psi' - K_3 = -\cos(\phi - \psi') \tag{3.14a}$$

where

$$\psi' = \psi - 180^\circ$$
$$K_1 = d/c$$
$$K_2 = d/a$$
$$K_3 = (a^2 - b^2 + c^2 + d^2)/2ac$$

Differentiating Eq. (3.14a) with respect to ϕ and solving for $d\psi/d\phi = n_1$ yields

$$n_1 = \frac{a}{c} \cdot \frac{d \sin \phi - c \sin(\psi' - \phi)}{d \sin \psi' - a \sin(\psi' - \phi)} \tag{3.15}$$

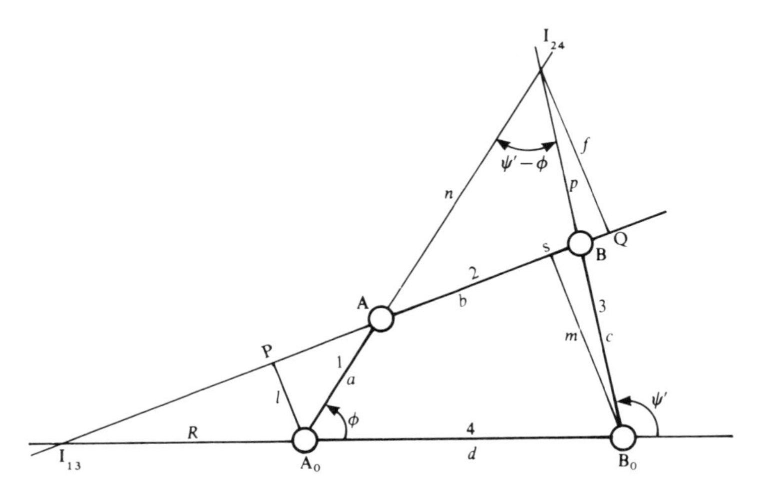

Figure 3.12 Determination of the frst derivative.

Referring to Fig. 3.12 we find:

1. From the triangle $A_0B_0I_{24}$,

$$\frac{p+c}{\sin\phi}=\frac{d}{\sin(\psi'-\phi)}$$

$$\frac{n+a}{\sin(180-\psi')}=\frac{\mathrm{d}}{\sin(\psi'-\phi)}$$

Solving for p and n and combining with eq. (3.15), we get

$$n_1=\frac{pa}{nc}$$

2. From triangles $AI_{24}Q$ and A_0PA,

$$\frac{l}{a}=\frac{f}{n}$$

3. From triangles B_0SB and BQI_{24},

$$\frac{m}{c}=\frac{f}{p}$$

Hence
$$\frac{p}{n}=\frac{l}{m}\cdot\frac{c}{a}$$

therefore
$$n_1=\frac{l}{m}$$

4. Finally from triangles $I_{13}A_0P$ and $I_{13}SB_0$,

$$\frac{l}{m}=\frac{R}{R+d}$$

It therefore follows that

$$n_1=\frac{R}{R+d}$$

This is identical to Eq. (2.4).

3.4.4 Proof that the Second Derivative $n_2=n_1(1-n_1)/\tan\lambda$

The following proof differs from that of Freudenstein of 1956. From Sec. 3.4.3 we know that

$$n_1=\frac{R}{R+d}$$

$$n_2=\frac{\mathrm{d}}{\mathrm{d}\phi}(n_1)=\frac{\mathrm{d}}{\mathrm{d}\phi}\left(\frac{R}{R+d}\right)$$

$$=\frac{\mathrm{d}}{\mathrm{d}R}\left(\frac{R}{R+d}\right)\frac{\mathrm{d}R}{\mathrm{d}\phi}$$

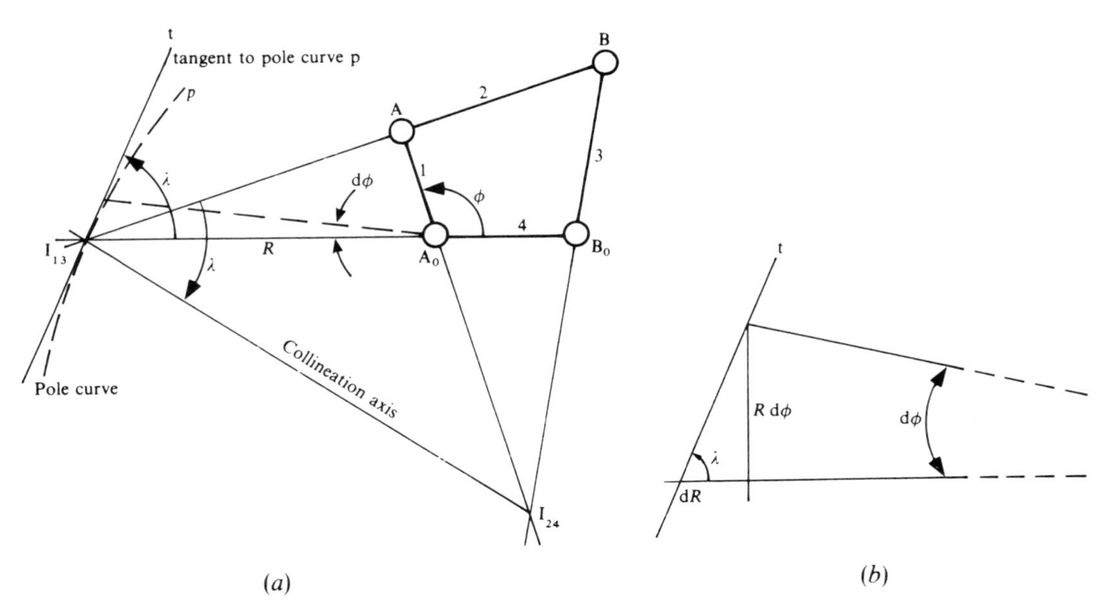

Figure 3.13 Determination of the second derivative.

$$n_2 = \frac{d}{(R+d)^2}\frac{\mathrm{d}R}{\mathrm{d}\phi} \tag{3.16}$$

In Fig. 3.13(*a*) p is the pole curve, i.e., the curve traced by I_{13} as ϕ varies, and t is the tangent to the pole curve. By a theorem due to Bobillier (1880), the angle $\angle A_0I_{13}t$ = the angle $\angle AI_{13}I_{24} = \lambda$. Let ϕ change by an amount $\mathrm{d}\phi$ and R by an amount $\mathrm{d}R$; then from Fig. 3.11(*b*) we have

$$\tan\lambda = \frac{R\cdot\mathrm{d}\phi}{\mathrm{d}R} \qquad \text{hence} \qquad \frac{\mathrm{d}R}{\mathrm{d}\phi} = \frac{R}{\tan\lambda}$$

Substituting in Eq. (3.16) yields

$$n_2 = \frac{\mathrm{R}d}{(R+d)^2}\Big/\tan\lambda$$

but

$$\frac{Rd}{(R+d)^2} = n_1\frac{d}{(R+d)} = n_1\left(1 - \frac{R}{R+d}\right) = n_1(1-n_1)$$

It therefore follows that

$$n_2 = n_1(1-n_1)/\tan\lambda$$

3.4.5 Sign Convention

When using the graphical method just described it is important to observe a sign convention. The six cases shown in Fig. 3.14 should meet those encountered in practice. The Carter–Hall circles have been drawn in two particular cases to illustrate the construction.

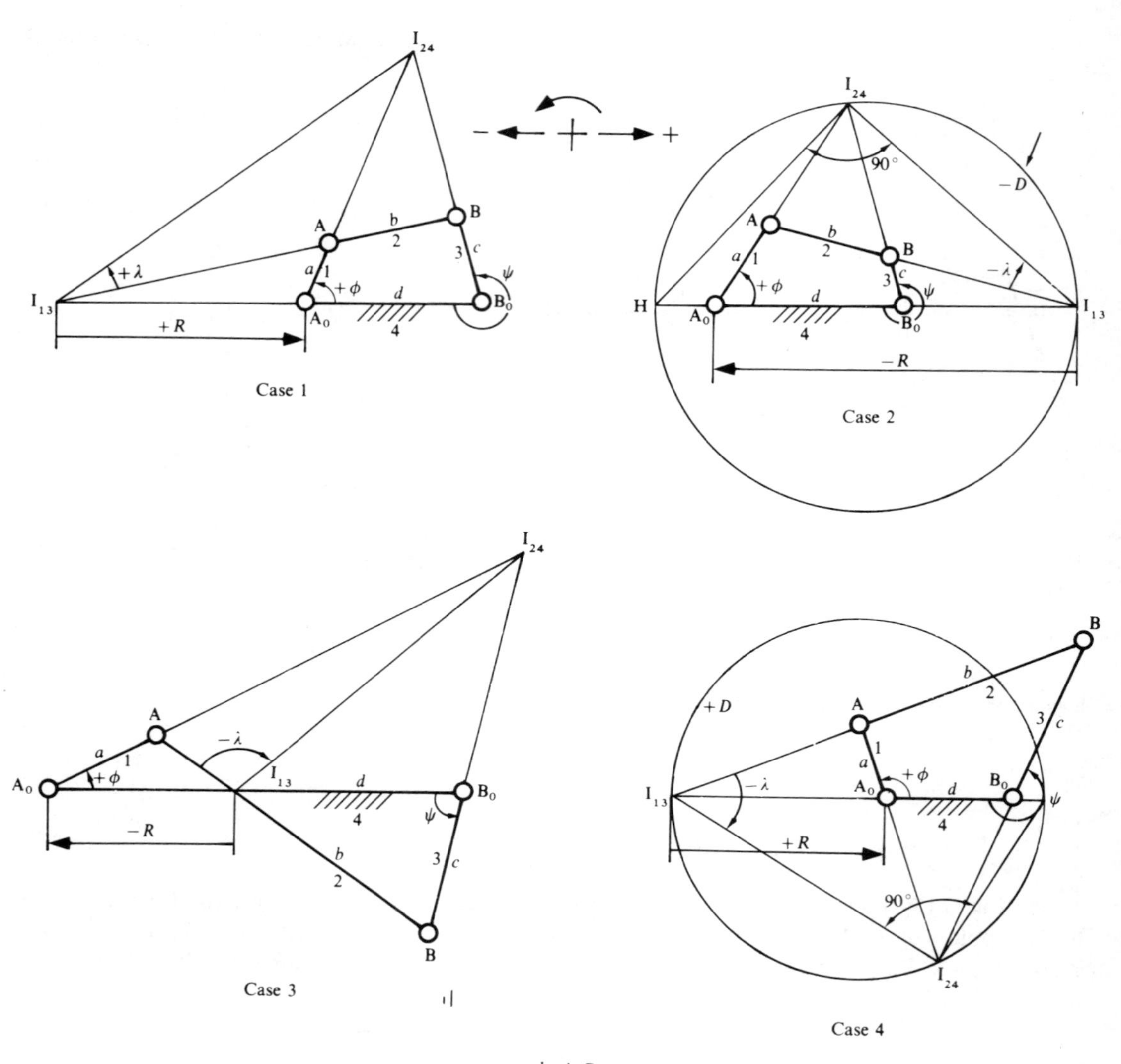

$d = A_0B_0$

Fixed link in all cases

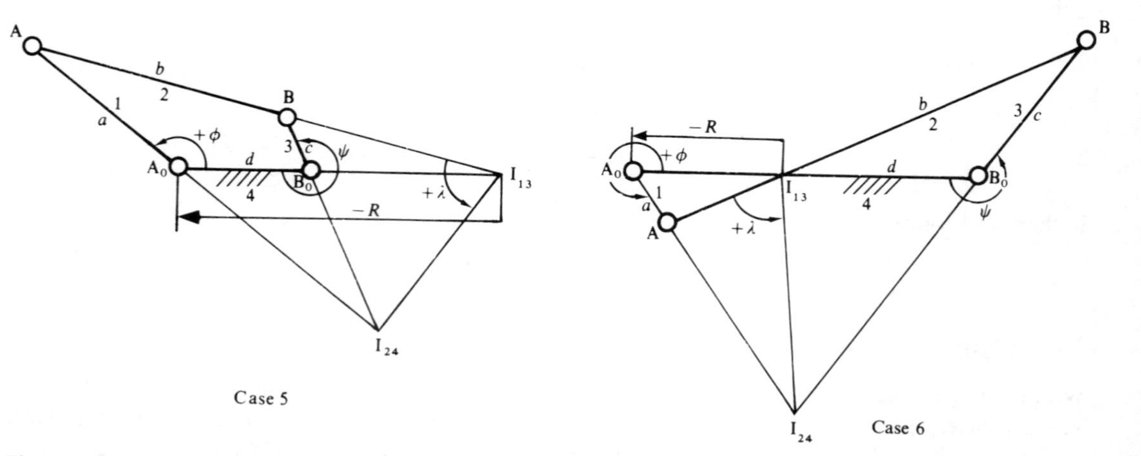

Figure 3.14 Carter–Hall construction: sign convention.

1. The angle λ is measured from the coupler line BAI_{13} towards the collineation axis $I_{13}I_{24}$ and *positive* in an anticlockwise direction.
2. $R = A_0I_{13}$ is *positive* as shown.
3. In all cases,

$$n_1 = \frac{R}{R+d};$$

d is always a positive number;

n_1 will be either positive or negative.

4. In all cases, $n_2 = n_1(1 - n_1)/\tan\lambda$, therefore n_2 will be either positive or negative.

Example 3.3 Calculate the velocity, acceleration, and jerk of the output link of the stitching machine (Example 3.1) using the Carter–Hall construction. $A_0A = 40$ mm, $AB = 100$ mm, $B_0B = 50$ mm, $A_0B_0 = 100$ mm, $\phi = -25°$.

SOLUTION

$$\omega_1 = \frac{2\pi \cdot 500}{60} = 52.36 \text{ rad/s}$$

The linkage is drawn full-size (Fig. 3.15) and the two instant centres I_{24} and I_{13} located. The collineation axis $I_{13}I_{24}$ is drawn and the angle λ measured, positive in this case. The distance $R = A_0I_{13}$ is read off, negative in this instance. To locate the point H for the Carter–Hall circle the line $I_{24}H$ is drawn at 90° to the collineation axis and the diameter of the circle measured, i.e., $D = I_{13}H$.

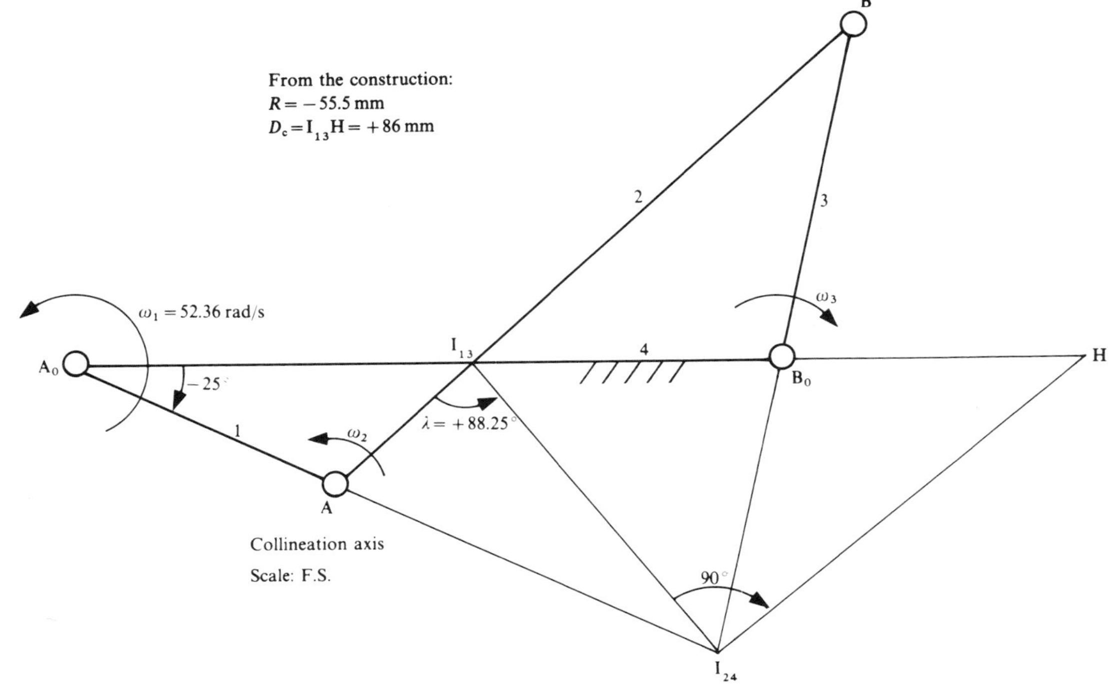

Figure 3.15

The characteristic derivatives are:

$$n_1 = \frac{R}{R+d} = \frac{-55.5}{-55.5+100} = -1.247 \qquad \text{(Eq. (3.11))}$$

$$n_2 = n_1(1-n_1)/\tan\lambda = 1.247(1+1.247)/\tan 88 = -0.0978 \qquad \text{(Eq. (3.12))}$$

$$n_3 = \{3n^2\, d/D_c - n_1(1-n_1)[3n_1 + (1-2n_1)\sin^2\lambda]\}/\sin^2\lambda \qquad \text{(Eq. (3.13))}$$

$$= \{3\times 1.247^2 \times 100/86 + 1.247\times 2.247[-3\times 1.247 + (1+2\times 1.247)\sin^2 88]\}/\sin^2 88$$

$$= 4.7262$$

Hence using (3.10) we find:

$$\text{Output angular velocity} = n_1\omega_1 = -1.247\times 52.36 = \underline{-65.3\text{ rad/s}}$$

$$\text{Output angular acceleration} = n_2\omega_1^2 = 0.0978\times 52.36^2 = \underline{-268\text{ rad/s}^2}$$

$$\text{Output angular jerk} = n_3\omega_1^3 = 4.7262\times 52.36^3 = \underline{6.8\times 10^5\text{rad/s}^3}$$

Note: Using the author's computer program the values obtained were:

-66 rad/s $\quad -247$ rad/s^2 and 6.9×10^5 rad/s^3 respectively

When the angle λ exceeds 80° great care must be taken in measuring it because the tangents of angles greater than 80° change very rapidly; e.g., tan 87.5 = 22.9, tan 88 = 28.636, and tan 88.5 = 38.188. If the linkage has been drawn to a small scale first and the angle λ when measured is 80° or greater, it may be wiser to redraw it to a larger scale in order to improve the accuracy in the measurement of λ.

Example 3.4 Figure 3.16 shows a six-bar linkage $A_0AB_0BCDD_0$ (Watt type 2) in which A_0B_0 and B_0D_0 are the fixed links. B_0BC is a ternary link with pin joints at B_0, B, and C. The input is at A_0 where a shaft rotates at 75 rad/s and, at the instant when $\phi = 100°$, is accelerating at 200 rad/s^2. Calculate the velocity, acceleration, and jerk of the output shaft at D_0 in magnitude and direction.

SOLUTION If ω_2, α_2, and J_2 are the angular velocity, acceleration, and jerk of the link B_0BC, then these values become the input to the four-bar linkage B_0CDD_0.

The first thing to do is to find the best scale so that the linkage and the necessary construction lines will fit within whatever size of paper is selected. The construction in Fig. 3.16 was drawn to a scale of $\frac{1}{10}$ FS on a sheet of A4. An A3 sheet would most likely have been better, nevertheless the accuracy achieved on the A4 size was quite acceptable. The values of λ, R and D_c are measured in accordance with our sign convention as shown.

1. *Linkage A_0ABB_0*

$$R = +390\text{ mm} \qquad D_c = I_{13}H = +1730\text{ mm} \qquad \lambda = -84°$$

Hence

$$n_1 = R/(R+d) = 390/(390+260) = 0.60$$

$$\omega_2 = n_1\omega_1 = 0.6\times 75 = \underline{45\text{ rad/s}} \qquad (44.7\text{ rad/s})$$

$$n_2 = n_1(1-n_1)/\tan\lambda = 0.6\times 0.4/\tan -84 = -0.025$$

$$\alpha_2 = n_1\ddot{\phi} + n_2\dot{\phi}^2 = 0.06\times 200 - 0.025\times 75^2$$

$$= \underline{20.6\text{ rad/s}^2} \qquad \text{i.e. } 20.6\text{ rad/s}^2 \circlearrowright \quad (19.92\text{ rad/s}^2)$$

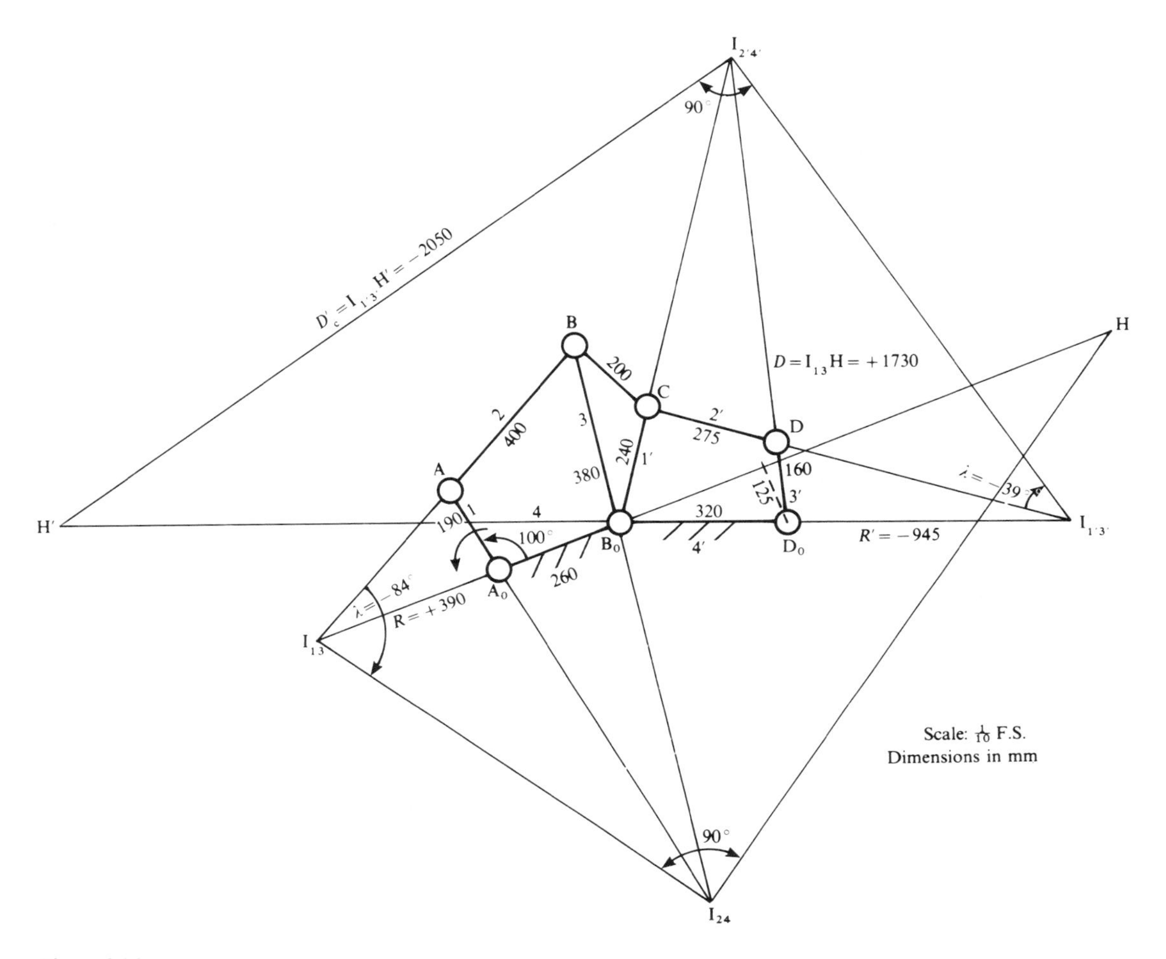

Figure 3.16

Substituting in Eq. (3.13), we find

$$n_3 = -0.2247$$

and, from Eq. (3.9),

$$J_3 = 3 \times -0.25 \times 75 \times 200 - 0.2247 \times 75^3$$
$$= \underline{-9.59 \times 10^4 \text{ rad/s}^3} \qquad \text{i.e. } 9.59 \times 10^4 \text{ rad/s}^3 \curvearrowright \quad (-9.44 \times 10^4 \text{ rad/s}^3)$$

2. *Linkage* B_0CDD_0

$$R' = -925 \text{ mm} \qquad D'_c = -2050 \text{ mm} \qquad \lambda' = -39^\circ$$

Hence with ω_2, α_2 and J_2 as input we find

$$n'_1 = -R'/(-R' + d') = -925/(-925 + 343.5) = 1.59$$
$$\omega_3 = n'_1 \omega_2 = 1.57 \times 45 = \underline{71.6 \text{ rad/s}} \circlearrowleft \quad (73)$$
$$n'_2 = n'_1(1 - n'_1)/\tan \lambda' = 1.59(1 - 1.59)/\tan -39 = 1.16$$

$$\alpha_3 = n'_1\alpha'_2 + n'_2\omega_2^2 = 1.59(-20.63) + 1.16 \times 45^2 = \underline{2317 \text{ rad/s}^2} \qquad (2630)$$
$$n_3 = 5.59$$
$$J_3 = 1.59 \times (-9.59 \times 10^4) + 3 \times 1.16 \times 45 \times -20.63 + 5.59 \times 45^3$$
$$= 3.53 \times 10^5 \text{ rad/s}^3 \qquad (4.93 \times 10^5)$$

Note: The figures in brackets are the values obtained using the author's computer program; in view of the small scale employed the answers obtained graphically are quite acceptable.

3.4.6 Linkages with Equal Properties

All linkages such as $A_0A'B'B_0$ and A_0ABB_0 having the same instant centre I_{13}, as shown in Fig. 3.17, and hence the same R and d, have equal velocity ratios.

All linkages such as A_0ABB_0 and $A_0A'B'B_0$, as shown in Fig. 3.18, having the same I and λ have equal accelerations as well as equal velocity ratios.

In Fig. 3.19, all linkages such as A_0ABB_0 and $A_0A'B'B_0$ having the same Carter–Hall circle, and the same λ, R, and d, have equal jerks as well as equal velocities and accelerations.

These properties are useful in the design of linkages since, given particular desired properties, there exists an infinite number of solutions.

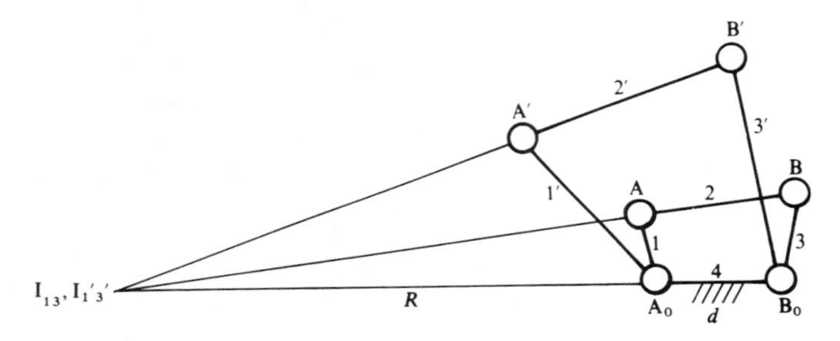

Figure 3.17 Linkages having equal velocity ratio.

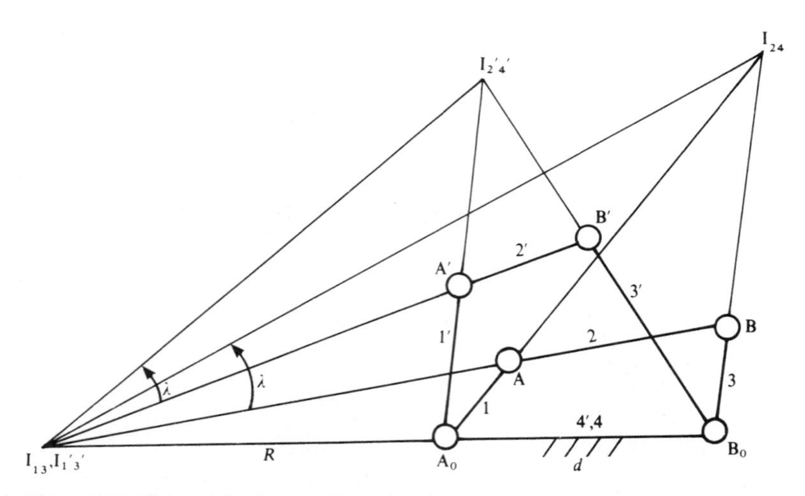

Figure 3.18 Linkages having equal accelerations.

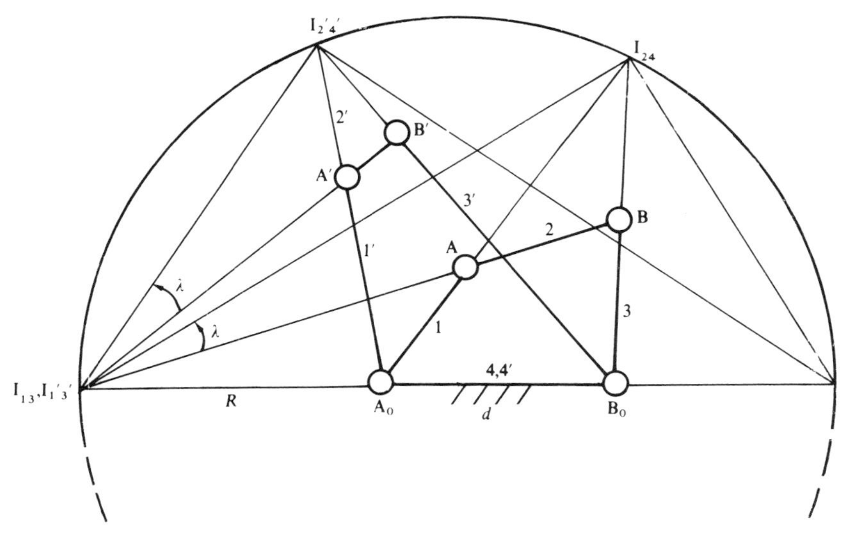

Figure 3.19 Linkages having equal jerks.

3.4.7 Freudenstein's Theorem for Extreme Values of the Velocity Ratio

Figure 3.20 shows a four-bar linkage A_0ABB_0 and the two instant centres I_{13} and I_{24} for a particular position ϕ of the input link A_0A.

Theorem When the angle λ is equal to a right angle the velocity ratio is either a maximum or a minimum.

PROOF From Eq. (3.7),

$$n_1 = \dot{\psi}/\dot{\phi}$$

For n_1 to be a maximum or a minimum,

$$\frac{\mathrm{d}n_1}{\mathrm{d}t} = 0$$

i.e.,

$$\frac{\mathrm{d}n_1}{\mathrm{d}\phi}\frac{\mathrm{d}\phi}{\mathrm{d}t} = n_2\dot{\phi} = 0$$

But from Eq. (3.12)

$$n_2 = n_1(1 - n_1)/\tan\lambda$$

hence

$$\frac{\mathrm{d}n_1}{\mathrm{d}t} = 0 \qquad \text{when} \quad n_2 = 0$$

i.e.,

$$\text{when} \qquad \tan\lambda = \infty \longrightarrow \lambda = \pi/2$$

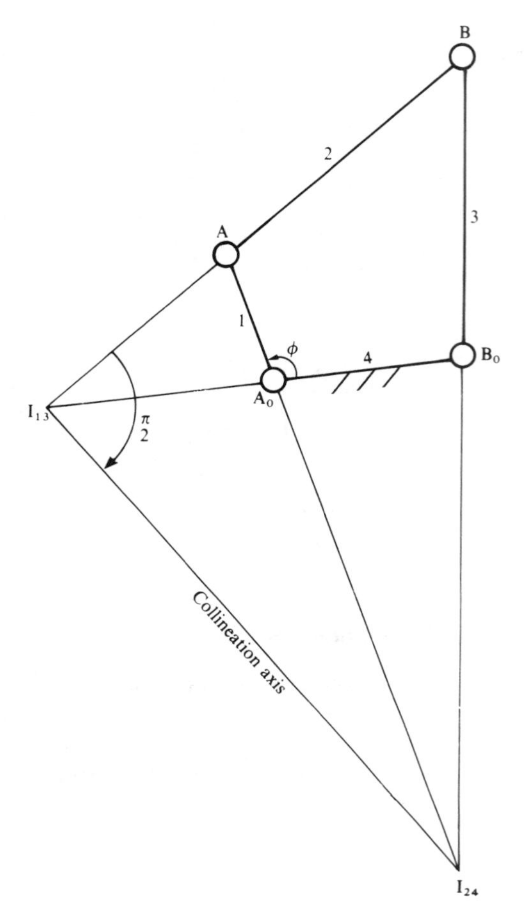

Figure 3.20 Extreme velocity ratio of the four-bar linkage.

Thus the velocity ratio is a maximum or a minimum when the collineation axis is perpendicular to the coupler line $I_{13}AB$.

3.4.8 Application of the Carter–Hall Construction to the Design of four-bar Linkages

A four-bar linkage which is required to co-ordinate input and output can be designed quite quickly, using the Carter–Hall construction in reverse order to that of the analysis discussed in the previous sections. This method is not only simple but visual at every step, enabling the engineer to select the best linkage which will satisfy a desired performance from a large number of possible solutions. Like all graphical methods, however, the accuracy is limited by the size of paper used and the skill of the engineer, but in many designs the solutions obtained are generally satisfactory.

The method is particularly useful in the case of the generation of functions locally, i.e., around a particular position as for example when replacing geared segments by a four-bar linkage. Linkages used to replace geared segments or cams offer a number of technical advantages such as reduced friction, the ability to transmit large forces or torques, reduced play between members, lower cost etc.; they are especially useful in the construction of robotic devices or dedicated manipulators in manufacturing processes.

The purpose of a mechanism is to transform motion as illustrated in Sec. 1.2; we are thus concerned with the analogue generation, mechanically, of a function $\psi = f(\phi)$ between an input angular rotation ϕ and an output angular rotation ψ. One thing we should realize is that it is always possible to introduce, mechanically, an offset or a constant phase angle to modify the input and output angles, since what matters in practice are the relations between the variations of the input and the output; thus the real variables from the point of view of practical applications are $\Delta\phi = \phi - \phi_0$ and $\Delta\psi = \psi - \psi_0$ where ϕ_0 and ψ_0 can be chosen arbitrarily and incorporated physically in the mechanism; hence the functional relations to be considered will, in reality, be of the form

$$\Delta\psi = f(\Delta\phi).$$

Thus coordination between input and output will, in general, be defined as follows:

1. n_1, n_2, and n_3 which specify the velocity, acceleration and jerk relationship; or
2. $\psi = f(\phi)$ or $\Delta\psi = f(\Delta\phi)$ which specify the position relationships from which n_1, n_2 and n_3 can be calculated.

In each case an initial value for ϕ_0 or for ψ_0 may or may not be specified.

To carry out the construction we need the values of R, λ, and D_c. From Eqs (3.11), (3.12), and (3.13) we find

$$R = \frac{n_1}{1 - n_1} d \tag{3.17}$$

$$\lambda = \arctan[n_1(1 - n_1)/n_2] \tag{3.18}$$

and

$$D_c = 3n_1^2 d / \{[(1 - 2n_1)s^2 + 3n_1](1 - n_1)n_1 + n_3 s^2\} \tag{3.19}$$

where $s = \sin\lambda$.

For the construction there are two cases to consider:

case 1, when ϕ_0 (or ψ_0) is specified; and
case 2, when ϕ_0 (or ψ_0) is not specified.

The first case leads to two unique solutions whereas the second case leads to an infinite number of solutions.

To illustrate the procedure in case 1, we consider the situations where R is positive and where R is negative, i.e., corresponding to an 'open' and a 'crossed' linkage as shown in Fig. 3.21(*a*) and (*b*) respectively.

From the specifications we first calculate the values of R, λ, and D_c and carry out the following steps:

Step 1 Set $A_0B_0 = d$, the fixed link to an arbitrary value.
Step 2 Select a suitable sheet of paper; the value of D_c can be used as a guide for the scale to be used in the case of the 'open' linkage and the values of D_c and R for the 'crossed' linkage. In either case part of the required mechanism may lie outside the circle, in which case a smaller scale will have to be selected or a larger size of paper.
Step 3 Set $A_0I_{13} = R$ to locate the instant centre I_{13} between the input and output links.
Step 4 Set $I_{13}H = D_c$ to locate the point H.

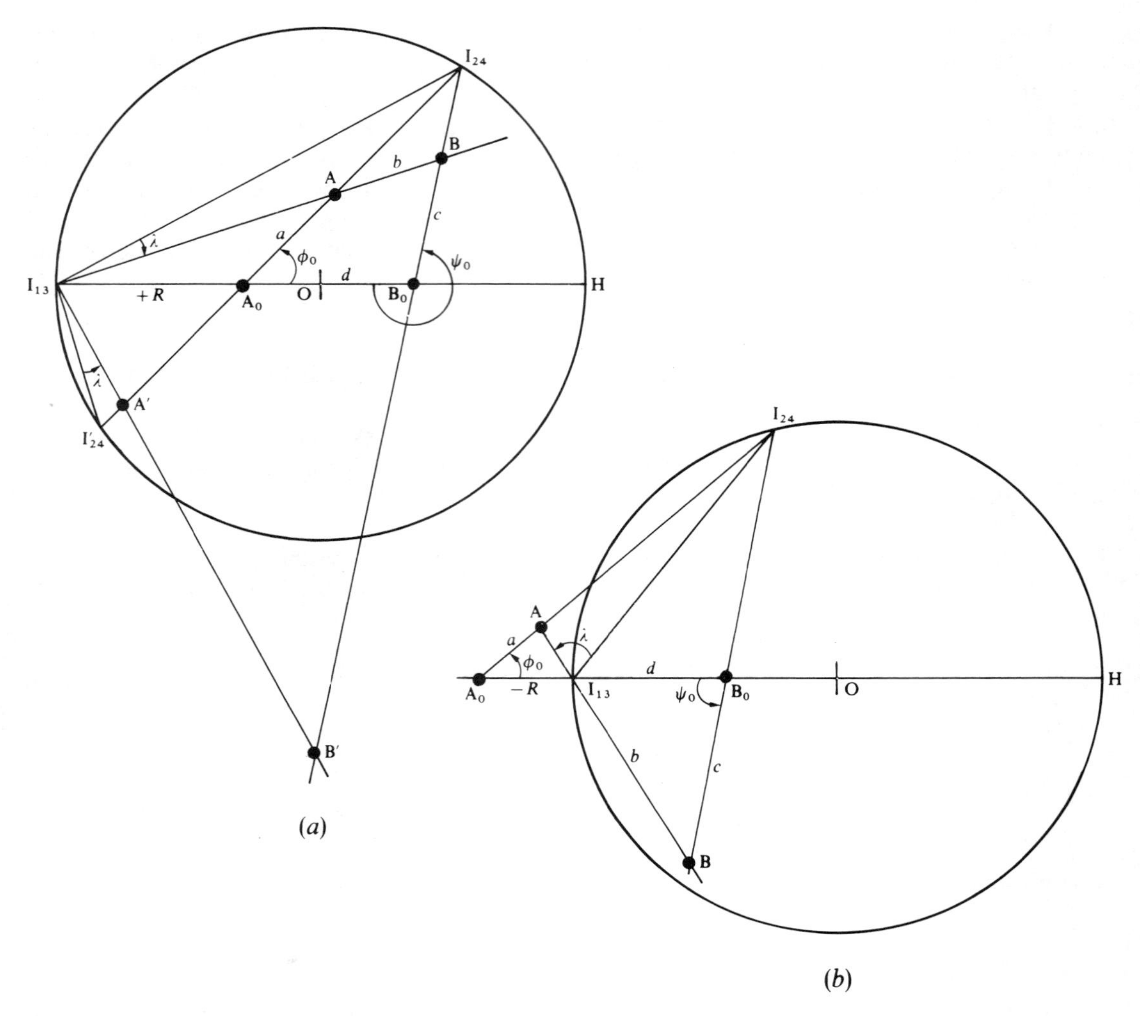

Figure 3.21 Carter–Hall construction for the design of a four-bar linkage.

Step 5 Locate O, the centre of the Carter–Hall circle, and draw a circle to a radius of $D_c/2$.

Step 6 Draw $I_{24}AA_0I'_{24}$ at an angle ϕ_0 (or B_0BI_{24} at an angle ψ_0) to locate I_{24} and I'_{24} on the circumference of the circle; these two points will lead to the two solutions for the size of the linkage.

Step 7 Consider the point I_{24}: join $I_{24}I_{13}$.

Step 8 Draw a line through I_{13} at an angle λ to $I_{13}I_{24}$, observing the sign convention defined in Sec. 3.4.5, cutting A_0I_{24} and B_0I_{24} at A and B.

Step 9 The required mechanism is defined by A_0ABB_0 and its dimensions are $a = A_0A$, $b = B_0B$, $c = BB_0$ to the scale of the diagram.

Step 10 Repeat step 7 to 9 with I'_{24}. The resulting linkage will be $A_0A'B'B_0$.

Example 3.5 Design a four-bar linkage to coordinate the input and output of two shafts

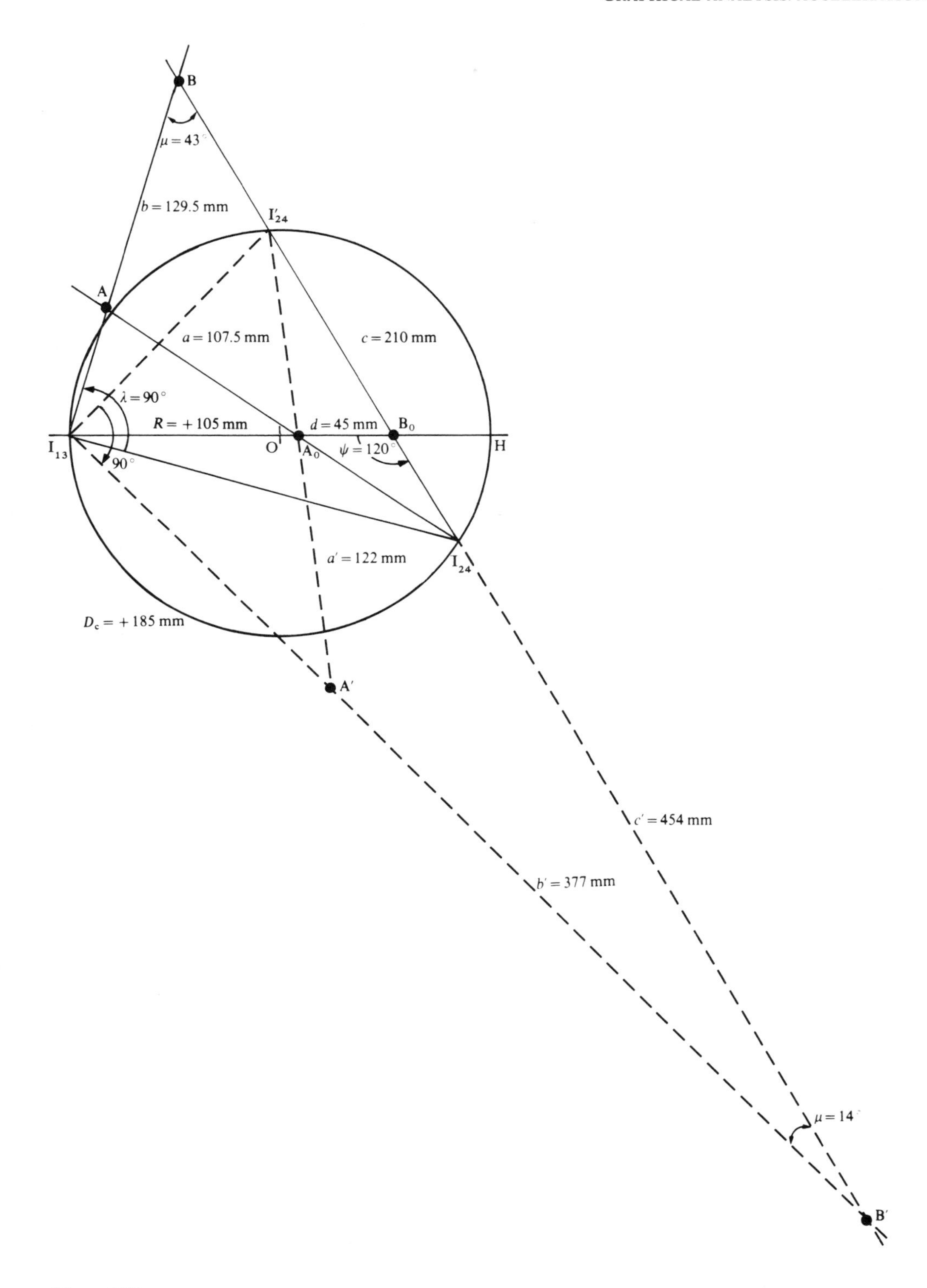
B
μ = 43°
b = 129.5 mm
I'24
A
a = 107.5 mm
c = 210 mm
λ = 90°
R = +105 mm
d = 45 mm
B0
I13
O
A0
ψ = 120°
H
90°
a' = 122 mm
I24
Dc = +185 mm
A'
c' = 454 mm
b' = 377 mm
μ = 14°
B'

Figure 3.22

such that

$$n_1 = 0.7 \qquad n_2 = n_3 = 0 \qquad \text{for} \quad \psi_0 = 120^\circ$$

i.e. there is a constant velocity ratio locally.

SOLUTION We first calculate the values of R, λ, and D_c. Substituting in Eqs (3.17), (3.18), and (3.19), we find

$$R = \frac{0.7}{1-0.7} d = 2.33d$$

$$\lambda = 90^\circ \qquad \text{since} \quad n_2 = 0$$

$$D_c = \frac{3 \times 0.7^2 \times d}{[(1-2\times 0.7)+3\times 0.7](1-0.7)0.7} = 4.118d$$

Next we need to select a suitable value for d, the length of the fixed link, so that the circle will fit within the size of paper available (Fig. 3.22). Using, for instance, an A3 size sheet of paper (420 mm × 297 mm) requires that d should not exceed 45 mm.

Thus with $d = 45$ mm we find $R = 105$ mm, $D_c = 185$ mm.

Following the steps mentioned above, we obtain from the construction the linkage A_0ABB_0 shown for $\psi_0 = 120°$. The lengths of each link are found by measuring the lengths of the lines A_0A, AB, and BB_0, giving:

$$a = A_0A = 107.5\text{ mm} \qquad b = AB = 129.5\text{ mm} \qquad c = BB_0 = 210\text{ mm}$$

and the link ratios for this linkage are:

$$2.39,\ 2.88,\ \text{and}\ 4.67 \qquad \text{when dividing by } d = 45\text{ mm}$$

Using a protractor, we find the value of the transmission angle:

$$\mu = 43^\circ$$

From what has been said previously, we see that this linkage is just about acceptable from the point of view of quality of transmission. It can be shown that we can improve on this by introducing an offset or phase angle on ϕ_0(or ψ_0).

We now consider the second solution by taking the point I'_{24} on the circumference as shown in Fig. 3.22. By joining I'_{24} to I_{13} and drawing a line at $\lambda = 90^\circ$ to $I_{13}I'_{24}$ we obtain the second linkage $A_0A'B'B_0$ (shown dotted). It is obvious that this second linkage would be quite impractical. Hence the only acceptable linkage in this case is A_0ABB_0.

It is left to the reader to show that when ϕ_0 (or ψ_0) is not specified (case 2) there are an infinite number of solutions, some of which would be quite impractical. It is however possible to obtain a satisfactory solution fairly quickly and with reasonable accuracy.

EXERCISES

3.1 Figure 3.23 shows a mechanism (Stephenson type 1) used in a wrapping machine. Dimensions: $A_0A = 50$ mm; AC = 140 mm; CB = CD = $DC_0 = 65$ mm; $BB_0 = 100$ mm; $C_0E = 165$ mm; $\angle DC_0E = 90°$; $x_{C_0} = 100$ mm; $y_{C_0} = 50$ mm; $x_{B_0} = 200$ mm; $y_{B_0} = 75$ mm. The oscillations of the point E feed the paper to the machine while the crank A_0A rotates at a constant angular velocity of 50 rad/s anticlockwise.

When $\phi = 320°$ calculate graphically:

(*a*) the angular acceleration of the link BB_0;

(*b*) the angular acceleration of the coupler AB;
(*c*) the angular acceleration of the crank DC_0E;
(*d*) the total acceleration of the feed point E.

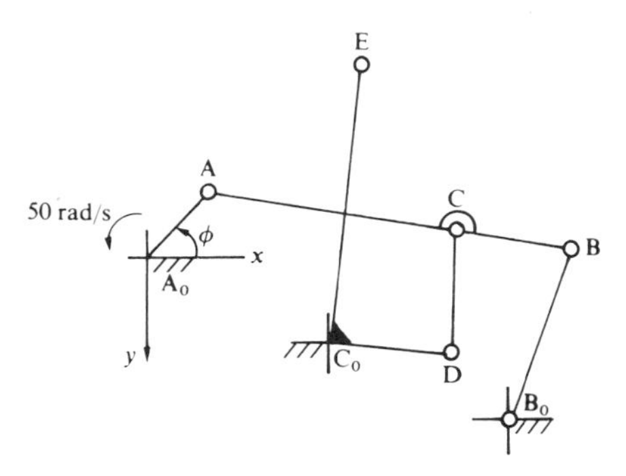

Figure 3.23

3.2 For the linkage shown in Fig. 3.24 calculate graphically ω_0 and α_0, when $\phi = 70°$ and $\omega = 720$ rev/min. Dimensions are: $a = 25$ mm; $b = 95$ mm; $c = 90$ mm; $d = 100$ mm; $m = 55$ mm; $n = 50$ mm; $e = 50$ mm; $f = 102$ mm; $L = 25$ mm; $h = 75$ mm.

Note. f and h are measured to E.

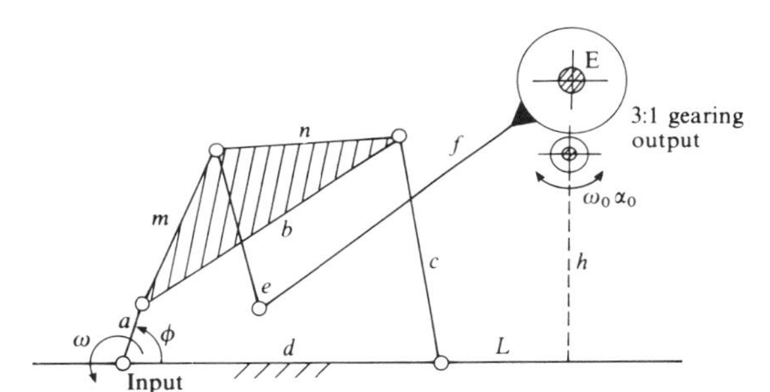

Figure 3.24

3.3 In the mechanism shown in Fig. 3.25 the crank A_0A rotates at a constant speed of 120 rev/min in the direction indicated. A vertical motion of the rack is imparted by means of the geared segment. A_0 and B_0 are fixed centres. Calculate the velocity and acceleration of the rack when the angle B_0A_0A is 135°.

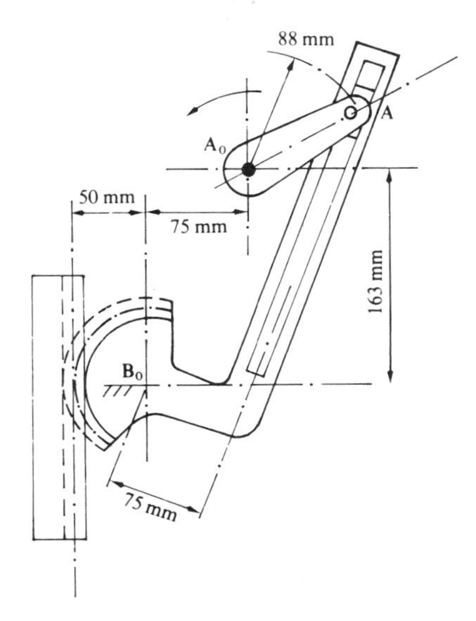

Figure 3.25

3.4 Figure 3.26 shows a quick-return mechanism used to save time in a machine which cuts in one direction only (time is money!). At the end of the crank BE, rotating at 60 rev/min, is a pin E sliding in the link DF, which slides on a fixed pin A at one end. At the other end a pin D slides in a fixed slot CC′. Obtain an analytical expression for the velocity of D in terms of θ, ω, and the lengths a, b, and r. If $a = 178$ mm, $b = 165$ mm, $r = 57$ mm, and $\theta = 60°$, calculate (*a*) the total stroke of D, (*b*) the velocities of D as it passes the mid-stroke position in each direction, and (*c*) the acceleration of D when $\theta = 60°$.

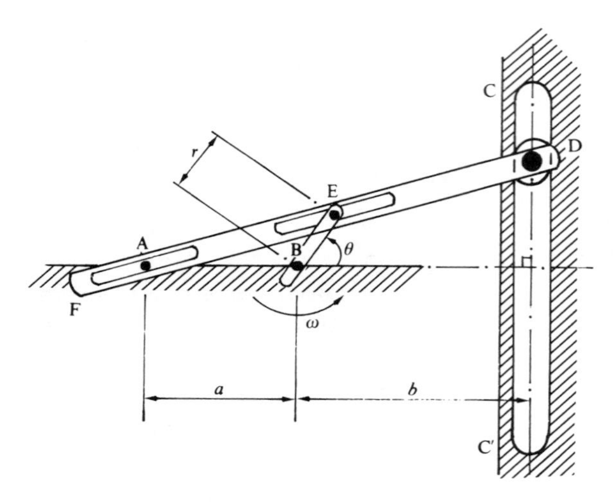

Figure 3.26

3.5 Part of a mechanism consists of a spring-loaded follower driven by a cam as shown in Fig. 3.27. The cam has an angular velocity of 10 rad/s anticlockwise and an angular acceleration of 5 rad/s² clockwise.

Replace the cam follower arrangement by an equivalent four-bar linkage and hence compute the angular velocity and acceleration of the follower in magnitude and direction.

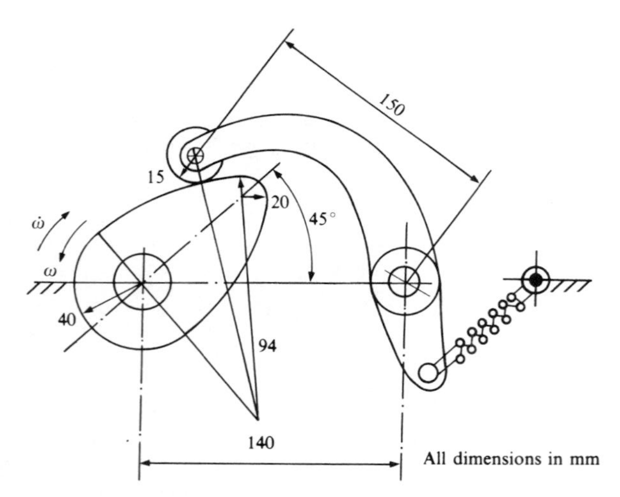

Figure 3.27

3.6 In the crank-and-slotted-lever mechanism shown in Fig. 3.28, the crank A_0A is driven at constant angular velocity ω measured in rad/s. If A_0Q is perpendicular to the centreline B_0P show that the angular acceleration of the lever B_0P is given by

$$\ddot{\phi} = \omega^2 \frac{A_0Q}{B_0A}\left(\frac{2 \times AQ}{B_0A} - 1\right)$$

Calculate graphically the acceleration of point P in magnitude and direction when $\theta = 60°$, $\omega = 10$ rad/s, $A_0A = 75$ mm, $B_0P = 450$ mm.

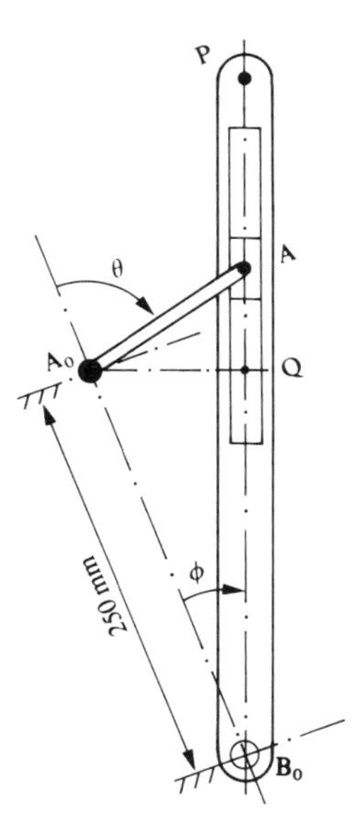

Figure 3.28

3.7 For the four-bar linkage shown in Fig. 3.29 calculate, using the Carter–Hall construction, the velocity, acceleration, and jerk of the output shaft B_0B when the input crank A_0A makes an angle of 30° with A_0B_0 and $\omega = 50$ rad/s, $\dot{\omega} = 350$ rad/s^2 in that position and the directions indicated. Data: $A_0A = 25$ mm, $AB = 80$ mm, $B_0B = 36$ mm, $A_0B_0 = 101$ mm.

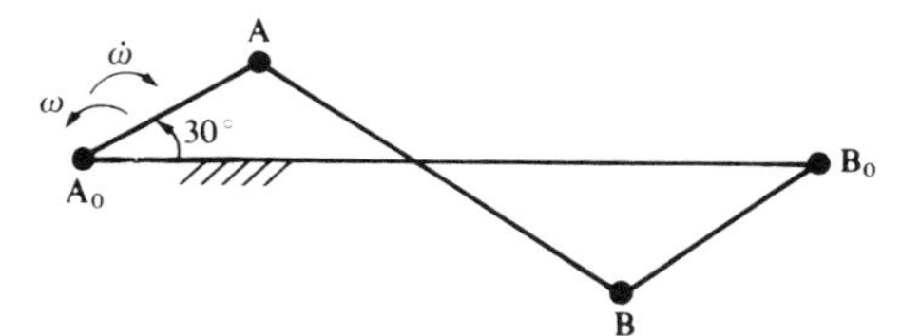

Figure 3.29

3.8 A four-bar linkage has the following dimensions: $A_0A = -108$ mm, $AB = 99$ mm, $B_0B = -172$ mm, $A_0B_0 = 90$ mm, $\phi = 25°$ when using the convention for angles and signs. In that particular position the input link A_0A is rotating at a speed of 500 rev/min and accelerating at the rate of 750 rev/min/s, both in an anticlockwise direction.

Using the Carter–Hall construction calculate the velocity, acceleration, and jerk of the output link B_0B_0.

3.9 Referring to Exercise 3.5, calculate the velocity, acceleration, and jerk of the follower using the Carter–Hall construction.

3.10 Figure 3.30 shows a six-bar linkage used in a shoe-stitching machine. Link A_0A, the input, is driven at a constant

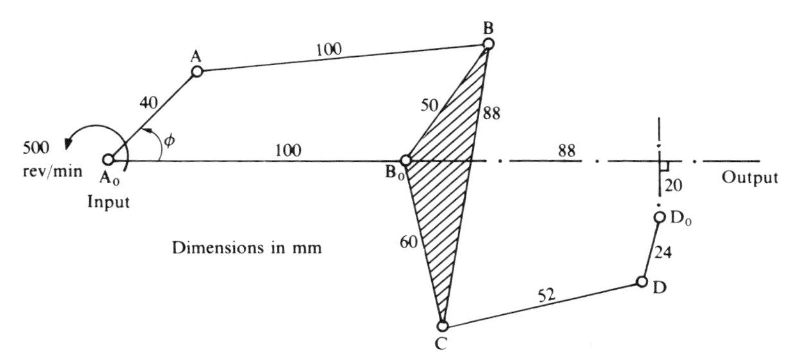

Figure 3.30

speed of 500 rev/min in the direction shown, and the output is taken at the link D_0D to which the needle, not shown, is attached.

When $\phi = 30°$ calculate the angular velocity and angular acceleration of the output link D_0D in magnitude and direction using the Carter–Hall construction.

3.11 Figure 3.31 is a diagram of the linkages which operate the bay doors on a space shuttle. Using the Carter–Hall construction, calculate the angular velocity and angular acceleration of the bay door BB_0, when the input link DD_0 is in the position shown and rotating at a constant speed of 10 rev/min. Use an A3 sheet of paper and a scale of $\frac{1}{2}$ F.S.

A_0, B_0 and D_0 are fixed pivots. Dimensions: $A_0B_1 = 65$ mm, $B_1B_0 = 60$ mm, $BB_0 = 120$ mm, $AB = 135$ mm, $AC = 210$ mm, $CA_0 = 90$ mm, $CD = 200$ mm, $DD_0 = 80$ mm, $D_0D_1 = 90$ mm, $D_1A_0 = 210$ mm.

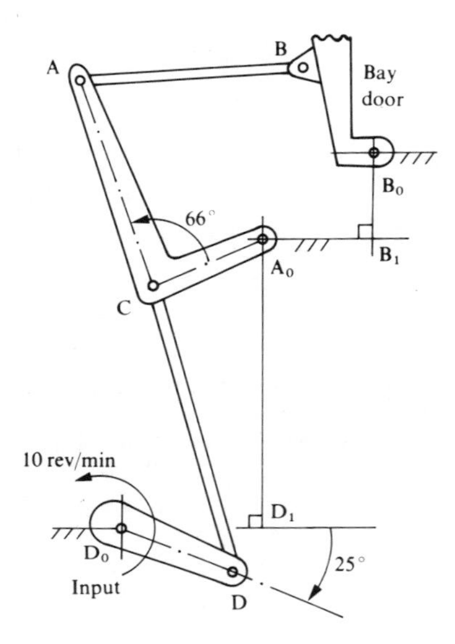

Figure 3.31

3.12 A four-bar linkage is required to replace geared segments to coordinate the output to input velocities of two shafts over small finite rotations in the neighbourhood of the output position ψ_0 of 120°, given that the velocity ratio should be equal to 0.75 with zero acceleration and zero jerk. The fixed link should not exceed 75 mm in length.

CHAPTER
FOUR
KINEMATIC OF CAMS

4.1 INTRODUCTION: CAMS AND FOLLOWERS

A brief reference to cams was made in Sec. 1.2 in connection with motion transformation; in fact a cam is a very versatile mechanism with very many applications. Its purpose is to impart an oscillatory, rectilinear or rotary, motion to an element referred to as a follower which is in contact with the cam face as shown in Fig. 4.1. Many kinds of motion of the follower are possible because the cam face can have any desired profile, within certain limitations such as acceptable dynamic forces.

The cam, fixed to a shaft, usually rotates at constant speed, although in some cases its motion may be oscillatory. The cam face determines the kind of motion the follower will have, e.g., a slow rise to a given value, then a dwell followed by a rapid descent or fall.

By proper design the motion of the follower can be accurately controlled in all situations where precise positioning of a machine element is required, e.g., the valves in an internal-combustion engine as in Fig. 1.11(*a*), or the position of a bar to hold down a rubber tube for a fraction of a second during the manufacture of rubber bands.

Cams are usually divided into two main groups:

1. disc or radial cams which impart motion to the follower in a plane perpendicular to the cam face as in Fig. 4.1; and

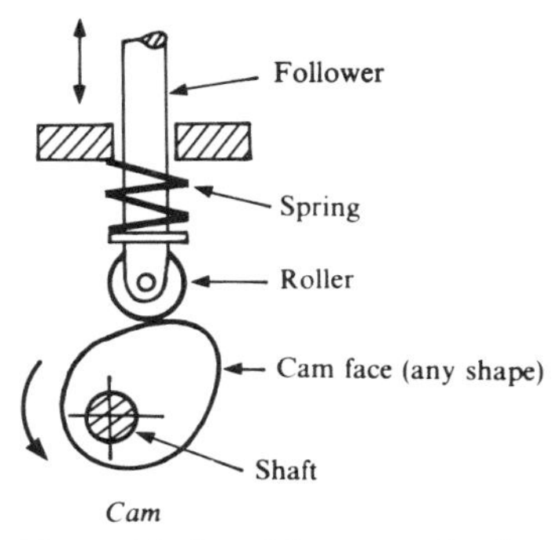

Figure 4.1 Cam follower mechanism.

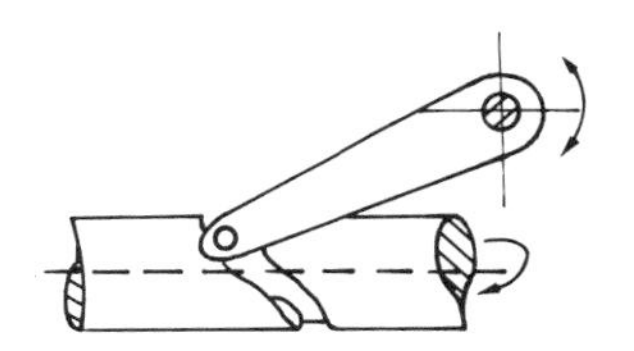
Figure 4.2 Cylindrical cam and follower.

2. cylindrical cams which impart motion to the follower in a plane parallel to the axis of the cam as illustrated in Fig. 4.2.

There are also three-dimensional cams but these will not be considered in this text. Since the majority of cams fall into group 1 this text will limit its exposition to that type.

Followers have three possible shapes:

1. knife edge: seldom employed because of high stresses and rapid wear (Fig. 4.3(*a*));
2. roller: eliminates sliding between the cam face and the follower and is in common use (Fig. 4.3(*b*));
3. flat face: except for friction it eliminates side thrust between the contact surfaces (Fig. 4.3(*c*)).

In some cases the follower centreline is offset to that of the cam axis as indicated in Fig. 4.3(*d*).

Figures 4.3(*e*) and (*f*) show the rotary–oscillatory types of followers.

In all situations it is most important to ensure that the follower maintains contact with the cam face and adequate means must be provided such as the spring shown in Fig. 4.1. This is of particular importance in high-speed cams since inertia effects become very large; also, high values of the jerk (rate of change of the acceleration) lead to wear, vibrations, and noise. In slow speeds of rotation and with smooth changes in contour of the cam profile these dynamic effects are not so important.

As for linkages, two cases need consideration:

1. *Synthesis.* For a desired motion of the follower, the problem is to determine the profile of the cam either graphically or analytically.
2. *Analysis.* Given a cam profile, we need to calculate either graphically or analytically the follower displacement, velocity, acceleration, and jerk.

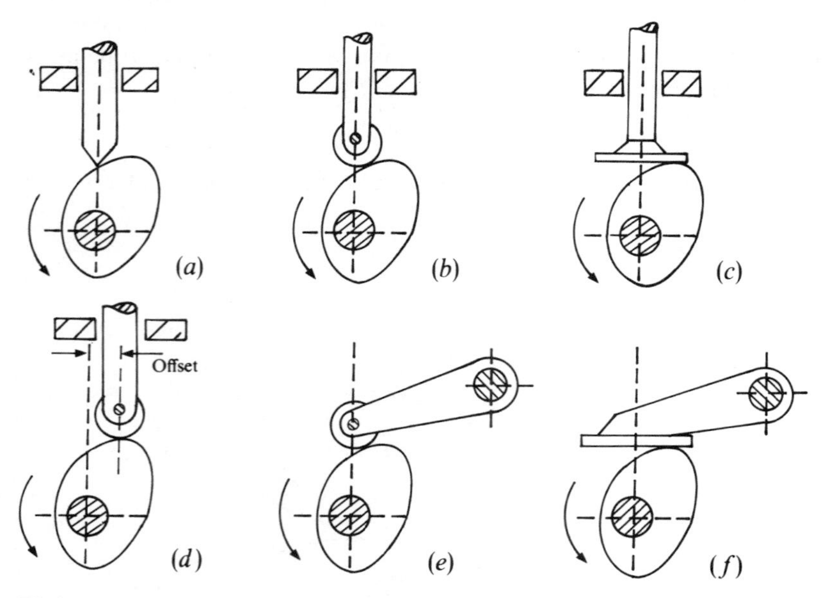

Figure 4.3 Typical cam follower arrangements.

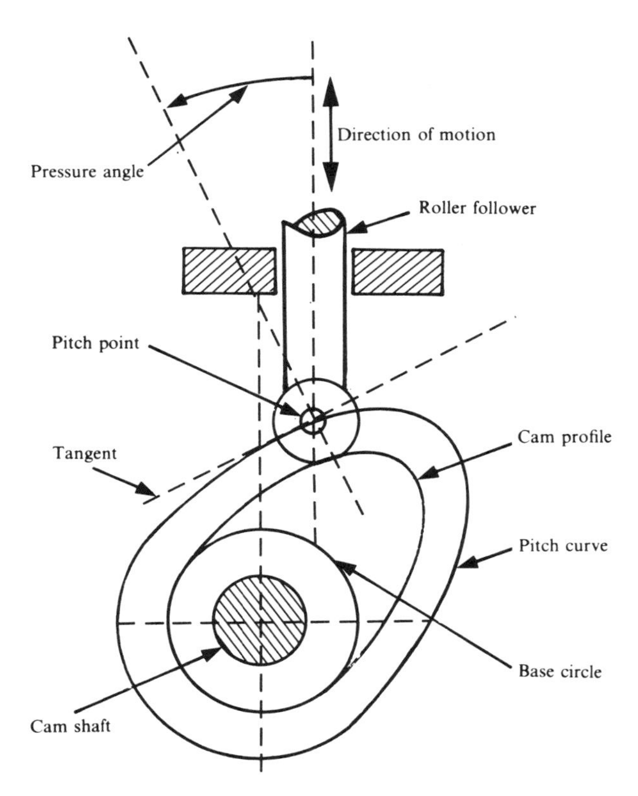

Figure 4.4 Definitions of cam parameters.

A number of terms in common usage to define various geometrical aspects of cam and follower motion are shown in Fig. 4.4 for reference.

4.2 GRAPHICAL DESIGN OF CAMS

The problem in the graphical method of synthesis is to determine the profile of the cam face that will transmit the desired motion to the follower. The construction is simplified by using the principle of inversion we have met in a previous chapter. This is achieved by holding the cam stationary and rotating the follower in the opposite direction to the true cam rotation, thus preserving the correct sequence when motion of the cam takes place.

By placing the follower in a number of successive positions the profile of the cam can be drawn. In order to obtain a 'decent' curve many positions are needed; however, the process involved is relatively simple but does require a great deal of care.

To illustrate the construction, as shown in Fig. 4.5, let us design a cam that will impart a constant velocity to the follower along a straight path passing through the centre of the cam. One complete revolution of the cam is to give one rise and one fall to a roller follower.

Let O be the centre of the cam and AB the lift and BA the fall of the follower,

and R = radius of the shaft that carries the cam

L = minimum distance from the shaft to the roller

h = the total travel of the follower from A to B

r = radius of the roller

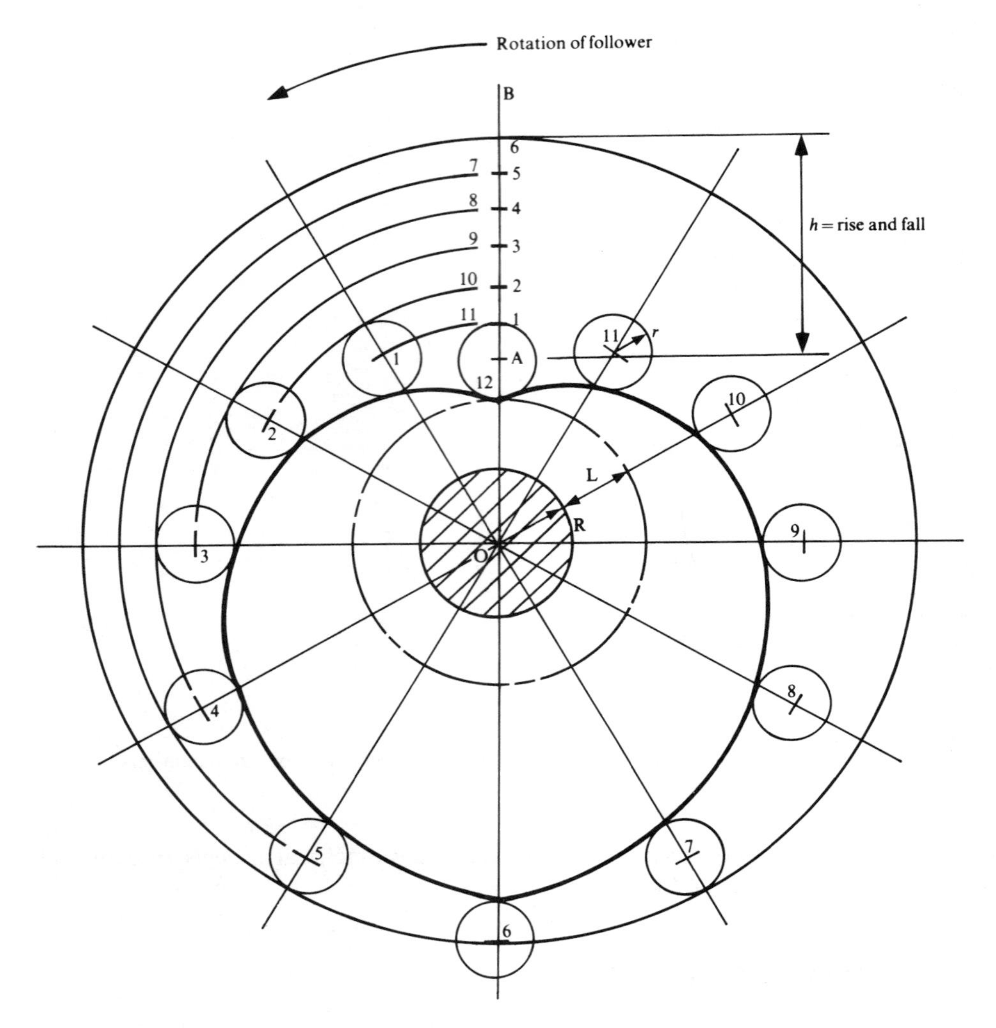

Figure 4.5 Graphical cam design: constant velocity of the follower.

Then $R+L$ = radius of the base circle

$R_{min}=R+L+r$ = least radius of the pitch curve

$R_{max}=R_{min}+h$ = maximum radius of the pitch circle

To determine the cam profile the following steps are required:

1. Draw a circle of radius = OB = R_{max}.
2. Divide AB into a number of equal intervals since the velocity is constant; six have been chosen in this case. In practice such a number would be a minimum.
3. From O draw radial lines O1, O2, O3 . . . O12 equally spaced, at 30° in this case.
4. The points 1, 2, 3 . . . 12 on these radial lines lie on the pitch curve of the cam.
5. With each of these points as centres draw circles having a radius equal to that of the roller.
6. With the aid of French curves draw a curve as accurately as posssible to touch those circles

tangentially. This curve is the required profile of the cam. We notice that in this case it is symmetrical about OAB.

In the above construction the cam profile was designed to give the roller follower a rise (and fall) of 30 mm, with a camshaft diameter of 20 mm, a base circle of 40 mm, and a roller diameter of 10 mm.

Example 4.1 Design a cam to give a swinging link with a roller follower a uniform angular velocity about a fixed pivot during its rise and fall. Data: shaft diameter = 10 mm; base circle diameter = 23 mm; roller diameter = 7 mm; position of pivot from the cam centre, $x = 30$ mm, $y = 15$ mm vertically; total angular movement of the link = 30°; rotation of the cam is to be clockwise.

SOLUTION Refer to Fig. 4.6. Draw the shaft diameter, centre O, of the base circle, the

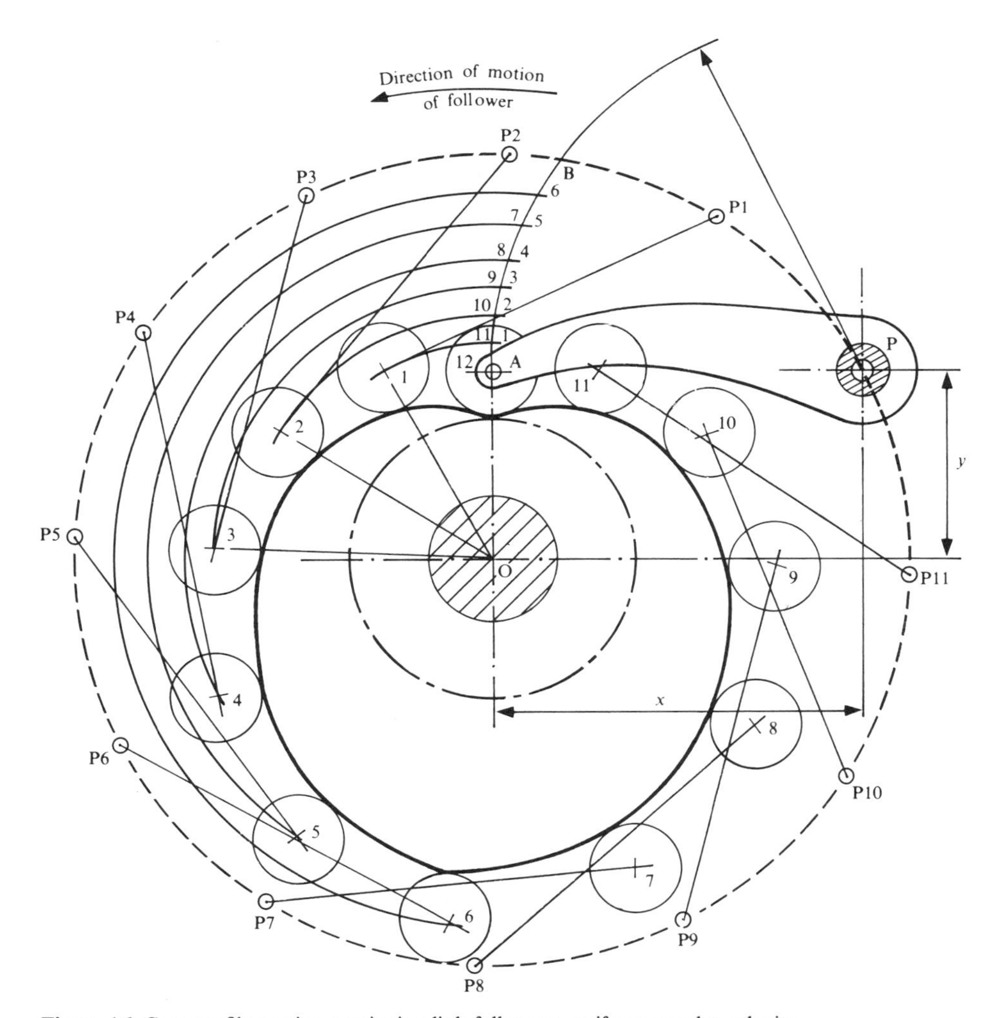

Figure 4.6 Cam profile to give a swinging link follower a uniform angular velocity.

position P of the pivot, and the length PA of the link to a suitable scale; a scale of twice full size was used in this design.

Draw a circle of radius OP in order to locate the successive positions of the pivot P1, P2, P3 ... of the swinging link.

The cam being held stationary, the follower is made to rotate in an anticlockwise direction.

From P draw an arc of a circle AB of radius PA such that the angle APB = 30° and divide it into six equal parts numbered 1, 2, 3, ...

Next draw radial lines O1, O2, O3 ... at 30° to each other and locate the points 1, 2, 3 ... on these lines.

With these points as centres draw the rollers; it is not always necessary to draw the complete roller; in many cases an arc of a circle will be sufficient.

The required cam profile is given by the curve drawn to touch the rollers tangentially.

4.3 ANALYSIS OF DIFFERENT TYPES OF FOLLOWER MOTIONS

In many cases in practice the follower is required to dwell once or several times during one rotation of the cam, as illustrated in Fig. 4.7. The rise, fall, and dwells may be expressed as a fraction of one cycle, such as 360°, or in terms of the times t_1, t_2, t_3, and t_4 measured in seconds.

The displacement shown in Fig. 4.7 corresponding to a uniform velocity is not desirable in practice unless the motion is a very slow one; this is because the accelerations at the transition points O, A, B, and C become infinitely large. To overcome the problem of high accelerations and hence large dynamic forces in high-speed machinery a cam profile must be chosen that will minimize these effects, as well as avoiding possible separation of the follower from the cam surface. This can be achieved by defining the motion of the follower in terms of the way it gets to its desired positions as functions of the cam rotation.

In order to analyse various follower motions we need to investigate the velocity, acceleration, and jerk.

The displacement y of the follower after a rotation θ of the cam can be expressed as follows (see Fig. 4.7):

$$y = f(\theta) \qquad \text{or} \qquad y = f(t)$$

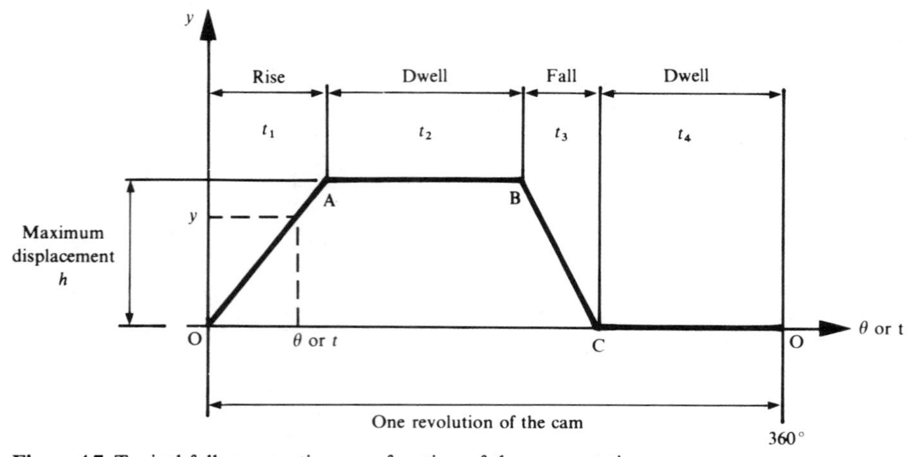

Figure 4.7 Typical follower motion as a function of the cam rotation.

Hence by successive differentiation we get:

Velocity of the follower, $$v=\frac{\mathrm{d}y}{\mathrm{d}t}=\dot{y}=\frac{\mathrm{d}f(t)}{\mathrm{d}t}=\frac{\mathrm{d}f(\theta)}{\mathrm{d}\theta}\frac{\mathrm{d}\theta}{\mathrm{d}t}=\omega\frac{\mathrm{d}f}{\mathrm{d}\theta}$$

where ω = angular velocity of the cam.

Acceleration of the follower, $$a=\frac{\mathrm{d}^2y}{\mathrm{d}t^2}=\ddot{y}=\omega^2\frac{\mathrm{d}^2f}{\mathrm{d}\theta^2}+\dot{\omega}\frac{\mathrm{d}f}{\mathrm{d}\theta}$$

Jerk of the follower, $$J=\frac{\mathrm{d}^3y}{\mathrm{d}t^3}=\dddot{y}=\omega\frac{\mathrm{d}f}{\mathrm{d}\theta}+3\omega\dot{\omega}\frac{\mathrm{d}^2f}{\mathrm{d}\theta^2}+\omega^3\frac{\mathrm{d}^3f}{\mathrm{d}\theta^3}$$

If the cam rotates at constant speed as in many practical situations then $\dot{\omega}=0$. Hence:

Acceleration, $$a=\omega^2\frac{\mathrm{d}^2f}{\mathrm{d}\theta^2}$$

Jerk, $$J=\omega^3\frac{\mathrm{d}^3f}{\mathrm{d}\theta^3}$$

Also, let T = time taken for the cam to rotate through an angle β

$$=\frac{\beta}{\omega}$$

β = rotation of the cam for a rise h.

In the next sections we analyse those motions that are useful in practice by considering the rise of the follower only; the fall or descent can be analysed in a similar fashion.

4.3.1 Constant Velocity of the Follower (Fig. 4.8)

The follower moves from O to A.

Displacement, $$y=h\frac{t}{T} \quad \text{or} \quad y=\frac{h\theta}{\beta}$$

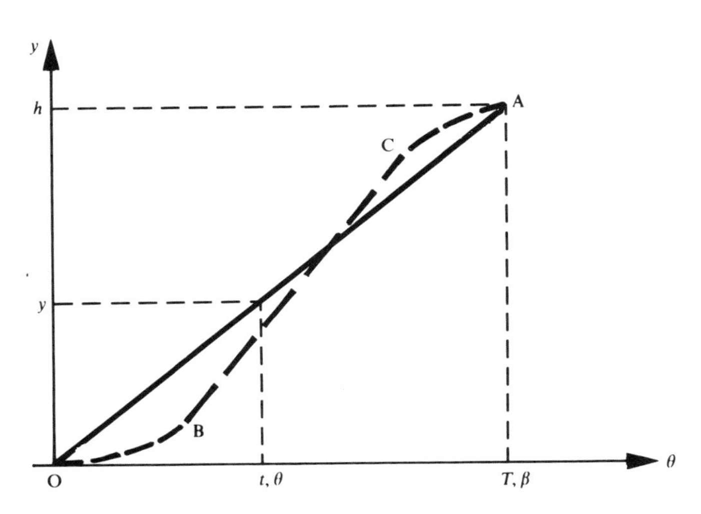

Figure 4.8 Constant velocity of a follower.

Velocity, $v = \dfrac{h}{T}$ or $v = \dfrac{h\omega}{\beta}$

Acceleration, $a = 0$ except at O and A where it is infinite.

We mentioned earlier that this type of motion is undesirable except at very low speeds. We can however modify this motion to reduce dynamic effects by introducing curved portions, e.g., portions of parabolas, to produce constant acceleration from O to B, followed by constant velocity from B to C and finally constant deceleration from C to A.

4.3.2 Parabolic Motion of the Follower (Fig. 4.9)

In this case we have:

Displacement, $y = 2h\left(\dfrac{t}{T}\right)^2$ or $y = 2h\left(\dfrac{\theta}{\beta}\right)^2$

Velocity $v = 4h\dfrac{t}{T^2}$ or $v = \dfrac{4h\omega\theta}{\beta^2}$

Acceleration, $a = \dfrac{4h}{T}$ or $a = \dfrac{4h\omega^2}{\beta^2}$

Jerk, $J = 0$ except at O, B, and A where it is infinite.

We see that with this type of motion the acceleration has a constant value from O to B where it is positive, and also constant from B to A where it is negative. There is one disadvantage, however, because impact will occur due to the effect of the jerk. Nevertheless, this kind of motion would be suitable in low-speed systems.

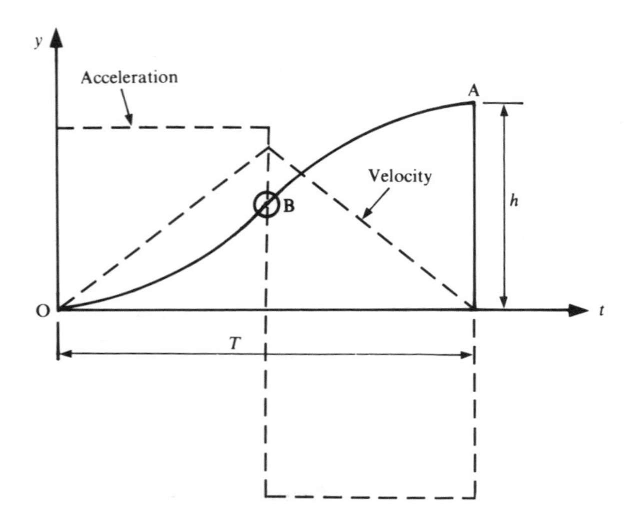

Figure 4.9 Parabolic motion of a follower.

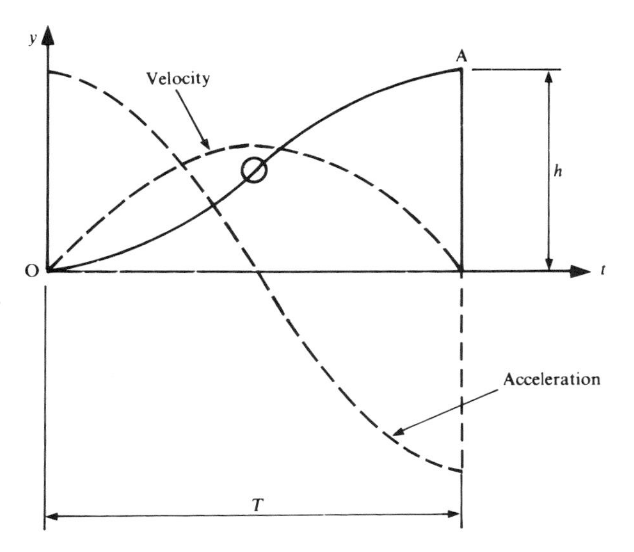

Figure 4.10 Simple harmonic motion of a follower.

4.3.3 Simple Harmonic Motion of the Follower (Fig. 4.10)

With this type of motion we have:

Displacement, $y=\frac{h}{2}\left(1-\cos\frac{\pi t}{T}\right)$ or $y=\frac{h}{2}\left(1-\cos\frac{\pi\theta}{\beta}\right)$

Velocity, $v=\frac{\pi h}{2T}\sin\frac{\pi t}{T}$ or $v=\frac{\pi h\omega}{2\beta}\sin\frac{\pi\theta}{\beta}$

Acceleration, $a=\frac{\pi^2 h}{2T^2}\cos\frac{\pi t}{T}$ or $a=\frac{\pi^2 h\omega^2}{2\beta^2}\cos\frac{\pi\theta}{\beta}$

Jerk, $J=-\frac{\pi^3 h}{2T^3}\sin\frac{\pi t}{T}$ or $J=-\frac{\pi^3 h\omega^3}{2\beta^3}\sin\frac{\pi\theta}{\beta}$

We observe from the plot of velocity and acceleration shown in the figure that they are both smooth with no discontinuities. Unfortunately there are abrupt changes in the acceleration at O and A, resulting in high jerks which will result in vibrations, noise, and wear. Also, the acceleration has maximum values at O and A, resulting in high inertia loads which will in general be much greater than the applied loads. Nevertheless, simple harmonic motion is useful when the speeds involved are not too high.

Figures 4.11 and 4.12 illustrate the constructions necessary to determine the cam profiles that will give different types of followers simple harmonic motion. During the cam cycle of 360° the follower motion for each type is as follows:

1. a rise from 0 to 120°;
2. a dwell from 120° to 180°;
3. a fall from 180° to 300°;
4. a dwell from 300° to 0, i.e., back to the start of a new cycle.

Figure 4.11 is for inline followers and Fig. 4.12 is for offset followers.

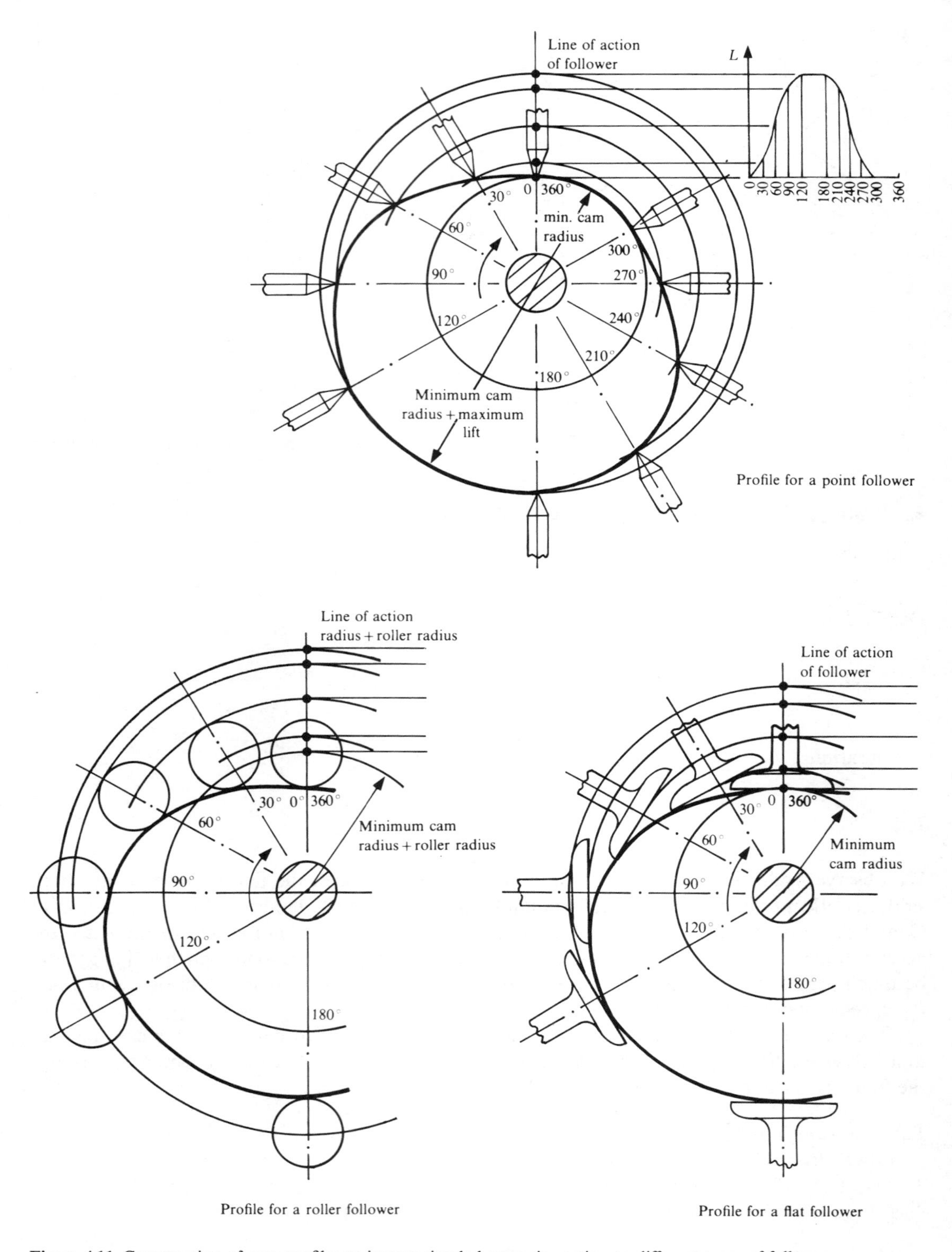

Figure 4.11 Construction of cam profiles to impart simple harmonic motion to different types of followers.

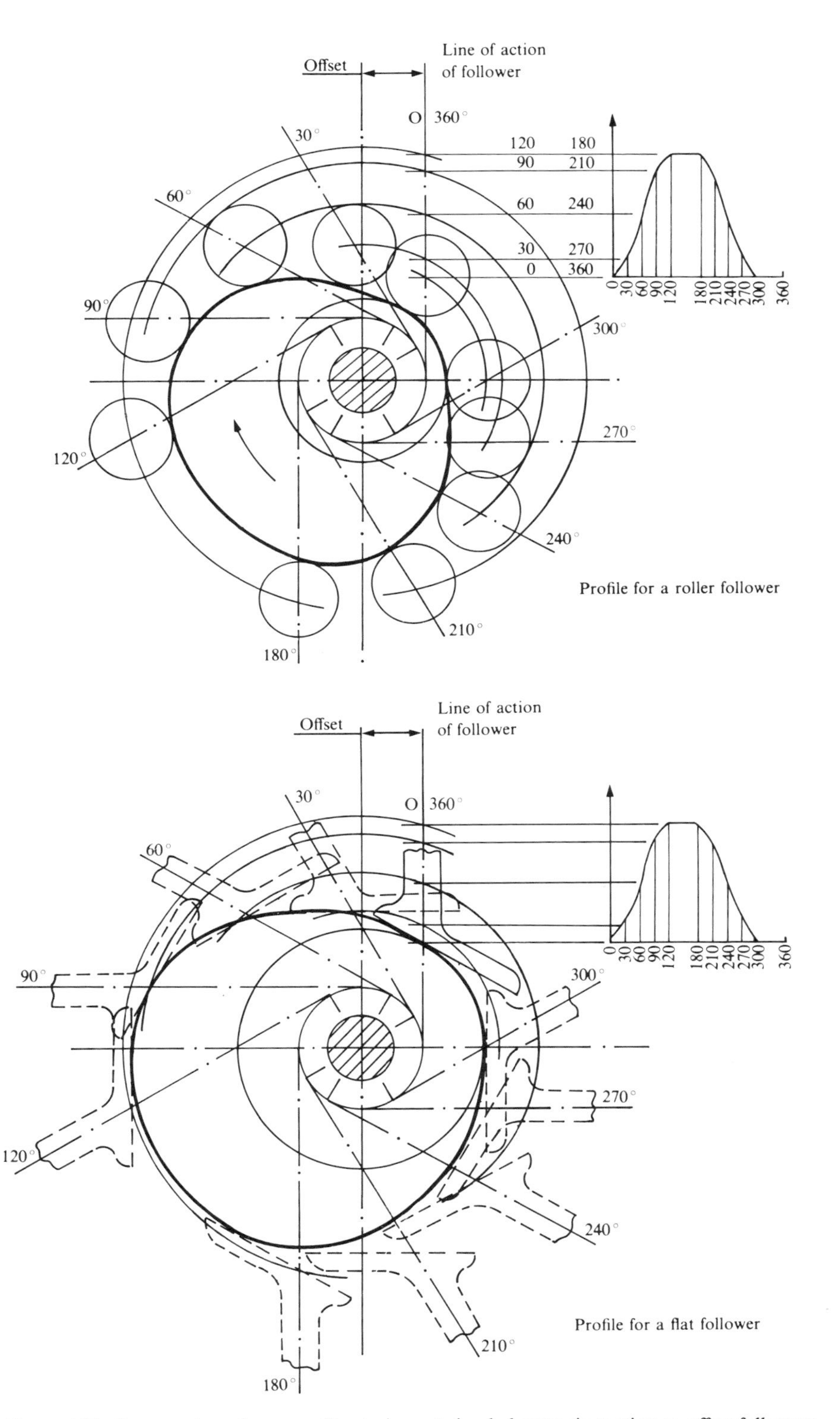

Figure 4.12 Construction of cam profiles to impart simple harmonic motion to offset followers.

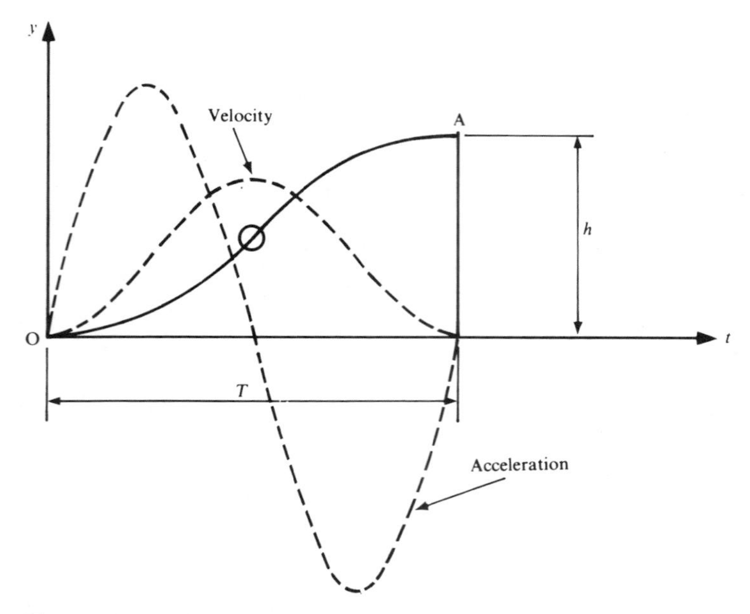

Figure 4.13 Cycloidal motion of a follower.

4.3.4 Cycloidal Motion of the Follower (Fig. 4.13)

For this type of motion we have the following:

Displacement, $y=h\left(\frac{t}{T}-\frac{1}{2\pi}\sin\frac{2\pi t}{T}\right)$ or $y=h\left(\frac{\theta}{\beta}-\frac{1}{2\pi}\sin\frac{2\pi\theta}{\beta}\right)$

Velocity, $v=\frac{h}{T}\left(1-\cos\frac{2\pi t}{T}\right)$ or $v=\frac{h\omega}{\beta}\left(1-\cos\frac{2\pi\theta}{\beta}\right)$

Acceleration, $a=\frac{2\pi h}{T^2}\sin\frac{2\pi t}{T}$ or $a=\frac{2\pi h\omega^2}{\beta^2}\sin\frac{2\pi\theta}{\beta}$

Jerk, $J=\frac{4\pi^2 h}{T^3}\cos\frac{2\pi t}{T}$ or $J=\frac{4\pi^2 h\omega^3}{\beta^3}\cos\frac{2\pi\theta}{\beta}$

We can see from the curves shown in the figure that this type has very good characteristics with no abrupt changes in the acceleration; the jerk has finite values, giving much lower levels of vibration and noise and hence less wear in high-speed machinery.

Although the acceleration has a greater value than in the case of simple harmonic motion for the same parameters, i.e., rise and time, this curve is often used in high-speed machinery.

Furthermore, since this curve has zero acceleration at the beginning and end of the rise it can easily be coupled to dwells at these points, but because the pressure angle (see Sec. 4.4) is high, coupling two cycloidal curves should be avoided.

In order to minimize the effects of the jerk, other curves have been devised by M. K. Kloomok and R. V. Muffley.[2] These curves use polynomials in θ/β up to the eight power.

4.4 PRESSURE ANGLE

The pressure angle at any point on the cam profile is the angle between the follower direction of motion and the normal to the tangent at the point of contact of the roller with the cam profile as shown in Fig. 4.4. The value of the pressure angle is important because as it increases the side thrust between the cam and follower also increases. Decreasing the pressure angle increases the size of the cam, which can be undesirable when space is at a premium; furthermore, large cams require greater care in manufacture. Also, larger cams mean more mass, hence greater inertia forces which may lead to unwanted vibrations when the speeds are high. It is generally accepted that for roller followers the pressure angle should not exceed 30°, and 45° in the case of oscillating or swinging followers.

4.5 RADIUS OF THE ROLLER FOLLOWER

The radius of a roller follower must be chosen to ensure that the desired motion of the follower will be transmitted smoothly. Three possible situations are shown in Fig. 4.14, in which r is the radius of the follower and ρ is the radius of curvature of the pitch curve.

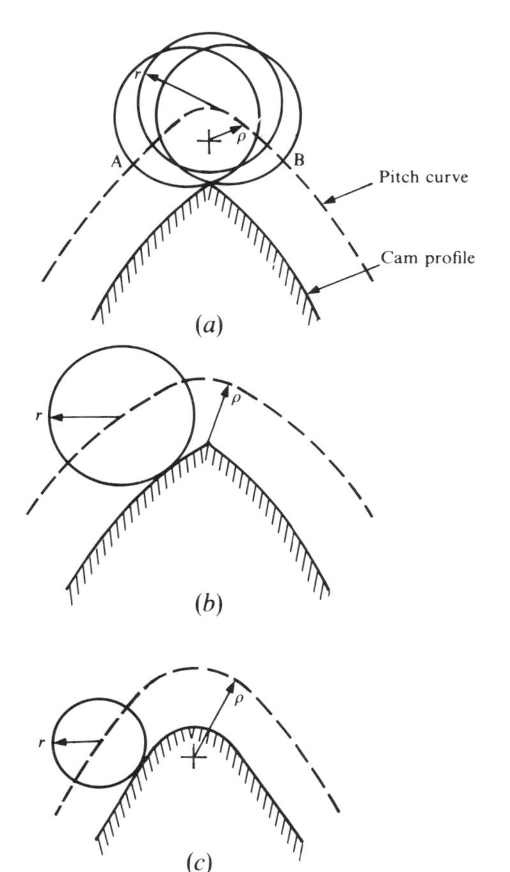

Figure 4.14 Effect of the radius of a roller follower on the profile of the cam.

(*a*) $r > \rho$: the cam profile crosses itself to allow the roller to move from A to B. This is obviously undesirable and is an impossible situation.

(*b*) $r = \rho$: this condition leads to a pointed cam profile, again an undesirable situation.

(*c*) $r < \rho$: In this case the cam profile has a finite value and the motion of the follower will be smooth.

EXERCISES

4.1 Draw the displacement, velocity, acceleration, and jerk as functions of the time for a follower having a lift of 20 mm in order to satisfy the following conditions: The rise takes place during 90° of cam rotation, and the fall during 75° of cam rotation. The follower is to dwell in the lifted position for 45° of cam rotation. The motion is parabolic and the cam rotates uniformly at 1440 rev/min.

4.2 If in Exercise 4.1 the motion is simple harmonic draw the displacement, velocity, acceleration, and jerk as functions of time and compare maximum values.

4.3 Repeat Exercise 4.2, but with the motion cycloidal, and draw the cam profile if the diameter of a roller follower is 15 mm and the base circle diameter is to have a minimum value of 40 mm. The follower is offset from the cam centreline by 15 mm.

4.4 A cam rotating with constant angular velocity operates an oscillating follower through a roller 19 mm diameter. The pivot of the follower is 50 mm from the axis of the cam and the distance from the centre of the roller to the pivot centre is 45 mm. The minimum radius of the cam is 25 mm and the angular displacement of the follower is 30°. The rise and fall of the follower each occupy 60° of cam rotation and there is no dwell in the lifted position. If the motion is simple harmonic, draw the cam profile and the velocity and acceleration diagrams of the follower when the cam rotates at 1500 rev/min.

4.5 A valve is operated by a cam having the following dimensions: base circle diameter = 19 mm, nose circle diameter = 13.5 mm, distance between the centres of these circles = 8.5 mm; the flanks of the cam are straight. The cam acts on a tappet roller of 25 mm diameter. The valve opens 25° before top dead centre and the camshaft rotates at half the engine speed of 3000 rev/min.

Calculate:

(*a*) the valve lift;
(*b*) the maximum velocity;
(*c*) the maximum acceleration of the valve.

4.6 To minimize the dynamic effects in a high-speed cam the rise and fall of a roller follower are required to follow a fifth-order polynomial in θ/β where θ is the cam position and β is the cam rotation corresponding to a rise h (refer to Sec. 4.3). The follower cycle for one revolution of the cam is as follows:

(*a*) a rise of 20 mm during 100° rotation of the cam;
(*b*) a dwell of 60°;
(*c*) a fall during the next 100° of cam rotation;
(*d*) a dwell for the remainder of cam rotation.

The roller is required to have a radius of 6.5 mm and the diameter of the base circle should not be less than 30 mm; the line of action of the follower is to have an offset of 12 mm.

Calculate the maximum velocity and acceleration of the follower when the cam rotates at a constant speed of 4500 rev/min. Draw the cam profile.

CHAPTER

FIVE

MATHEMATICAL ANALYSIS AND SYNTHESIS

5.1 INTRODUCTION

The graphical methods discussed previously are quite valuable in practice; they are lively in the sense that one can see the results at a glance and modifications can be undertaken with speed. These methods can still constitute powerful tools in any design office and their value should not be underestimated, e.g., rapid first solution and a check on the computer output. The computer, however, plays a very important role nowadays in the analysis and synthesis of mechanisms; this gives greater accuracy while interactive facilities save a considerable amount of time.

Conventional graphical methods have two main drawbacks, which should be clearly appreciated; they are as follows:

1. Accuracy is limited unless a large drawing board is used; it does to a large extent depend on the skill of the designer.
2. It is only possible to look at one position at a time. For example, in the case of a mechanism with a 360° cycle it will be necessary to draw as many as 24 positions in each case if the displacement, velocity, and acceleration are required; this is time-consuming.

Nevertheless, a great deal can be learnt graphically in the initial design of a mechanism before undertaking a complete analytical study.

The greater accuracy achievable using analytical techniques is particularly noticeable in the case of synthesis. With the use of programmable calculators or computers a satisfactory mechanism to achieve a given functional relationship between input and output can be obtained very quickly even if several iterations are necessary to arrive at a practical solution. The computer can also be used to draw the final mechanism if necessary. Furthermore, it is very much quicker to see the effect of a change in one or more dimensions on the output.

In this section we examine various mathematical methods for calculating displacement, velocities, accelerations, and jerks of the four-bar linkage. The methods can easily be extended to include N-bar linkages, e.g. the six-bar linkage of Example 3.4.

5.2 INPUT–OUTPUT RELATIONSHIP: FREUDENSTEIN'S EQUATION

When analysing a mechanism it is very important to define clearly a frame of reference for displacements, velocities, accelerations, jerks, forces, and moments before anything else. If this is not done mistakes will be made which may take a long time to discover; this is particularly important when using a computer.

Consider the four-bar linkage shown in Fig. 5.1; the frame of reference is A_0x, A_0y and the angular positions of the links are measured positively from the A_0x axis. The four-bar linkage may be regarded as being made up of four vectors A_0A, AB, BB_0 and A_0B_0 joined at A, B, B_0, and A_0 and of magnitude a, b, c, and d respectively. d is the frame or fixed link and μ is the transmission angle. In this case ϕ or link A_0A is the input and ψ or link B_0B the output.

The position for any value ϕ of the input can be expressed in three ways, as follows:

(a) By a vector equation

$$\overrightarrow{A_0A}+\overrightarrow{AB}=\overrightarrow{A_0B_0}+\overrightarrow{B_0B}$$

or more simply by

$$\mathbf{a}+\mathbf{b}=\mathbf{d}+\mathbf{c}$$

where

$$\begin{aligned}
\mathbf{a}&=a\cos\phi\mathbf{i}+a\sin\phi\mathbf{j}\\
\mathbf{b}&=b\cos\theta\mathbf{i}+b\sin\theta\mathbf{j}\\
\mathbf{c}&=c\cos(\psi-180)\mathbf{i}+c\sin(\psi-180)\mathbf{j}\\
&=-c\cos\psi\mathbf{i}-c\sin\psi\mathbf{j}\\
\mathbf{d}&=d\mathbf{i}
\end{aligned}$$

The angle θ is the angle the coupler link AB makes with an axis parallel to the A_0x axis as shown in Fig. 5.1.

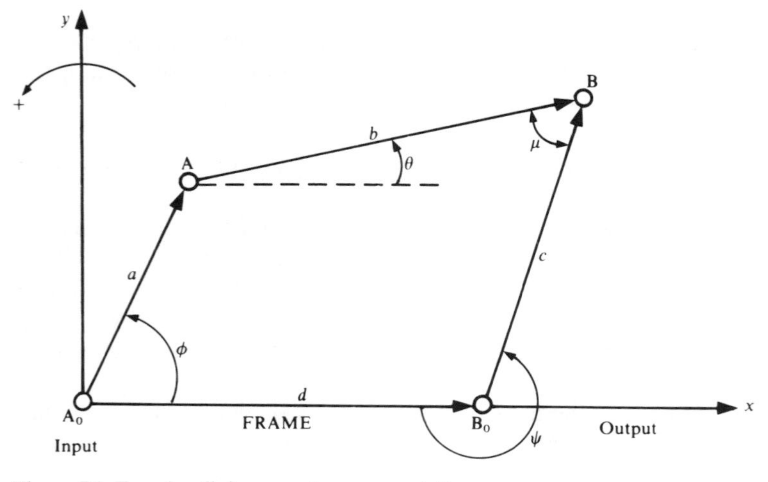

Figure 5.1 Four-bar linkage: vector representation.

(b) By use of complex numbers

$$a\,e^{j\phi}+b\,e^{j\theta}-c\,e^{j(\psi-180)}-d\,e^{j0}=0$$

where $e^{j\alpha}=\cos\alpha+j\sin\alpha$

This representation has certain advantages in calculating velocities and accelerations.

(c) By projection on the *x* and *y* axes (we adopt this method for the present)
Projection on the x axis gives

$$a\cos\phi+b\cos\theta=c\cos(\psi-180)+d \tag{5.1}$$

and projection on the y axis gives

$$a\sin\phi+b\sin\theta=c\sin(\psi-180) \tag{5.2}$$

Since we are interested in many cases in the output-to-input relationship, we can eliminate θ from Eqs 5.1 and 5.2. θ will however be required in considering the velocity and acceleration of the coupler link AB or when the output is taken from a point on the coupler or coupler extended, e.g., Sec. 1.2(f), Fig. 1.6 and Fig. 1.8. Equations (5.1) and (5.2) become

$$b\cos\theta=d-(a\cos\phi+c\cos\psi) \tag{5.3}$$

$$b\sin\theta=-(c\sin\psi+a\sin\theta) \tag{5.4}$$

Squaring and adding yields

$$b^2=a^2+c^2+d^2+2a\cos(\phi-\psi)-2ad\cos\phi-2cd\cos\psi$$

$$\Leftrightarrow\qquad \frac{b^2-(a^2+c^2+d^2)}{2ac}+\frac{d}{c}\cos\phi+\frac{d}{a}\cos\psi=\cos(\phi-\psi) \tag{5.5}$$

This equation may be written more simply as follows:

$$K_1\cos\phi+K_2\cos\psi-K_3=\cos(\phi-\psi) \tag{5.6}$$

where

$$\left.\begin{aligned}K_1&=d/c\\K_2&=d/a\\K_3&=(a^2-b^2+c^2+d^2)/2ac\end{aligned}\right\} \tag{5.7}$$

Equation (5.6) is known as *Freudenstein's equation.* It relates the input to output as a function of the size of the linkages and can be used in two ways.

For a given ϕ, the input, we can solve for ψ, the output, and hence, by differentiating with respect to time, obtain the velocity, acceleration, and jerk of the output in terms of the input velocity, acceleration, and jerk. In many practical situations the input has a constant velocity such as an electric motor rotating at constant speed or a hydraulic jack moving uniformly. It does not, however, follow that if the input moves uniformly the output will also have a uniform motion; this is due to the geometry of the mechanism, which transforms uniform motion into a non-uniform one (see Sec. 1.2).

The equation can also be used to design a mechanism from a required relationship between input and output, in which case the unknowns are the lengths of the links or, more frequently, the link ratios a/d, c/d, *and* b/d. If we fix three desired values of the output for three given values of the input (coordinating input and output positions) we then have three equations in three unknowns, and hence the size of the links for a given length of the fixed link d to suit whatever physical space is available. We shall see this in Sect. 5.7 when we consider synthesis.

To obtain an expression for the output ψ as a function of the input ϕ it is more convenient to rewrite Freudenstein's equation as follows:

$$K_1 \cos\phi + K_2 \cos\psi - K_3 = \cos\phi \cos\psi + \sin\phi \sin\psi$$

collecting terms involving ψ gives

$$\sin\psi \sin\phi + \cos\psi(\cos\phi - K_2) = K_1 \cos\phi - K_3$$

$$\Leftrightarrow \qquad A \sin\psi + B \cos\psi = C \tag{5.8}$$

where

$$\left.\begin{aligned} A &= \sin\phi \\ B &= (\cos\phi - K_2) \\ C &= (K_1 \cos\phi - K_3) \end{aligned}\right\} \tag{5.9}$$

To solve for ψ, we recall from trigonometry that

$$\sin\psi = \frac{2t}{1+t^2}$$

where $t = \tan\frac{1}{2}\psi$,

and

$$\cos\psi = \frac{1-t^2}{1+t^2}$$

Hence Eqn. (5.8) becomes

$$(B+C)t^2 - 2At + (C-B) = 0$$

This is a quadratic in t whose solution is

$$t = \frac{A \pm \sqrt{A^2 + B^2 - C^2}}{B+C} \tag{5.10}$$

There are two solutions, namely

$$\psi^+ = 2\tan^{-1}\left(\frac{A + \sqrt{A^2 + B^2 - C^2}}{B+C}\right) \tag{5.11}$$

corresponding to the positive sign, and

$$\psi^- = 2\tan^{-1}\left(\frac{A - \sqrt{A^2 + B^2 - C^2}}{B+C}\right) \tag{5.12}$$

corresponding to the negative sign.

The two values correspond to the two possible ways the linkage can be assembled as shown in Fig. 5.2 for given values of a, b, c, d, and ϕ. A_0ABB_0 will be referred to as the 'OPEN' linkage and $A_0AB'B_0$ as the 'CROSSED' linkage.

Example 5.1 A four-bar linkage has the following dimensions: crank = 50 mm, coupler = 200 mm, follower = 150 mm, fixed link = 205 mm, and the crank rotates uniformly at 240 rev/min.

Calculate the angular displacement, velocity and acceleration for one complete revol-

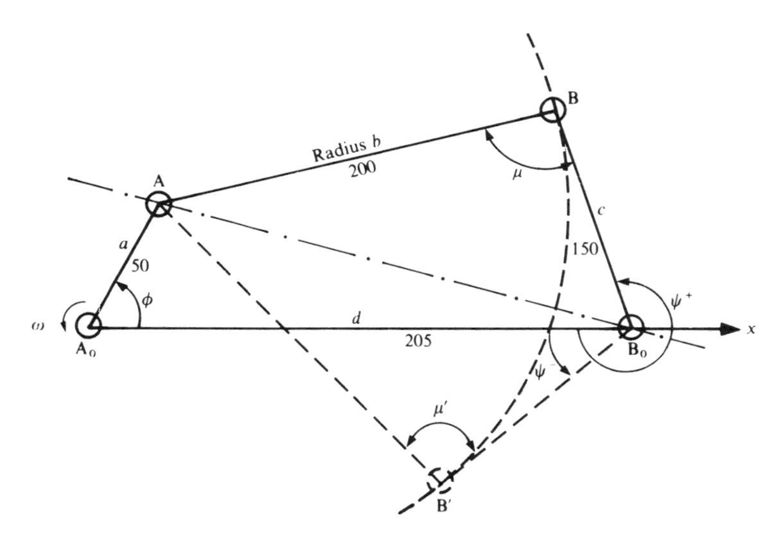

Figure 5.2 The two possible ways of assembling a four-bar linkage.

ution of the crank for each mechanism, i.e., corresponding to the two ways the mechanism can be assembled. Check that the mechanism obeys Grashof's criterion.

SOLUTION *Grashof's criterion*:

$$\text{Shortest} + \text{longest links} \leqslant \text{sum of the other two}$$

Shortest = 50, longest = 205.

Sum of the other two = 150 + 200 = 350.

Hence the conditions are satisfied and one complete revolution of the input link is possible.

Before we can calculate the angular displacement of each mechanism we need to know the values of K_1, K_2, K_3:

$$K_1 = d/c = 205/150 = 1.3667$$
$$K_2 = d/a = 205/50 = 4.1$$
$$K_3 = (50^2 - 200^2 + 150^2 + 205^2)/(2 \times 50 \times 150) = 1.8017$$

Using a programmable calculator or a computer we can give to the input ϕ values from 0 to 360° in steps of 15°, for example; substitution in Eqs (5.11) and (5.12) yields the results shown in Table 5.1.

From the table, total angular displacement of mechanism 1 is equal to 88.00–47.00 = 41° for a rotation of 225 − 30 = 195° of the input. Total angular displacement of mechanism 2 = 87 − 46 = 41° for a rotation of 330 − 135 = 195° of the input.

Thus we see that, in this case, both mechanisms have the same total angular displacement for the same rotation of the input crank but for different positions of the latter. The motion of the output is oscillatory and each mechanism is of the crank–rocker type.

Calculation of the velocity and acceleration will be discussed in sec. 5.4.

The two mechanisms in their extreme positions are shown in Fig. 5.3.

Table 5.1

Mechanism 1		Mechanism 2	Mechanism 1		Mechanism 2	
ϕ,°	ψ^+,°	ψ^-,°	$\dot{\psi}/\dot{\phi}$	$\ddot{\psi}/\dot{\phi}^2$	$\dot{\psi}/\dot{\phi}$	$\ddot{\psi}/\dot{\Phi}^2$
0	−82	81	−0.323	0.384	−0.323	−0.385
15	−86	76	−0.203	0.516	−0.397	−0.174
30	−88	70	−0.063	0.532	−0.415	0.032
45	−88	64	0.069	0.462	−0.386	0.172
60	−86	58	0.176	0.355	−0.33	0.242
75	−83	54	0.255	0.244	−0.263	0.264
90	−79	51	0.305	0.143	−0.194	0.264
105	−74	48	0.331	0.056	−0.125	0.258
120	−69	47	0.336	−0.02	−0.058	0.253
135	−64	46	0.322	−0.086	8E-03	0.25
150	−60	47	0.292	−0.142	0.073	0.245
165	−55	49	0.249	−0.186	0.136	0.236
180	−52	51	0.196	−0.218	0.196	0.217
195	−50	54	0.136	−0.237	0.249	0.185
210	−48	59	0.073	−0.246	0.292	0.141
225	−47	63	8E-03	−0.251	0.322	0.085
240	−48	68	−0.058	−0.254	0.336	0.019
255	−49	73	−0.125	−0.259	0.331	−0.057
270	−52	78	−0.194	−0.265	0.305	−0.144
285	−55	82	−0.263	−0.265	0.255	−0.245
300	−59	85	−0.33	−0.243	0.176	−0.356
315	−65	87	−0.386	−0.173	0.069	−0.463
330	−71	87	−0.415	−0.033	−0.063	−0.533
345	−77	85	−0.397	0.173	−0.203	−0.517
360	−82	81	−0.323	0.384	−0.323	−0.385

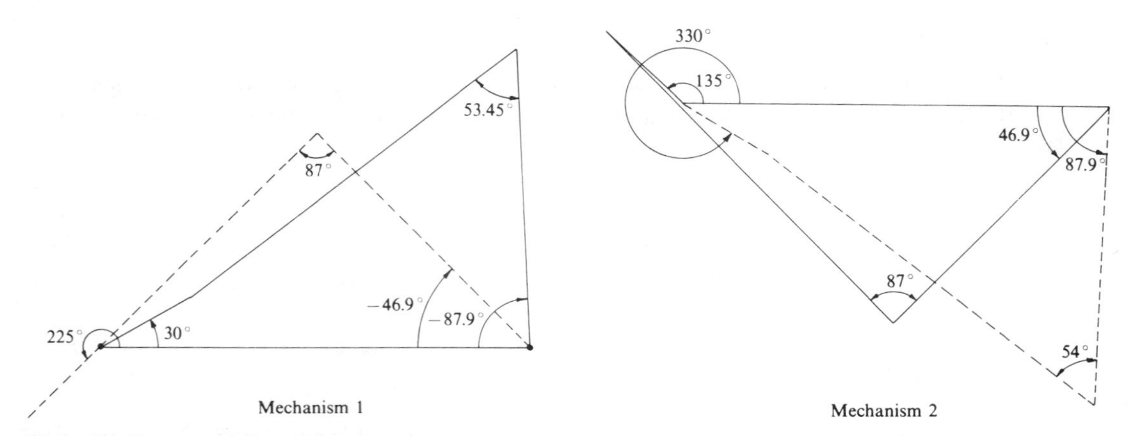

Figure 5.3 Extreme positions of the four-bar linkage of Example 5.1.

5.3 TRANSMISSION ANGLE

For the smooth operation of any mechanism without jerky movements it is important to ensure that the transmission angle μ (see Fig. 5.1) lies between 40° and 140°; the ideal value is 90° but this cannot remain constant during a cycle.

Referring to Fig. 5.1 or Fig. 5.2, we have

$$\begin{aligned} AB_0^2 &= a^2 + d^2 - 2ad \cos \phi \\ &= b^2 + c^2 - 2bc \cos \mu \end{aligned}$$

Solving for $\cos \mu$ yields

$$\cos \mu = \frac{(b^2 + c^2) - (a^2 + d^2) + 2ad \cos \phi}{2bc} \tag{5.13}$$

This equation is valid for all configurations of the mechanism, which readers should verify for themselves.

Example 5.2 Calculate the variation in the transmission angle for mechanisms 1 and 2 of Example 5.1.

SOLUTION Referring to Fig. 5.2, we see from the geometry that $\mu = \mu'$, so that substitution in Eq. (5.13) for $\phi = 0$ to $360°$ in steps of $30°$ gives the values for μ shown in Table 5.2.

Table 5.2

ϕ,°	μ,°
0,360	50.11
30	53.45
60	61.93
90	72.56
120	82.60
150	89.78
180	92.41
210	89.78
240	82.60
270	72.56
300	61.93
330	53.45

The results show that for both mechanisms the transmission angle is acceptable, being greater than 40° throughout one cycle of operation. It increases from 50.11° for $\phi = 0$ to 92.41° for the first half-cycle and decreases to 50.11° during the second half of the cycle. We conclude from this that both mechanisms should operate smoothly even when friction at the joints is taken into account.

5.4 VELOCITY AND ACCELERATION OF THE OUTPUT

Let us recall Freudenstein's equation (Eq. (5.6)):

$$K_1 \cos \phi + K_2 \cos \psi - K_3 = \cos(\phi - \psi)$$

To obtain a velocity relationship we differentiate Freudenstein's equation with respect to time; this leads to

$$-K_1 \sin \phi \dot{\phi} - K_2 \sin \psi \dot{\psi} = -\sin(\phi - \psi)(\dot{\phi} - \dot{\psi}) \tag{5.14}$$

Solving for $\dot{\psi}$, the output angular velocity, yields

$$\dot{\psi}=\frac{\sin(\phi-\psi)-K_1\sin\phi}{\sin(\phi-\psi)+K_2\sin\psi}\dot{\phi} \qquad \text{or} \qquad \dot{\psi}=G_1\dot{\phi} \tag{5.15}$$

where $\dot{\phi}$ is the angular velocity of the input and G is the velocity ratio. We see that G_1 is a function of ϕ and ψ for a given mechanism.

Let
$$\dot{\psi}=\frac{z_1}{z_2}\dot{\phi} \qquad \text{i.e.} \quad G_1=z_1/z_2 \tag{5.16}$$

where
$$z_1=\sin(\phi-\psi)-K_1\sin\phi$$
$$z_2=\sin(\phi-\psi)+K_2\sin\psi$$

The acceleration is obtained by differentiating Eq. (5.16); this leads to

$$\ddot{\psi}=\frac{\mathrm{d}}{\mathrm{d}t}(G_1\dot{\phi})=\dot{\phi}\frac{\mathrm{d}G_1}{\mathrm{d}t}+G_1\ddot{\phi}$$

$$=\dot{\phi}\frac{\mathrm{d}G_1}{\mathrm{d}\phi}\cdot\frac{\mathrm{d}\phi}{\mathrm{d}t}+G_1\ddot{\phi}$$

Hence
$$\ddot{\psi}=G_1'\dot{\phi}^2+G_1\ddot{\phi} \tag{5.17}$$
$$=\frac{z_2z_1'-z_1z_2'}{z_2^2}\dot{\phi}^2+\frac{z_1}{z_2}\ddot{\phi}$$

where a prime ($'$) represents differentiation with respect to the input ϕ.

Hence
$$z_1'=(1-G_1)\cos(\phi-\psi)-K_1\cos\phi \tag{5.18}$$

and
$$z_2'=(1-G_1)\cos(\phi-\psi)+G_1K_2\cos\psi \tag{5.19}$$

Equations (5.15) and (5.17) are best left in this form, as they are not easily simplified. In any case, when computing $\dot{\psi}$ or $\ddot{\psi}$ it is prudent to express these equations in terms of intermediate parameters such as z_1 and z_2 to avoid errors during calculations.

We notice that before computing the output angular velocity and acceleration it is necessary to calculate the value of the output angle ψ from Eq. (5.11) or (5.12), depending on the type of mechanism used.

It can also be seen from Eq. (5.17) that if the input has a constant angular velocity, so that $\ddot{\phi}=0$, the output does not rotate with constant angular velocity unless

$$z_2z_1'=z_1z_2' \qquad \Leftrightarrow \qquad \frac{\dot{\psi}}{\dot{\phi}}=\frac{z_1'}{z_2'}$$

which is very unlikely in practice. The only four-bar linkage with this property is the parallelogram where $c=a$, and hence $\psi-180=\phi$; by substituting in Eq. (5.15) we get

$$\dot{\psi}=\frac{0-d/c\sin\phi}{0+d/a\sin(180+\phi)}\dot{\phi}\Rightarrow\dot{\psi}=\dot{\phi}$$

The velocity ratio $\dot{\psi}/\dot{\phi}$ and acceleration ratio $\ddot{\psi}/\dot{\phi}^2$ for mechanisms 1 and 2 of Example 5.1 have been computed and the results are shown in Table 5.1.

The results for mechanism 1 have been plotted and are shown in Fig. 5.4(*a*). Such a

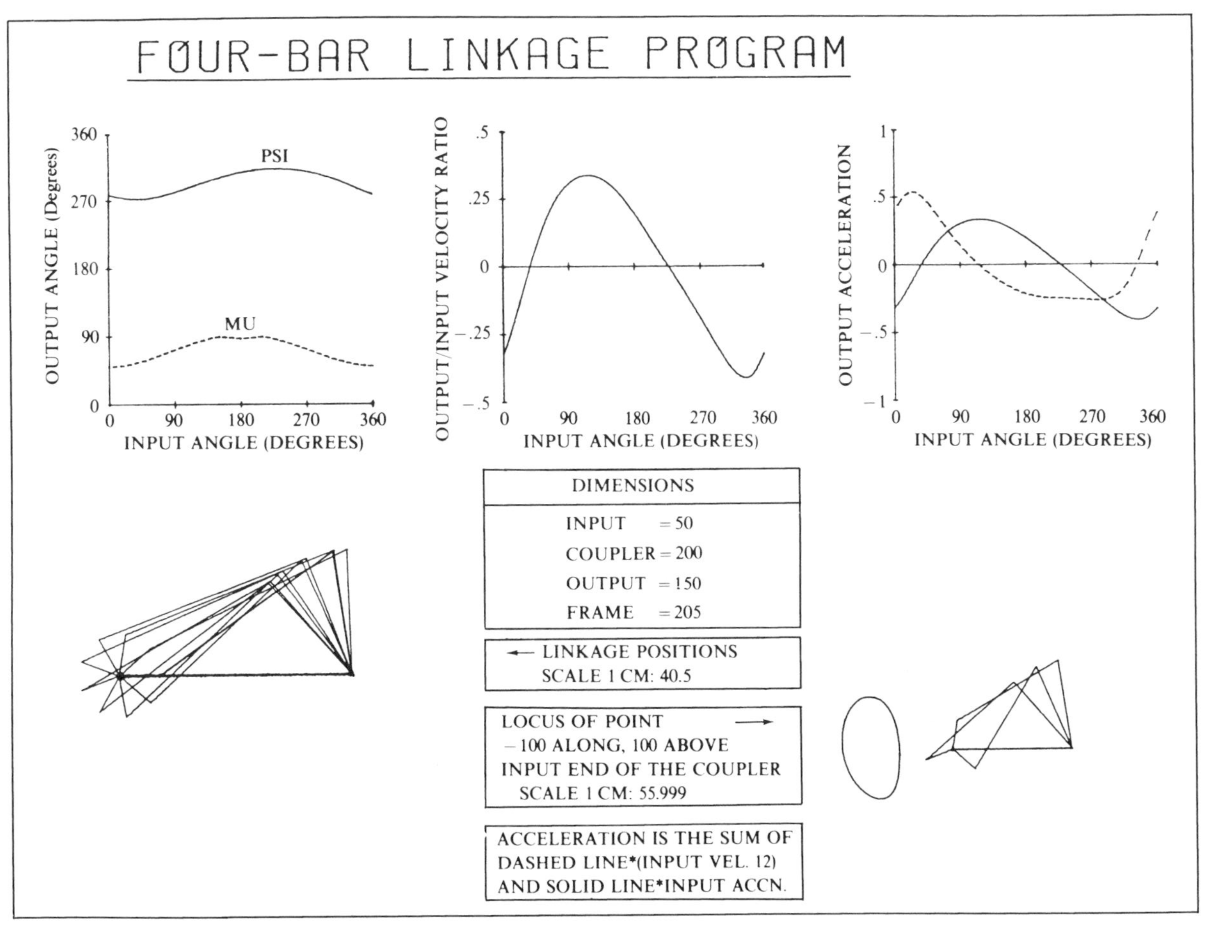

Figure 5.4a Computer printout for Example 5.1.

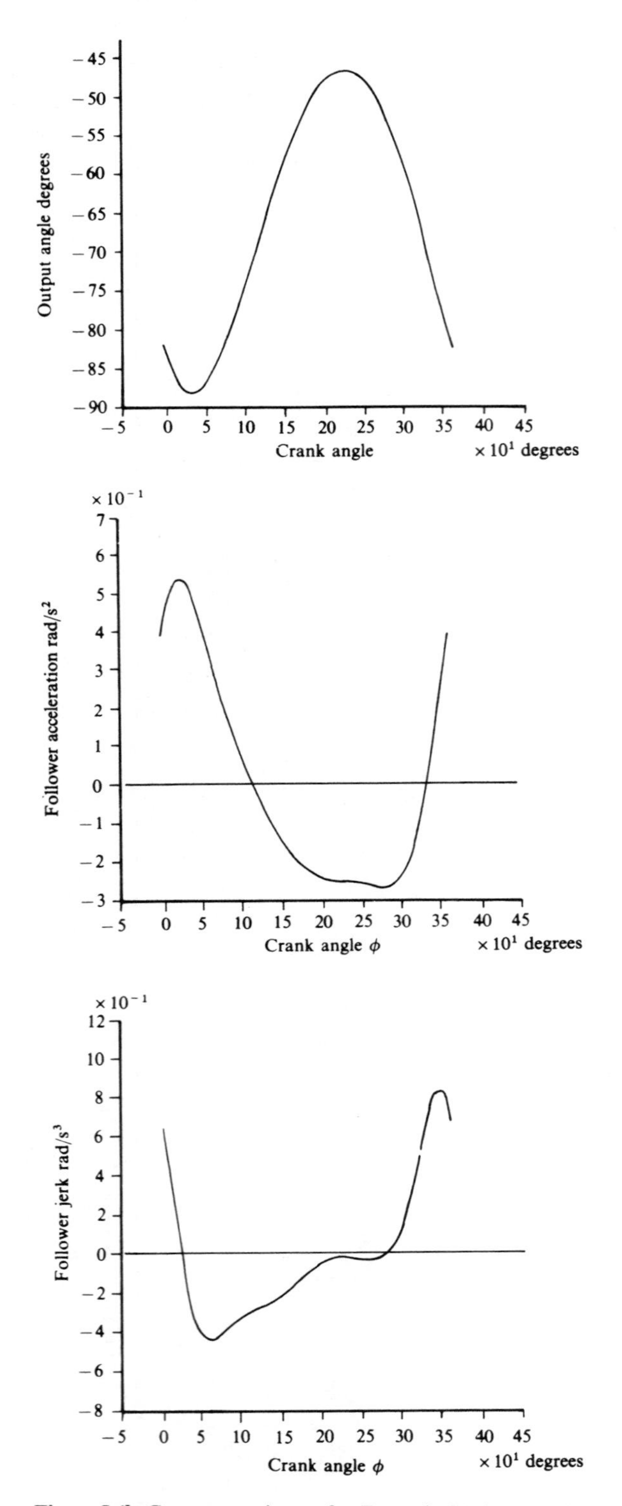

Figure 5.4b Computer printout for Example 5.1 (*continued*)

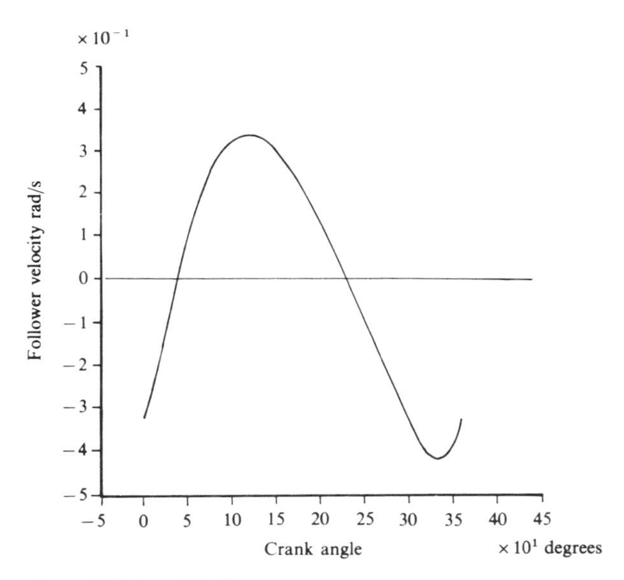

Figure 5.4b *(concluded)*

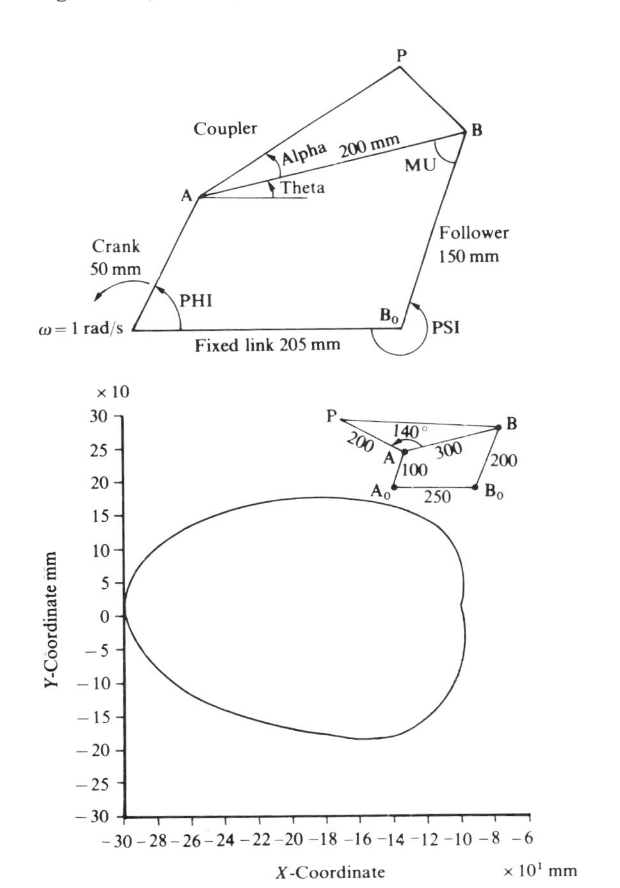

Figure 5.5 Computer plot of the motion of a coupler point.

presentation has obvious advantages over the tabular method because we can see at a glance all the maxima and minima.[3]

5.4.1 Derivative of the Acceleration: Jerk

To obtain an expression for the jerk we differentiate Eq. (5.17) with respect to the time and obtain

$$\dddot{\psi} = \frac{d}{dt}(G_1'\dot{\phi}^2) + \frac{d}{dt}(G_1\ddot{\phi})$$

$$= \dot{\phi}^2 \frac{d}{dt}G_1' + G_1' 2\dot{\phi}\ddot{\phi} + \ddot{\phi}\frac{d}{dt}G_1 + G_1\dddot{\phi}$$

$$= G_1''\dot{\phi}^3 + 3G_1'\dot{\phi}\ddot{\phi} + G_1\dddot{\phi} \tag{5.20}$$

since
$$\frac{d}{dt}G_1' = \frac{d}{d\phi}G_1' \cdot \frac{d\phi}{dt} = G_1''\dot{\phi}$$

$$G_1'' = \frac{z_2 z_1'' - z_1 z_2''}{z_2^2} - \frac{z_2 z_1' - z_1 z_2'}{z_2^2}\left(\frac{2z_2'}{z_2}\right)$$

where
$$z_1'' = -(1-G_1)^2 \sin(\phi-\psi) - G_1' \cos(\phi-\psi) + K_1 \sin\phi$$

$$z_2'' = -(1-G_1)^2 \sin(\phi-\psi) - G_1' \cos(\phi-\psi) - G_1^2 K_2 \sin\psi + G_1' K_2 \cos\psi$$

Figure 5.4(b) is another example of a computer printout[4] showing the output position, velocity, acceleration, and jerk of the linkage (Example 5.1) for 1 rad/s input. If the angular velocity of the input is ω rad/s the values shown on the graphs are to be multiplied by ω, ω^2, and ω^3 to obtain the actual velocity, acceleration, and jerk respectively for this particular linkage.

Figure 5.5 shows the path generated by point P on the coupler ABP for the linkage having the dimensions shown using the above computer program. The X and Y coordinates are measured relative to A_0.

These graphs illustrate clearly the importance of visual displays when we need to see quickly the performance of the linkage we are designing or that of an existing one we wish to analyse or modify to satisfy a particular requirement.

5.5 POSITION, VELOCITY AND ACCELERATION OF THE COUPLER

To obtain an expression for the position of the coupler as a function of the input angle ϕ we write Eqs (5.3) and (5.4) as follows:

$$c\cos\psi = d - (a\cos\phi + b\cos\theta)$$
$$c\sin\psi = -(a\sin\phi + b\sin\theta)$$

Squaring, adding and collecting terms to eliminate ψ leads to

$$(\sin\phi)\sin\theta+\left(\cos\phi-\frac{d}{a}\right)\cos\theta=\frac{c^2-(a^2+b^2+d^2)}{2ab}+\frac{d}{b}\cos\phi$$

$$\Leftrightarrow \qquad D\sin\theta+E\cos\theta=F \tag{5.21}$$

where

$$D=\sin\phi$$

$$E=\cos\phi-\frac{d}{a}$$

$$F=\frac{c^2-(a^2+b^2+d^2)}{2ab}+\frac{d}{b}\cos\phi$$

To solve for θ we note once again that

$$\sin\theta=\frac{2t}{1+t^2} \qquad \cos\theta=\frac{1-t^2}{1+t^2}$$

where $t=\tan\theta/2$.

Hence Eq. (5.21) becomes

$$(E+F)t^2-2Dt+(F-E)=0$$

Solving for θ yields

$$\theta=2\arctan\left(\frac{D\pm\sqrt{D^2+E^2-F^2}}{(E+F)}\right) \tag{5.22}$$

Thus there are two values for θ:
θ^+, corresponding to the positive sign;
θ^-, corresponding to the negative sign.

The positions of the coupler in Example 5.1 have been computed and are shown in Table 5.3.

Table 5.3

ϕ,°	Mechanism 1 θ^-,°	Mechanism 2 θ^+,°
0,360	47.95	−47.95
30	38.64	−56.22
60	32.11	−59.16
90	28.99	−56.41
120	28.80	−50.12
150	31.19	−42.69
180	35.99	−35.99
210	42.69	−31.19
240	50.12	−28.80
270	56.41	−28.99
300	59.16	−32.11
330	56.22	−38.64

To obtain expressions for the angular velocity and acceleration of the coupler we differentiate Eq. (5.21) implicitly with respect to time, getting

$$D \cos\theta\dot{\theta} + \sin\theta\dot{D} - E \sin\theta\dot{\theta} + \cos\theta\dot{E} = \dot{F},$$

where

$$\dot{D} = \cos\phi\dot{\phi}$$

$$\dot{E} = -\sin\phi\dot{\phi}$$

$$\dot{F} = -\frac{d}{b}\sin\phi\dot{\phi}$$

Substitution in the above equation yields the angular velocity of the coupler:

$$\dot{\theta} = \left(\frac{\sin(\phi-\theta) - \dfrac{d}{b}\sin\phi}{\sin(\phi-\theta) + \dfrac{d}{a}\sin\theta} \right)\dot{\phi} \qquad \text{or} \qquad \dot{\theta} = G_c\dot{\phi} \tag{5.23}$$

where G_c, the velocity ratio (the suffix 'c' stands for 'coupler'), is a function of ϕ and θ. The equation is similar to Eq. (5.15) for the output link B_0B.

Let

$$y_1 = \sin(\phi-\theta) - \frac{d}{b}\sin\phi$$

$$y_2 = \sin(\phi-\theta) + \frac{d}{a}\sin\theta$$

Differentiating Eq. (5.23) yields the angular acceleration of the coupler:

$$\ddot{\theta} = G_c'\dot{\phi}^2 + G_c\ddot{\phi} = \frac{y_2y_1' - y_1y_2'}{y_2^2}\dot{\phi}^2 + G_c\ddot{\phi} \tag{5.24}$$

where the dashes refer to differentiation with respect to ϕ, so that

$$y_1' = (1-G_c)\cos(\phi-\theta) - \frac{d}{b}\cos\phi$$

and

$$y_2' = (1-G_c)\cos(\phi-\theta) + G_c\frac{d}{a}\cos\theta \tag{5.24a}$$

The reader is advised to tabulate the angular velocity and acceleration for mechanism 1 of Example 5.1 for one cycle of the input to gain experience in handling complex expressions. The reader should also appreciate that once the equations have been programmed in a suitable 'computer' the influence on the output parameters and transmission angles due to changes in one or more dimensions of the mechanism takes very little time to compute, which cannot be said of any graphical techniques. It is, however, wise and good practice to draw the mechanism in one position and measure the output and transmission angles as a verification of the program, as shown in Fig. 5.6 for mechanism 1 of Example 5.1.

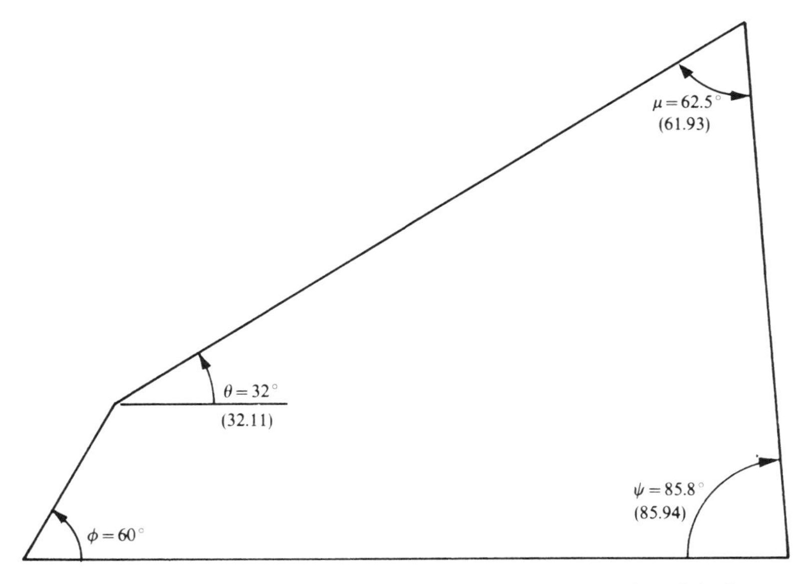

Figure 5.6 Graphical verification of computed values for one position of the input crank.

5.6 VELOCITY AND ACCELERATION USING COMPLEX NUMBERS

We return to the vector representation of the four-bar linkage mentioned in Sec. 5.2, i.e.,

$$\mathbf{a}+\mathbf{b}=\mathbf{c}+\mathbf{d} \tag{5.25}$$

and express each vector in its complex form thus:

$\mathbf{a}=a\,\mathrm{e}^{\mathrm{j}\phi}$
$\mathbf{b}=b\,\mathrm{e}^{\mathrm{j}\theta}$
$\mathbf{c}=-c\,\mathrm{e}^{\mathrm{j}\psi}$
$\mathbf{d}=d\,\mathrm{e}^{\mathrm{j}0}=d$ since this vector makes an angle of 0° with the x-axis

The above equation becomes

$$a\,\mathrm{e}^{\mathrm{j}\phi}+b\,\mathrm{e}^{\mathrm{j}\theta}+c\,\mathrm{e}^{\mathrm{j}\psi}-d=0 \tag{5.25a}$$

Differentiating twice with respect to time yields

$$\mathrm{j}\dot{\phi}a\,\mathrm{e}^{\mathrm{j}\phi}+\mathrm{j}\dot{\theta}b\,\mathrm{e}^{\mathrm{j}\theta}+\mathrm{j}\dot{\psi}c\,\mathrm{e}^{\mathrm{j}\psi}=0 \tag{5.26}$$

and

$$(\mathrm{j}\ddot{\phi}-\dot{\phi}^2)a\,\mathrm{e}^{\mathrm{j}\phi}+(\mathrm{j}\ddot{\theta}-\dot{\theta}^2)b\,\mathrm{e}^{\mathrm{j}\theta}+(\mathrm{j}\ddot{\psi}-\dot{\psi}^2)c\,\mathrm{e}^{\mathrm{j}\psi} \tag{5.27}$$

From Eq. (5.26) we observe that point A (Fig. 5.1) has a velocity equal to $a\dot{\phi}$ perpendicular to A_0A, since multiplication by j rotates a vector through 90°. Similarly, the velocity of B relative to A is $b\dot{\theta}$ at right angles to AB while that of B on B_0B is $c\dot{\psi}$.

In the same way an examinaton of Eq. (5.27) shows that the acceleration of point A on A_0A has a component equal to $a\ddot{\phi}$ at 90° to A_0A and another component equal to $a\dot{\phi}^2$ along AA_0, i.e., from A towards A_0. Similarly, the acceleration of B relative to A has a component of

$b\ddot{\theta}$ perpendicular to AB and a component of $b\dot{\theta}^2$ directed from B towards A. The point B on B_0B has a component equal to $c\ddot{\psi}$ perpendicular to B_0B and another equal to $c\dot{\psi}^2$ from B towards B_0.

Expressions for the angular velocities of the output and of the coupler are obtained by replacing each exponential term by its trigonometrical identity and equating real and imaginary parts. This results in the following two equations:

$$\dot{\theta}b \cos \theta + \dot{\psi}c \cos \psi = -\dot{\phi}a \cos \phi$$
$$\dot{\theta}b \sin \theta + \dot{\psi}c \sin \psi = -\dot{\phi}a \sin \phi$$

Solving for $\dot{\psi}$, the output velocity, and $\dot{\theta}$, the coupler velocity, yields

$$\dot{\psi} = \frac{a}{c}\frac{\sin(\theta - \phi)}{\sin(\psi - \theta)}\dot{\phi} \tag{5.28}$$

and

$$\dot{\theta} = \frac{a}{b}\frac{\sin(\phi - \psi)}{\sin(\psi - \theta)}\dot{\phi} \tag{5.29}$$

We notice that when these expressions are used it is necessary to calculate the angular position of the coupler, unlike Eqs (5.15) and (5.23) which are independent of θ. In the case of mechanism 1 of Example 5.1 for $\phi = 60°$, the corresponding values of ψ and θ are $-85.94°$ and $32.11°$ from Tables 5.1 and 5.3.

Substitution in Eq. (5.28) yields

$$\frac{\dot{\psi}}{\dot{\phi}} = \frac{50}{150} \times \frac{\sin(32.11 - 60)}{\sin(-85.99 - 32.11)} = 0.17$$

which is the value computed using Eq. 5.15 and shown in Table 5.1.

To derive expressions for the angular accelerations of the output and of the coupler we replace the exponentials by their trigonometrical identities in Eq. (5.27) and equate real and imaginary parts and get

$$-\ddot{\theta}b \sin \theta - \ddot{\psi}c \sin \psi = \ddot{\phi}a \sin \phi + \dot{\phi}^2 a \cos \phi + \dot{\theta}^2 b \cos \theta + \dot{\psi}^2 c \cos \psi$$
$$\ddot{\theta}b \cos \theta + \ddot{\psi}c \cos \psi = -\ddot{\phi}a \cos \phi + \dot{\phi}^2 a \sin \phi + \dot{\theta}^2 b \sin \theta + \dot{\psi}^2 c \sin \psi$$

Let

$$\Delta = bc(\sin \psi \cos \theta - \cos \psi \sin \theta) = bc \sin(\psi - \theta)$$

The angular acceleration of the output is then given by

$$\ddot{\psi} = \frac{AF - CD}{\Delta} \tag{5.30}$$

and that of the coupler by

$$\ddot{\theta} = \frac{CE - BF}{\Delta} \tag{5.31}$$

where

$$A = -b \sin \theta$$
$$B = -c \sin \psi$$
$$C = \ddot{\phi}a \sin \phi + \dot{\phi}^2 a \cos \phi + \dot{\theta}^2 b \cos \theta + \dot{\psi}^2 c \cos \psi$$
$$D = b \cos \theta$$
$$E = c \cos \psi$$
$$F = -\ddot{\phi}a \cos \phi + \dot{\phi}^2 a \sin \phi + \dot{\theta}^2 b \sin \theta + \dot{\psi}^2 c \sin \psi$$

As for velocity we see that the acceleration depends on θ and ψ as well as on $\dot{\theta}$ and $\dot{\psi}$. In practice if only the output angular velocity and acceleration are required it will be preferable to use Eqs (5.15) and (5.17) which are independent of θ and $\dot{\theta}$.

Example 5.2 Figure 5.7 shows a floating dockside crane whose design consists of the four-bar linkage A_0ABCB_0 where C is a point on the coupler ABC, i.e., the jib. Link AA_0 is the driving member which rotates about an axis through A_0, and is driven by an electric motor whose speed is constant at 720 rev/min, via a quadruple-reduction gear box of 1430 ratio.

Calculate analytically the luffing speed, the change in the elevation of the point C, and the sinking speed as the input link AA_o rotates from $\phi = 60°$ to $\phi = 140°$ in steps of 10°.

Solution To solve this problem we could use Eq. (5.23) for the angular velocity $\dot{\theta}$ of the coupler (the jib of the crane in this case), remembering that the fixed link is $A_oB_o = d$ and that the angles ϕ, θ, and ψ are measured relative to this link. Instead we propose to develop the necessary equations using the angles ϕ, θ, and ψ shown in Fig. 5.8 and the dimensions e and f since these are as defined by the manufacturer. ($a = 22.05$ m; $b = 9.75$ m; $c = 28.95$ m; $e = 7.95$ m; $f = 9.60$ m; $L = 33.45$ m; CB = 24 m.)

Let Oxy be a set of Cartesian axes as shown with Ox horizontal through B_o and Oy vertical through A_o. Furthermore, let ϕ, ψ, and θ be measured in the directions indicated relative to the x axis.

Resolving along Ox and Oy yields

$$a\cos\phi + b\cos\theta - c\cos\psi = f \tag{5.32}$$

$$a\sin\phi + b\sin\theta + e = c\sin\psi \tag{5.33}$$

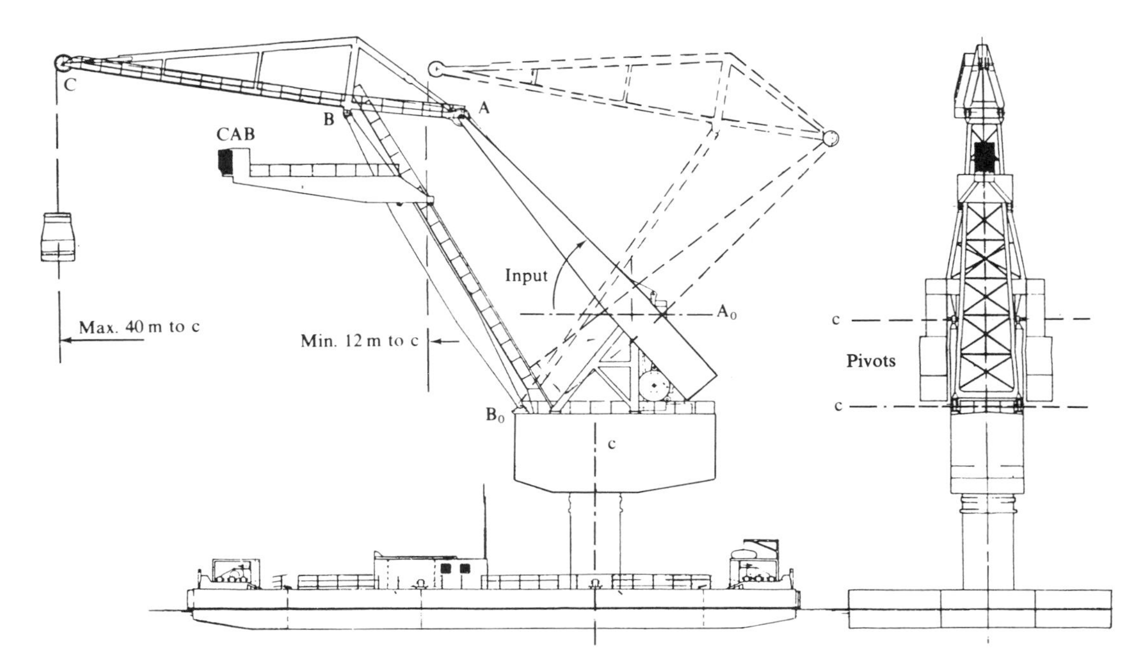

Figure 5.7

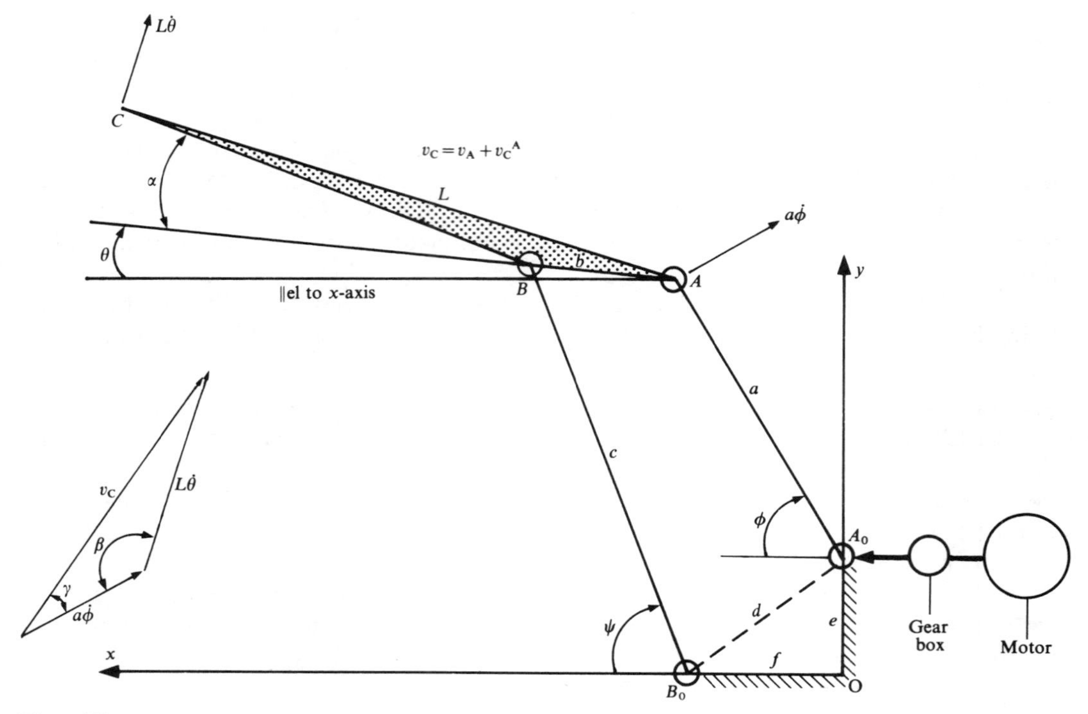

Figure 5.8

Since we will not require the angle ψ to calculate the velocity of C we can eliminate it from these two equations.

Rearranging Eqs (5.32) and (5.33) we get

$$c\cos\psi = a\cos\phi + b\cos\theta - f$$
$$c\sin\psi = a\sin\phi + b\sin\theta + e$$

Squaring and adding and a little algebra yields the following equation:

$$X\cos\theta + Y\sin\theta + Z = 0 \tag{5.34}$$

where
$$X = \cos\phi - \frac{f}{a}$$
$$Y = \sin\phi + \frac{e}{a}$$
$$Z = \frac{e}{b}\sin\phi - \frac{f}{b}\cos\phi + \frac{a^2+b^2+e^2+f^2-c^2}{2ab}$$

By replacing $\sin\theta$ by $2t/1+t^2$ and $\cos\theta$ by $(1-t^2)/(1+t^2)$, where $t=\tan(\theta/2)$, we obtain a quadratic in t, i.e.,

$$(Z-X)t^2 + 2Yt + (Z+X) = 0$$

Solving for t yields

$$t=\left(\frac{-Y\pm\sqrt{Y^2+X^2-Z^2}}{Z-X}\right)$$

and θ, the angle the jib makes with the horizontal, is given by

$$\theta=2\arctan\left(\frac{-Y\pm\sqrt{Y^2+X^2-Z^2}}{Z-X}\right) \tag{5.35}$$

Before we can calculate the velocity V_c of the point C on the jib we require the angular velocity $\dot{\theta}$ of the jib. This is obtained by differentiating Eq. (5.34) implicitly with respect to time so that:

$$-X\sin\theta\dot{\theta}+\dot{X}\cos\theta+Y\sin\theta\dot{\theta}+\dot{Y}\sin\theta+\dot{Z}=0$$

Solving for $\dot{\theta}$ we get

$$\dot{\theta}=\frac{\dot{Z}+\dot{X}\cos\theta+\dot{Y}\sin\theta}{X\sin\theta-Y\cos\theta}=\frac{Z'+X'\cos\theta+Y'\sin\theta}{X\sin\theta-Y\cos\theta}\dot{\phi} \tag{5.36}$$

where the dashes refer to differentiation with respect to ϕ, so that

$$X'=-\sin\phi$$

$$Y'=\cos\phi$$

$$Z'=\frac{e}{b}\cos\phi+\frac{f}{b}\sin\phi$$

since $$\dot{X}=\frac{dX}{dt}=\frac{dX}{d\phi}\cdot\frac{d\phi}{dt} \quad \text{etc.}$$

The absolute velocity V_c of C is made up of two terms: its velocity $V_{CA}=L\dot{\theta}$ relative to A and the velocity $V_A=a\dot{\phi}$ of A.

Hence $$\mathbf{V}_c=\mathbf{V}_A+\mathbf{V}_{cA},$$

and the magnitude of V_c is given by (see inset in Fig. 5.8)

$$V_c=\sqrt{(L\dot{\theta})^2+(a\dot{\phi})^2-2L\dot{\theta}a\dot{\phi}\cos\beta}$$

where $\beta=180-(\phi-\theta-\alpha)$, by the cosine rule

or $$V_c=\dot{\phi}\sqrt{k^2L^2+a^2+2kLa\cos(\phi-\theta-\alpha)} \tag{5.37}$$

where $k=\dot{\theta}/\dot{\phi}$.

The *luffing speed* is the horizontal speed of C given by

$$V_{cx}=V_c\cos(90-\phi+\gamma)=V_c\sin(\phi-\gamma)=K\dot{\phi}\sin(\phi-\gamma) \tag{5.38}$$

where $$K=\sqrt{k^2L^2+a^2+2kLa\cos(\phi-\theta-\alpha)}$$

and the *sinking speed* is the vertical speed of C, given by

$$V_{cy}=K\dot{\phi}\cos(\phi-\gamma) \tag{5.39}$$

where $$\gamma=\arctan\left(\frac{Lk\sin(\phi-\theta+\alpha)}{a+Lk\cos(\phi-\theta+\alpha)}\right) \tag{5.40}$$

Table 5.4

ϕ,°	θ,°	k	$V_c/\dot{\phi}$	V_{cx}, m/min	V_{cy}, m/min
60	−1.98	−0.308	17.32	54.7	2.8
70	−4.43	−0.187	19.94	62.8	4.3
80	−5.79	−0.087	21.42	68.5	3.0
90	−6.18	0.007	22.08	70.7	0.7
100	−5.63	0.105	22.11	70.8	−1.1
110	−4.04	0.214	21.72	69.5	−1.4
120	−1.28	0.339	21.21	67.9	−0.4
130	2.81	0.480	21.04	67.2	4.3
140	8.38	0.633	21.76	69.0	9.5

The angular velocity $\dot{\phi}$ of the link AA_o is

$$\dot{\phi} = 2\pi N/G$$

where N = motor speed and G = gear ratio.
The change in the elevation of C is given by

$$\Delta y_c = [a \sin \phi_2 + L\sin(\theta_2 + \alpha)] - [a \sin \phi_1 + L\sin(\theta_1 + \alpha)] \tag{5.41}$$

We now have all the required equations; hence we can calculate the following:

1. the luffing speed, V_{cx};
2. the sinking speed, V_{cy};
3. the change in the elevation of C, Δy_c.

Using the given data,

$$\dot{\phi} = 2\pi \times 720/1430 = 3.16 \text{ rad/min} \qquad \alpha = 12.02^\circ$$

The results of the computation are shown in Table 5.4.
Also

$$\Delta y_c = [22.05 \sin 140 + 33.45 \sin(8.38 + 12.02)] - [22.05 \sin 60 + 33.45 \sin(-1.98 + 12.02)]$$
$$= 25.83 - 24.93 = 0.9 \text{ m}$$

We notice from these values that the luffing speed is not constant throughout the range of operation of the crane; this is due to the fact that the path followed by the point C of the jib lies on a lemniscate, which also explains the variation in the sinking speed. From a practical point of view, however, this variation is not terribly significant.

The change in elevation of C means that the load W carried by the crane will be lifted a further 0.9 m above its initial position when the jib is fully extended; this corresponds to an increase in potential energy of 0.9 W. The reader should verify that the total horizontal travel of this crane is 29.5 m, a very useful practical value when handling cargo.

Note: A simple check on the above computation at $\phi = 60^\circ$ and 130° using the instantaneous centre construction gave $\mathbf{V}_c/\dot{\phi}$ of 16.5 and 21.9 respectively.

5.7 SYNTHESIS: FREUDENSTEIN'S METHOD

Synthesis is concerned with the problem of selecting the size of the linkage to perform a given function. In what follows we consider three common situations using a four-bar linkage:

1. the coordination of input and output positions;
2. the generation of functions;
3. the coordination of input and output velocities and input and output accelerations.

5.7.1 Coordination of Positions

We recall Freudenstein's equation, Eq. (5.6):

$$K_1 \cos \phi + K_2 \cos \psi - K_3 = \cos(\phi - \psi) \tag{5.6)*$$

where

$$K_1 = d/c$$
$$K_2 = d/a$$
$$K_3 = (a^2 - b^2 + c^2 + d^2)/2ac$$

The design of a four-bar linkage to coordinate input and output positions ϕ_i and ψ_i respectively, $i = 1, 2, \ldots, n$, requires calculating the link lengths a, b, c, and d. An examination of Eq. (5.5) shows that these would not be easy to compute due to the non-linear nature of the equation. Freudenstein's equation, on the other hand, is linear in K_1, K_2, and K_3, the link ratios, and since linkages having the same link ratios are geometrically and kinematically equivalent the problem of designing a linkage is much simplified. If we take $n = 3$, i.e., coordinating three input and three output positions, we have a set of three simultaneous linear equations in K_1, K_2 and K_3 to solve.

Solutions with $n > 3$ have been obtained but the process is more complicated; such cases will be considered later using the method of least squares.

Let us consider the case of coordinating three output positions ψ_1, ψ_2, and ψ_3 with three input positions ϕ_1, ϕ_2, and ϕ_3. Substitution in Eq. 5.6 yields the following three simultaneous equations in K_1, K_2, and K_3:

$$K_1 \cos \phi_1 + K_2 \cos \psi_1 - K_3 = \cos(\phi_1 - \psi_1) \tag{5.42}$$
$$K_1 \cos \phi_2 + K_2 \cos \psi_2 - K_3 = \cos(\phi_2 - \psi_2) \tag{5.43}$$
$$K_1 \cos \phi_3 + K_2 \cos \psi_3 - K_3 = \cos(\phi_3 - \psi_3) \tag{5.44}$$

These equations may be solved by Gaussian elimination as shown, in the example that follows, or by any other convenient method.

Example 5.3 Design a four-bar linkage to coordinate the rotation of two shafts whose input and output angles are given below. The linkage should fit in a space 220 mm by 250 mm.

$$\phi_1 = 30^\circ \qquad \psi_1 = 195^\circ$$
$$\phi_2 = 45^\circ \qquad \psi_2 = 220^\circ$$
$$\phi_3 = 60^\circ \qquad \psi_3 = 245^\circ$$

SOLUTION Substituting in Eqs (5.32), (5.33), and (5.34) gives

$$K_1 \cos 30 + K_2 \cos 195 - K_3 = \cos(-165) = \cos 165$$
$$K_1 \cos 45 + K_2 \cos 220 - K_3 = \cos(-175) = \cos 175$$
$$K_1 \cos 60 + K_2 \cos 245 - K_3 = \cos(-185) = \cos 185$$

$$\Leftrightarrow \qquad \begin{aligned} 0.8660\,K_1 - 0.9659\,K_2 - K_3 &= -0.9659 \\ 0.7071\,K_1 - 0.7660\,K_2 - K_3 &= -0.9962 \\ 0.5000\,K_1 - 0.4226\,K_2 - K_3 &= -0.9962 \end{aligned}$$

to four decimal places.

Using the Gauss elimination process leads to the following:

1. *First step.* Elimination of K_1 from the second and third equation by subtracting

$$\frac{0.7071}{0.8660} = 0.8165 \text{ times the first equation from the second}$$

and

$$\frac{0.5000}{0.8660} = 0.5774 \text{ times the first equation from the third}$$

This gives three new equations:

$$\begin{aligned} 0.8660K_1 - 0.9659K_2 - K_3 &= -0.9659 \\ 0.0227K_2 - 0.1835K_3 &= -0.2075 \\ 0.1351K_2 - 0.4226K_3 &= -0.4385 \end{aligned}$$

2. *Second step.* Elimination of K_2 from the third equation by subtracting

$$\frac{0.1351}{0.0227} = 5.9515 \text{ times the second equation from the third}$$

This gives the following new set of equations:

$$\begin{aligned} 0.8660K_1 - 0.9695K_2 - K_3 &= -0.9659 \\ 0.0227K_2 - 0.1835K_3 &= -0.2075 \\ 0.6695K_3 &= 0.7964 \end{aligned}$$

Hence starting with the last equation we have

$$K_3 = \frac{0.7964}{0.6695} = 1.1896$$

From the second equation,

$$K_2 = 0.4755$$

and from the first equation

$$K_1 = 0.7887$$

We are now in a position to calculate the length of each link by taking $d = 1$ initially, for the length of the fixed link. We could equally choose $a = 1$, the input link, if this is more convenient.

Since $\qquad K_1 = d/c \quad \Rightarrow \quad c = 1/K_1 = 1.2679$

also $\qquad K_2 = d/a \quad \Rightarrow \quad a = 1/K_2 = 2.1030$

and $\qquad b = \sqrt{a^2 + c^2 + d^2 - 2acK_3} = 0.8286$

working to four decimal places in order not to lose accuracy.

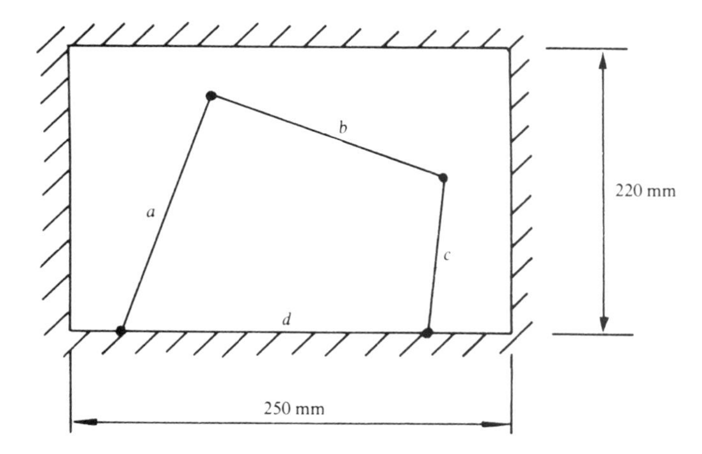

Figure 5.9

Before proceeding further it is strongly recommended that a check be carried out using one of the equations to ensure that the solution is correct. Thus from the first equation we have

$$0.8660 \times 0.7887 - 0.9659 \times 0.4755 - 1.1896 = -0.9659$$

which is exactly the right-hand side of that equation.

We can now decide on the actual size of the mechanism to fit in the space shown in Fig. 5.9.

In this case the conditions to satisfy are

$$d + c\cos(\psi_1 - 180) \leqslant 250\ \text{mm} \quad \Leftrightarrow \quad d\left(1 + \frac{1}{K_1}\cos 15\right) \leqslant 250$$

$$d \leqslant 250 \bigg/ \left(1 + \frac{1}{1.1896}\cos 15\right) \quad \Rightarrow \quad d \leqslant 138\ \text{mm}$$

and $\quad a\sin 60 \leqslant 220\ \text{mm} \quad \Leftrightarrow \quad 0.866a \leqslant 220 \quad \Rightarrow \quad a \leqslant 254\ \text{mm}$

Let us try $d = 100$ mm to allow for the fact that in practice all the links will have width.

Then $$c = \frac{d}{K_1} = \frac{100}{0.7887} = 126.79\ \text{mm}$$

This value satisfies the above conditions; and

$$a = \frac{d}{K_2} = \frac{100}{0.4755} = 210.30\ \text{mm}$$

this value satisfies the previous conditions.

Hence $$b = 100 \times 0.8286 = 82.86\ \text{mm}$$

It is also important to check the value of the transmission angle at each position. Substitution in Eq. (5.13) yields the following:

φ, °	μ, °
30	76
45	95
60	122

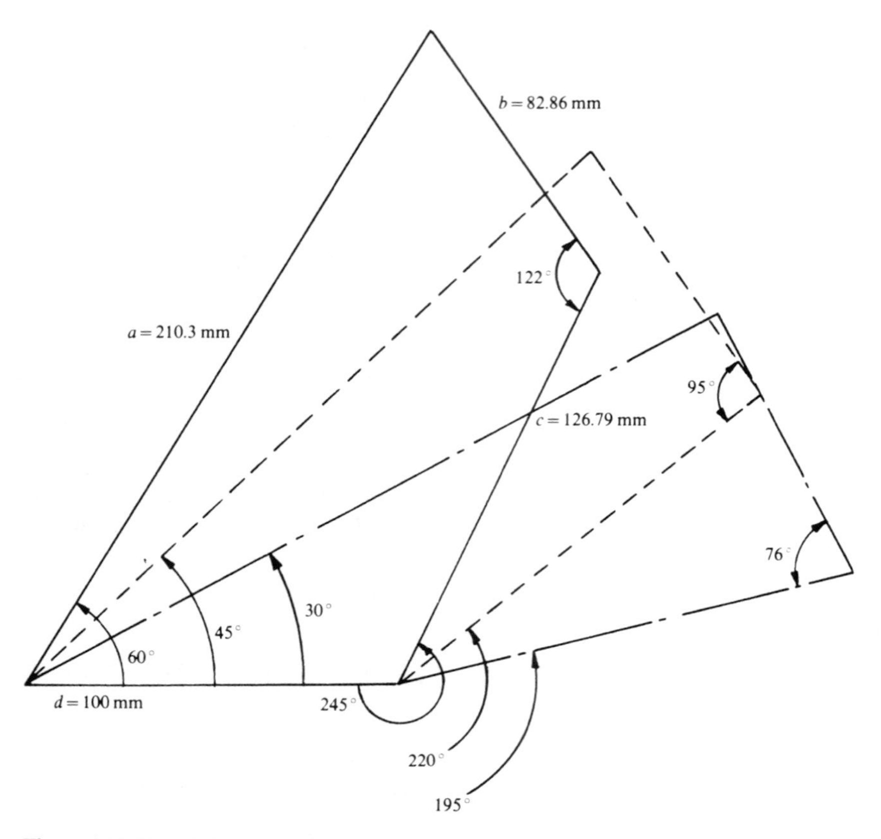

Figure 5.10 Size of the linkage showing good quality of transmission.

Since $40 < \mu < 140$ the quality of transmission should be good and the mechanism will operate smoothly even allowing for the inevitable presence of friction.

The resulting linkage is shown in Fig. 5.10. If the mechanism is a precision one manufacturing tolerances will have to be imposed on the dimensions, for example the 126.79 mm dimension for c could be specified as 126.80 ± 0.10 as typical of precision manufacture, but tolerances imposed on each link will, in general, have an influence on the performance of a mechanism, as we will see in Chapter 10.

The solution of the above problem was carried out using a pocket calculator, but for repeated calculations it is advisable to program a suitable machine such as a desk computer or a programmable calculator. Pocket computers are currently available that use BASIC as a language and are quite convenient for use in a drawing or design office.

The best way of handling Eqs (5.42), (5.43), and (5.44) for solution by a computer is to write them as follows:

$$\left.\begin{aligned} A_{11}K_1 + A_{12}K_2 + A_{13}K_3 = B_1 \\ A_{21}K_1 + A_{22}K_2 + A_{23}K_3 = B_2 \\ A_{31}K_1 + A_{32}K_2 + A_{33}K_3 = B_3 \end{aligned}\right\} \tag{5.45}$$

where
$$A_{11}=\cos\phi_1 \qquad A_{21}=\cos\phi_2 \qquad A_{31}=\cos\phi_3$$
$$A_{12}=\cos\psi_1 \qquad A_{22}=\cos\psi_2 \qquad A_{32}=\cos\psi_3$$
$$A_{13}=-1 \qquad A_{33}=-1 \qquad A_{33}=-1$$
$$\mathrm{B}_1=\cos(\phi_1-\psi_1) \qquad \mathrm{B}_2=\cos(\phi_2-\psi_2) \qquad \mathrm{B}_3=\cos(\phi_3-\psi_3)$$

We can also express these equations in matrix form, i.e.,

$$\mathbf{A}\cdot\mathbf{K}=\mathbf{B} \qquad \text{whose solution is } \mathbf{K}=\mathbf{A}^{-1}\cdot\mathbf{B}$$

where **A** is a 3×3 matrix of coefficients on the left-hand side, **K** is the vector of unknowns K_i, and **B** is the vector of constants on the right hand side, B_i.

Using Gaussian elimination the following equations are obtained for the unknowns K_i:

$$A_{11}K_1+A_{12}K_2+A_{13}K_3=B_1$$
$$(A_{22}-A_{12}A_{21}/A_{11})K_2+(A_{23}-A_{13}A_{21}/A_{11})K_3=(B_2-B_1A_{21}/A_{11})$$
$$(C_{22}-C_{12}C_{21}/C_{11})K_3=(D_2-D_1C_{21}/C_{11})$$

where
$$C_{11}=A_{22}-A_{12}A_{21}/A_{11}$$
$$C_{12}=A_{23}-A_{13}A_{21}/A_{11}$$
$$C_{21}=A_{32}-A_{12}A_{31}/A_{11}$$
$$C_{22}=A_{33}-A_{13}A_{31}/A_{11}$$
$$D_1=B_2-B_1A_{21}/A_{11}$$
$$D_2=B_3-B_1A_{31}/A_{11}$$

The unknowns K_1, K_2, and K_3 are then given by

$$\left.\begin{aligned} K_3&=(D_2-D_1C_{21}/C_{11})/(C_{22}-C_{12}C_{21}/C_{11})\\ K_2&=(D_1-C_{12}K_3)/C_{11}\\ K_1&=(B_1+K_3-A_{12}K_2)/A_{11} \qquad \text{since } A_{13}=-1 \end{aligned}\right\} \tag{5.46}$$

The values of a, b, and c can then be obtained for a given value of d.

Equations (5.46) were programmed on a pocket computer and Example 5.3 was solved again; the following values for the K's were obtained:

$$K_3=1.1895 \qquad K_2=0.4756 \qquad K_1=0.7887$$

and with $d=100$ mm we get

$$a=210.24\text{ mm} \qquad c=126.79\text{ mm} \qquad b=82.82\text{ mm}$$

The differences from the previous manual solution are seen to be very small and hence quite acceptable from a practical point of view.

5.7.2 Function Generation

The basic four-bar linkage has numerous applications in practice as mentioned earlier including the generation of functions where the output ψ is required to be some function of the input ϕ, i.e.,

$$\psi=f(\phi)$$

as for example in the case of a special purpose machine tool in which a certain profile on a

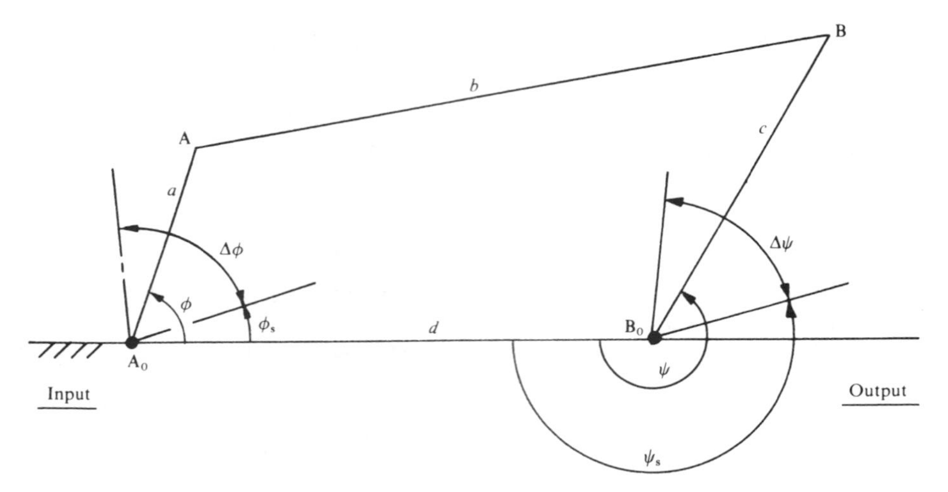

Figure 5.11 Design of a four-bar linkage to generate a given function: definition of the parameters involved.

workpiece had to be generated requiring the following relationship when using a four-bar linkage:

$$\psi = 240 + 0.095\phi^{1.5} \text{ degrees}$$

In practice the output is not always taken at the follower, as in the case of the dockside crane where the output is at the extension of the coupler forming the jib, but for the present we will consider the case where the output is taken at the follower of a four-bar linkage.

Let $\psi = f(\phi)$ be the required function; then referring to Fig. 5.11, the unknowns are: the link ratios a/d, c/d, and b/d, if d is taken as datum, the initial and final positions of the input and output links ϕ_s, ϕ_f, ψ_s, and ψ_f or ϕ_s, $\Delta\phi$, ψ_s, and $\Delta\psi$ where $\Delta\phi$ and $\Delta\psi$ are the total angular movements of the input and output respectively; thus we see that there are seven unknowns.

One of the simplest ways to calculate the link ratios is to give ϕ_s, $\Delta\phi$, ψ_s, and $\Delta\psi$ arbitrary values and solve for K_1, K_2, and K_3 using Freudenstein's equation. The process being repeated until a practical linkage is obtained; an optimum solution can be arrived at fairly quickly with a pocket or desk computer.

Recalling Freudenstein's equation, namely

$$K_1 \cos \phi_i + K_2 \cos \psi_i - K_3 = \cos(\phi_i - \psi_i) \qquad i = 1, 2, 3 \qquad (5.6)^*$$

We need three values for ϕ_i and ψ_i to solve for K_1, K_2, and K_3 which should ideally be chosen to minimize the error between the desired function and the function actually generated by the resulting mechanism throughout the operating ranges $\Delta\phi$ and $\Delta\psi$; it is only at the values ϕ_i and ψ_i that the error between the desired function and the generated function will be zero.

Thus if ψ_d is the desired function and ψ_g the generated function obtained from Freudenstein's equation then, except at the three chosen points referred to as the *accuracy* or *precision* points, there will be an error ε given by

$$\varepsilon = \psi_d - \psi_g = f_d(\phi) - f_g(\phi)$$

ε is referred to as the *structural error*. Figure 5.12(*a*) shows the desired and generated functions and Fig. 5.12(*b*) indicates the way in which the error varies as a function of the input angle ϕ. ϕ_1, ϕ_2, and ϕ_3 are the accuracy or precision points, and ϕ_s and ϕ_f are the starting and final

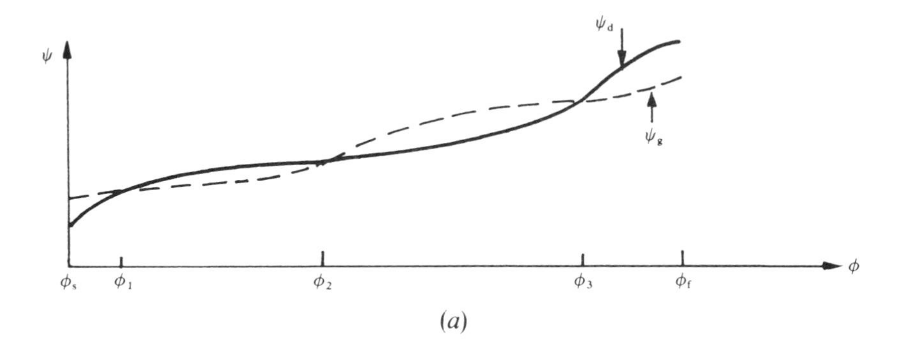

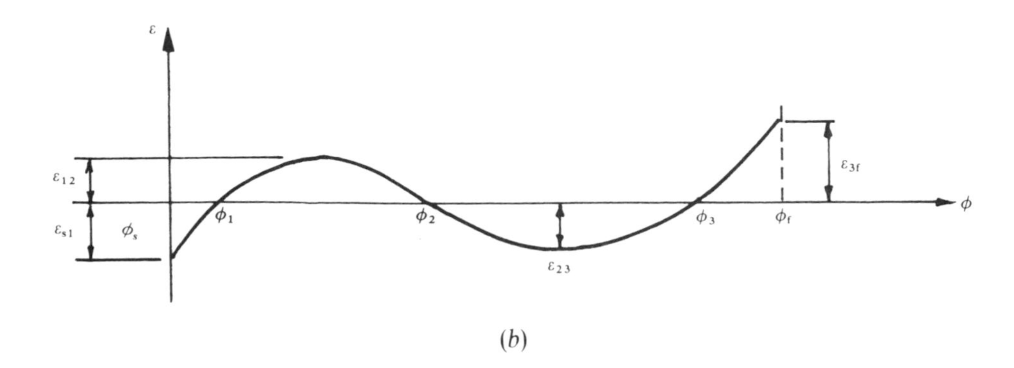

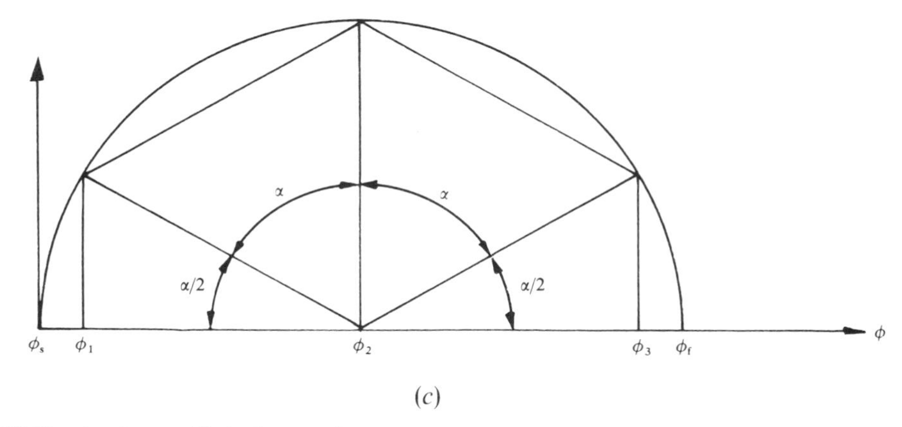

Figure 5.12 Structural error: Chebyshev spacings.

positions respectively. If ϕ_i are chosen arbitrarily then the maximum errors in the intervals will all have different values,

i.e.,
$$\varepsilon_{s1} \neq \varepsilon_{12} \neq \varepsilon_{23} \neq \varepsilon_{3f}$$

It is, however, possible to position ϕ_i by a method due to Chebyshev such that

$$\varepsilon_{s1} = \varepsilon_{12} = \varepsilon_{23} = \varepsilon_{3f} = \varepsilon_m \qquad \text{ideally}$$

in which case it can be demonstrated that the structural error ε_m is a minimum. In practice this is not completely achieved unless the function is a polynomial.[5]

For three accuracy points the method consists of inscribing a hexagon in a circle of diameter equal to the range of ϕ, i.e., $\Delta\Phi = \Phi_f - \Phi_s$; and the required values of ϕ_i located where the perpendiculars from the vertices of the hexagon cut the ϕ-axis as shown in Fig. 5.12(*c*). If four accuracy points were required an octogon would be needed, and hence for N accuracy points the polygon should have $2N$ sides.

The values ϕ_i are however best calculated using the following equation, which follows directly from the above construction:

$$\phi_i = \phi_s + \frac{\Delta\phi}{2}\left[1 - \cos\left(i\alpha - \frac{\alpha}{2}\right)\right] \qquad i = 1, 2, 3, \ldots, N \tag{5.47}$$

where $\Delta\phi = \phi_f - \phi_s$

N = number of accuracy points

$\alpha = 180/N$ degrees as shown in Fig. 5.12(*c*)

Thus for three accuracy points we have

$$N = 3 \qquad \alpha = 180/3 = 60°$$

giving

$$\phi_1 = \phi_s + 0.066\,99\Delta\phi$$
$$\phi_2 = \phi_s + 0.500\,00\Delta\phi$$
$$\phi_3 = \phi_s + 0.933\,01\Delta\phi$$

To illustrate the influence of the positions of the accuracy points on the structural error we propose to investigate the following example.

Example 5.4 Design a four-bar linkage to generate the function

$$\psi = 70 - \frac{18\,000}{\phi} \text{ degrees}$$

using the following accuracy points:

1. the end points and centre point;
2. Chebyshev's spacings;
3. any other arbitrary spacing;

and given that $\phi_s = 180°$ $\phi_f = 120°$ and with the crank as datum, i.e., $a = 1$.

SOLUTION Since $\psi = 70 - 18000/\phi$ degrees it follows that when

$$\phi_s = 180° \qquad \psi_s = -30° \qquad \text{and} \qquad \phi_f = 120° \qquad \psi_f = -80°$$

we get

$$\Delta\phi = \phi_f - \phi_s = 120 - 180 = -60°$$
$$\Delta\psi = \psi_f - \psi_s = -80 - (-30) = -50°$$

Case 1. End points and centre points as accuracy points

$$\phi_1 = 180° \qquad \psi_1 = -30°$$
$$\phi_2 = 150° \qquad \psi_2 = -50°$$
$$\phi_3 = 120° \qquad \psi_3 = -80°$$

Substitution in Eqs (5.42), (5.43), and (5.44) and solving yields

$$K_1 = 1.8326 \qquad K_2 = 1.4298 \qquad K_3 = 0.2716$$

and since
$$K_1 = d/c \qquad K_2 = d/a \qquad K_3 = \frac{a^2 - b^2 + c^2 + d^2}{2ac}$$

then with $a = 1$ we find that

$$b = 1.7970 \qquad c = 0.7802 \qquad d = 1.4298$$

Case 2. Chebyshev's spacings (Eq. (5.47))

$$\phi_1 = 180 - 60 \times 0.06699 = 175.98 \qquad \psi_1 = -32.2844°$$
$$\phi_2 = 180 - 0.5 \times 60 = 150.00 \qquad \psi_2 = -50.0000°$$
$$\phi_3 = 180 - 0.93301 \times 60 = 124.02 \qquad \psi_3 = -75.1379°$$

Substituting in Eqs (5.42), (5.43), and (5.44) and solving, we obtain

$$K_1 = 1.9244 \qquad K_2 = 1.5399 \qquad K_3 = 0.2628$$

Hence with $a = 1$, $b = 1.8950$, $c = 0.8002$, and $d = 1.5399$.

Case 3. Arbitrary spacings For comparison we will choose values that are not too far removed from cases (1) and (2).

Let
$$\phi_1 = 170.00° \qquad \psi_1 = -35.8824°$$
$$\phi_2 = 150.00° \qquad \psi_2 = -50.0000°$$
$$\phi_3 = 130.00° \qquad \psi_3 = -68.4615°$$

Solving Eqs (5.42), (5.43), and (5.44) yields

$$K_1 = 2.0599 \qquad K_2 = 1.7003 \qquad K_3 = 0.2486$$

Hence with $a = 1$, $b = 2.0401$, $c = 0.8254$, and $d = 1.7003$.

The structural error is now obtained by calculating the value of the generated function ψ_g from Eq. (5.11) or (5.12) with the values of a, b, c, and d obtained and subtracting it from the desired function $\psi_d = 70 - 18\,000/\phi$. Figure 5.13 shows the variation of the structural error as a function of the input angle for the three cases.

An examination of the results shows quite clearly that Chebyshev's spacing leads to nearly equal maxima for the error and whose values are less than the greatest error generated by other spacings, i.e., 0.55° compared with 0.87° for case 1 and 1.25° for case 3.

To illustrate and interpret the method of Freudenstein further we propose to examine in detail the following problem which arose in the design of a four-bar linkage, mentioned earlier, incorporated in a special-purpose machine.

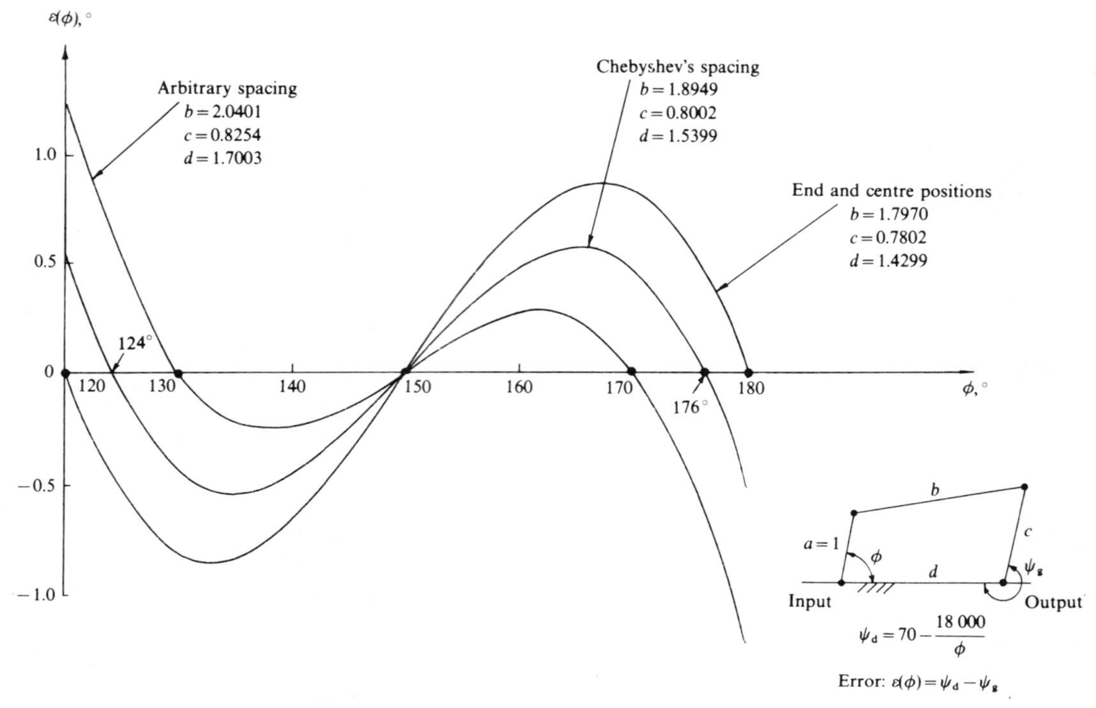

Figure 5.13 Comparison of the effect of different spacings on the value of the structural error.

Example 5.5 Design a four-bar linkage to generate the function

$$\psi = (240 + 0.095\phi^{1.5})^\circ$$

The linkage should fit easily in a space of 300 mm by 300 mm. ϕ_s should not be less than 20° and $45^\circ \leqslant \Delta\phi \leqslant 120^\circ$.

It is important that the structural error is kept as small as possible and the linkage operates smoothly throughout its range of operation; this means a good quality of transmission.

SOLUTION In this problem there are five unknowns, the link ratios a/d, b/d, and c/d, the starting position ϕ_s, and the angular movement $\Delta\phi$ of the input; the angular position and movement of the output are defined by the desired function. Hence the choice of ϕ_s and $\Delta\phi$ is arbitrary; the values selected, however, must be compatible with smooth operation, which implies a transmission angle in the range 40° to 140°. We should also bear in mind that the maximum error will depend on the precision with which the linkage can be manufactured; tolerances and clearances lead to mechanical errors which will be discussed in Chapter 10.

To solve this problem we proposed to employ Chebyshev spacings using three accuracy points with values of ϕ_s between 20° and 90° and values of $\Delta\phi$ between 45° and 120°. The best way to obtain a rapid solution is to use a programmable calculator or a desk computer. Some of the results of the computation are shown in the Table 5.5 and the variation in the maximum value of the structural error is shown in Fig. 5.14.

Table 5.5

		Size of fixed link: $d = 1.0$							
		ϕ_s, °							
$\Delta\phi$, °		20	30	40	50	60	70	80	90
45	a	$0.6\bar{6}01$	$0.9\bar{1}73$	$1.\bar{3}451$	$1.\bar{8}987$	$2.\bar{4}481$	$2.\bar{8}848$	$3.1\bar{6}32$	$3.\bar{2}787$
	b	1.4553	1.6339	1.9291	2.3060	2.6756	2.9689	3.1553	3.2302
	c	$0.5\bar{0}42$	$0.7\bar{0}45$	$1.\bar{0}127$	$1.3\bar{5}57$	$1.\bar{6}241$	$1.7\bar{7}86$	$1.\bar{8}417$	$1.\bar{8}517$
	$\Delta\psi$	41.28	46.09	50.41	54.37	58.06	61.51	64.79	67.90
	μ	75 102	68 97	64 92	63 89	62 88	61 86	59 84	56 80
	ε_{max}	0.063	0.083	0.091	0.083	0.066	0.046	0.033	0.019
60	a	$0.8\bar{3}10$	$1.\bar{2}236$	$1.\bar{7}601$	$2.\bar{3}212$	$2.\bar{7}854$	$3.\bar{0}936$	$3.2\bar{3}62$	$3.\bar{2}247$
	b	1.5748	1.8476	2.2158	2.5949	2.9064	3.1124	3.2048	3.1938
	c	−0.6458	−0.9407	−1.2930	−1.5870	−1.7644	−1.8404	−1.8540	−1.8441
	$\Delta\psi$	59.47	64.50	70.96	76.01	80.72	85.17	89.39	93.41
	μ	63 100	59 95	57 92	57 92	57 90	56 88	53 85	49 79
	ε_{max}	0.215	0.242	0.223	0.175	0.121	0.086	0.056	0.124
90	a	$1.\bar{4}718$	$2.\bar{0}452$	$2.\bar{5}600$	$2.\bar{9}255$	$3.\bar{1}173$	$3.\bar{1}450$	$3.\bar{0}299$	$2.\bar{7}998$
	b	2.0289	2.4229	2.7692	3.0131	3.1358	3.1463	3.0663	2.9442
	c	−1.1595	−1.5084	−1.7378	−1.8422	−1.8638	−1.8441	−1.8228	−1.8460
	$\Delta\psi$	101.1	109.27	116.77	123.78	130.37	136.37	142.59	148.30
	μ	45 96	47 95	48 93	48 94	47 93	44 89	40 82	36 73
	ε_{max}	1.015	0.77	0.497	0.326	0.242	0.45	0.97	1.71
120	a	$2.\bar{2}823$	$2.\bar{7}132$	$2.\bar{9}572$	$3.\bar{0}208$	$2.9\bar{2}90$	$2.\bar{7}110$	$2.\bar{4}011$	$2.\bar{0}436$
	b	2.6092	2.8928	3.0473	3.0773	3.0026	2.8600	2.7239	2.7958
	c	−1.7125	−1.8538	−1.8820	−1.8522	−1.805	−1.7778	−1.8295	−2.1303
	$\Delta\psi$	148.87	158.91	168.23	176.98	185.26	193.16	200.72	207.99
	μ	38 95	41 97	41 97	40 96	38 91	35 84	31 71	26 53
	ε_{max}	1.57	0.913	0.688	1.18	2.41	4.04	6.17	10.08

An examination of Table 5.5 shows that:

1. For a given range $\Delta\phi$ of the input:

   (*a*) the link ratios, i.e., a/d, b/c, and c/d, rise to a maximum value as the starting position ϕ_s of the input angle is increased;
   (*b*) the transmission angles μ corresponding to the beginning and end of travel of the linkage decrease as the starting position ϕ_s of the input angle is increased;
   (*c*) the maximum structural error ε_m decreases to a minimum value as ϕ_s increases before increasing to very large values with further increases in ϕ_s; this is particularly pronounced with large movements of the input;

2. For a given value of the starting position ϕ_s of the input angle:

   (*a*) the transmission angles decrease as the range $\Delta\phi$ of the input increases;
   (*b*) the maximum structural error ε_m increases as $\Delta\phi$ increases and, hence, as the range of the output $\Delta\psi$ increases.

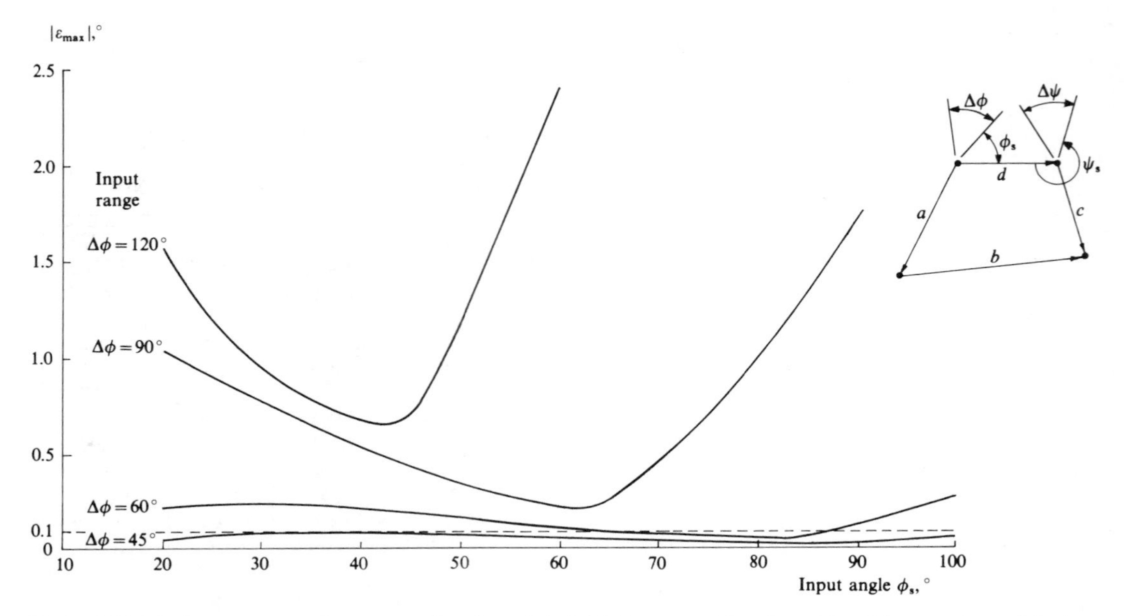

Figure 5.14 Maximum structural error as a function of the initial input position ϕ_s and for different values of the input range $\Delta\phi$.

From 2(*b*) we conclude that large angular movements should be avoided if we wish to limit the structural error and improve the quality of transmission. This will be particularly important in the case of precision mechanisms.

In our case we observe from 1(*c*) that, up to a point, the greater the starting angle ϕ_s the better from the point of view of the structural error.

In the present case if we can accept 0.10° as the maximum structural error, a typical value in practice for precision mechanisms, then we have a choice between all the mechanisms corresponding to $\Delta\phi = 45°$ and $20° \leqslant \phi_s \leqslant 90°$ (after 90° the error increases and the transmission angles get worse (not shown in the table)) and those corresponding to $\Delta\phi = 60°$, $65° \leqslant \phi_s \leqslant 87°$ (see Fig. 5.14).

However, since we are looking for a linkage with a good quality of transmission we can eliminate those corresponding to $\Delta\phi \geqslant 60°$.

We should also bear in mind the link ratios; it is considered good practice to avoid large values and ratios greater than 4 should be avoided whenever possible. A closer examination of Table 5.5 shows clearly that with larger ratios the quality of transmission deteriorates.

It would therefore appear that the mechanism with the best transmission angles and a structural error of less than 0.1° corresponds to

$$\underline{\phi_s = 20° \text{ and } \Delta\phi = 45°}$$

The negative signs for *a* and *c* in Table 5.5 mean that the lengths are to be drawn opposite to that shown in Fig. 5.1 since *a* and *c* are vectors as mentioned previously; we should note, however, that although *b* is also a vector its value is always positive, i.e., from *A* to *B*.

The steps necessary to calculate the size of the linkage, the transmission angle and the structural error are shown below for the linkage selected.

Solution with $\phi_s = 20°$ and $\Delta\phi = 45°$. For three accuracy points substitution in Eq. (5.37) yields

$$\phi_1 = 20 + 0.066\,99 \times 45 = 23.014\ 55°$$
$$\phi_2 = 20 + 0.050 \times 45 = 42.500\,00°$$
$$\phi_3 = 20 + 0.933.01 \times 45 = 61.985\,45°$$

using five decimal places throughout the computation.

Since the function to be generated is

$$\psi = (240 + 0.095\phi^{1.5})°$$

it follows that

$$\psi_1 = 240 + 0.095\,(23.014\,55)^{1.5} = 250.48\,884°$$
$$\psi_2 = 240 + 0.095\,(42.50)^{1.5} \qquad = 266.32\,128°$$
$$\psi_3 = 240 + 0.095\,(61.985\,45)^{1.5} = 286.36\,158°$$

for the output angles.

Substituting in Freudenstein's equation with three accuracy points and by Eq. (5.35) we find

$$K_1 = -1.983\,57 \qquad K_2 = -1.515\,15 \qquad K_3 = -0.643\,73$$

Hence with $d = 1.0$, the size of the required linkage is

$$a = -0.6601 \qquad b = 1.4553 \qquad c = -0.5042$$

The transmission angles corresponding to the extreme positions of the linkage are obtained by substitution in Eq. (5.13), giving

$$\phi_s = 20° \qquad \mu_s = 101.94°$$
$$\phi_f = 65° \qquad \mu_f = 75.03°$$

To compute the structural error we require the value of ψ_g generated by the linkage constructed with the link ratios obtained above. We saw earlier that there are two values of ψ given by Eqs (5.11) and (5.12). In our case we require ψ^- only; Fig. 5.15 which shows the linkage drawn to scale reveals quite clearly that the linkage corresponding to ψ^+ is impractical; the coupler crosses over the input link and the total movement of the output link is just over 7°. With ψ^- we see from Table 5.5 that the total movement of the output $\Delta\psi$ is 41.28°. If this movement is too small or too large in a particular application it can always be modified by introducing a geared system at the output of the linkage.

Calculation of the structural error. The structural error is given by

$$\varepsilon(\phi) = \psi_d^- - \psi_g$$
$$= (240 + 0.095\phi^{1.5}) - (360 + \psi^-)$$

We have to add 360° because ψ^- is negative as shown in Table 5.6, hence

$$\varepsilon(\phi) = 0.095\phi^{1.5} - 120 - \psi^-$$

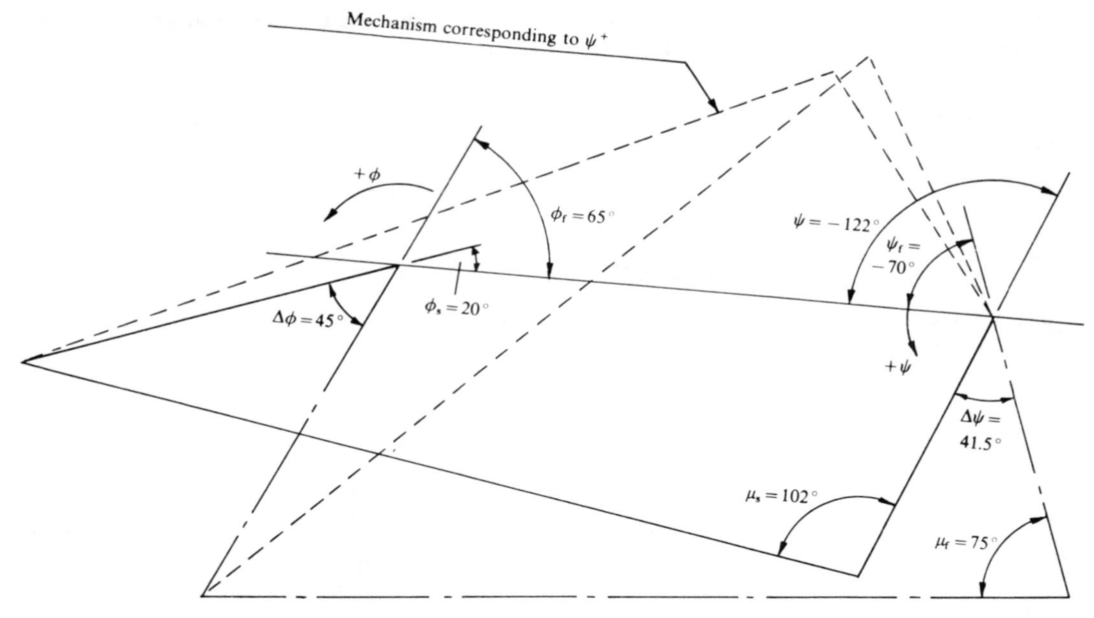

Figure 5.15 Optimum design having good quality of transmission.

Table 5.6

ϕ,°	ψ^-,°	ε,°	ψ^+,°
20	−111.57	−0.059	127.42
25	−108.11	0.024	127.90
30	−104.34	0.052	128.05
35	−100.29	0.044	127.89
40	−95.95	0.017	127.42
45	−91.34	−0.018	126.64
50	−86.46	−0.047	125.54
55	−81.31	−0.056	124.13
60	−75.88	−0.027	122.38
65	−70.16	0.063	120.28

Table 5.6 shows the values of the structural error, the generated output angle ψ^- and, for interest, the other generated output angle ψ^+ in steps of 5° of the input angle ϕ.

It now remains for us to fit the linkage in the space available, i.e., 300 mm × 300 mm. An examination of Fig. 5.15 shows that the overall length of the linkage will be the criterion; we must, therefore, satisfy the relationship.

$$a \cos \phi_s + d \leqslant 300$$

Since $a = d/K_2$ it follows that

$$d\left(\frac{\cos 20}{1.515\,15}\right) + 1 \leqslant 300$$

giving

$$d \leqslant 185$$

We could therefore choose $d = 150.000$ mm to allow for the fact that the links will have definite physical sizes. With this value for d the lengths of the link will be:

$$a = 0.6601 \qquad d = 99.015 \text{ mm}$$
$$b = 1.4553 \qquad d = 218.295 \text{ mm}$$
$$c = 0.5042 \qquad d = 75.63 \text{ mm}$$

In practice these dimensions would be modified so that the links can be manufactured to acceptable tolerances. For example the manufacturing drawings might specify that these dimensions should be

$$d = 150.00 \mp 0.15 \text{ mm}$$
$$a = 99.00 \pm 0.10 \text{ mm}$$
$$b = 218.25 \pm 0.15 \text{ mm}$$
$$c = 75.50 \pm 0.10 \text{ mm}$$

where the $\pm$ values are the tolerances imposed. In practice these values will depend very much on the manufacturing process employed. The associated mechanical errors will be functions of the tolerances imposed; in the present case we will see in Chapter 6 that the maximum mechanical error in the output ψ is 0.44°, that is 0.44/0.063, which is nearly seven times the maximum structural error.

5.8 METHOD OF LEAST SQUARES

We have seen that the four-bar linkage can be used to generate functions, i.e., $\psi = f(\phi)$, but that there will always be a structural error between the desired and generated function except at the precision points. In the preceding section these precision points were three in number because of the three unknowns in Freudenstein's equation, namely K_1, K_2, and K_3, and the maximum error between precision points were equalized by selecting the positions of these points in accordance with Chebyshev's spacing.

To reduce the structural error throughout the range of operation we now propose to consider more than three precision points. This will result in more equations than unknowns, which can be solved by a method due to Gauss known as the principle of least squares.

If $\psi_d(\phi)$ is the desired function and $\psi_g(\phi)$ the generated function using a four-bar linkage having link ratios a/d, b/d, and c/d then the structural error ε_i for any value ϕ_i of the input is given by:

$$\varepsilon_i = \psi_d(\phi_i) - \psi_g(\phi_i) \qquad i = 1, 2, 3, \ldots, N$$

This error is clearly a function of the link ratios and hence of the three design parameters K_1, K_2, and K_3 previously defined.

The sum of the squares of the errors is given by

$$S = \sum_{i=1}^{N} \varepsilon_i^2 = \sum_{i=1}^{N} [\psi_d(\phi_i) - \psi_g(\phi_i)]^2$$

The values of K_1, K_2, and K_3 which will make S a minimum are give by the following conditions:

$$\frac{\partial S}{\partial K_1} = 0 \qquad \frac{\partial S}{\partial K_2} = 0 \qquad \frac{\partial S}{\partial K_3} = 0$$

resulting in a set of three linear simultaneous equations in the three unknowns.

The simplest way of applying this principle to the design of a four-bar linkage to generate functions is to minimize Freudenstein's equation which for any ϕ_i and ψ_i is

$$K_1 \cos \phi_i + K_2 \cos \psi_i - K_3 = \cos(\phi_i - \psi_i)$$

A measure of the structural error is then given by the expression

$$e_i = K_1 \cos \phi_i + K_2 \cos \psi_i - K_3 - \cos(\phi_i - \psi_i)$$

Hence the square of the error for N points is

$$F = \sum e_i^2 = \sum_{i=1}^{N} [K_1 \cos \phi_i + K_2 \cos \psi_i - K_3 - \cos(\phi_i - \psi_i)]^2$$

For F to be minimum we must have

$$\frac{\partial F}{\partial K_1} = 0 \qquad \frac{\partial F}{\partial K_2} = 0 \qquad \frac{\partial F}{\partial F_3} = 0$$

This leads to the following equations:

$$\sum_{i=1}^{N} [K_1 \cos \phi_i + K_2 \cos \psi_i - K_3 - \cos(\phi_i - \psi_i)] \cos \phi_i = 0$$

$$\sum_{i=1}^{N} [K_1 \cos \phi_i + K_2 \cos \psi_i - K_3 - \cos(\phi_i - \psi_i)] \cos \psi_i = 0$$

$$\sum_{i=1}^{N} [K_1 \cos \phi_i + K_2 \cos \psi_i - K_3 - \cos(\phi_i - \psi_i)] = 0$$

Expanding each equation yields

$$\left.\begin{aligned} K_1 \sum \cos^2 \phi_i + K_2 \sum \cos \phi_i \cos \psi_i - K_3 \sum \cos \phi_i &= \sum \cos \phi_i \cos(\phi_i - \psi_i) \\ K_1 \sum \cos \phi_i \cos \psi_i + K_2 \sum \cos^2 \psi_i - K_3 \sum \cos \psi_i &= \sum \cos \psi_i \cos(\phi_i - \psi_i) \\ K_1 \sum \cos \phi_i + K_2 \sum \cos \psi_i - K_3 N &= \sum \cos(\phi_i - \psi_i) \end{aligned}\right\} \tag{5.48}$$

Equations (5.48) can be put in the form of Eq. (5.45) and solved using the Gaussian elimination technique explained in Sec. 5.7.1

To illustrate the results obtained by this method Example 5.5 has been analysed for three cases:

Case (1): $\phi = 20°$ $\quad \Delta\phi = 45°$
Case (2): $\phi = 20°$ $\quad \Delta\phi = 90°$
Case (3): $\phi = 20°$ $\quad \Delta\phi = 120°$

There are two ways of looking at the structural error, depending on the application. The first one is to consider the maximum value ε_m which occurs somewhere in the range of operation; this could be the criterion in the case of precision mechanisms such as computing mechanisms where positional error is to be kept as small as possible. The second is the root mean square value ε_{rms} given by

$$\varepsilon_{rms} = \sqrt{\left(\frac{1}{\phi_f - \phi_s}\int_{\phi_s}^{\phi_f} \varepsilon^2 d\phi\right)}$$

$$\varepsilon_{rms} = \sqrt{\frac{\Sigma\varepsilon^2 j}{N_1}} \qquad j = 1, 2, 3, \ldots, N_1$$

where N_1 is the number of discrete points along the total movement $\Delta\phi = \phi_f - \phi_s$ of the input. Thus ε_{rms} represents the overall error throughout the operating range of the mechanism and may be more significant in practice than the maximum error, particularly when combined with the mechanical errors due to manufacturing tolerances and clearances.

In Fig. 5.16 the maximum structural error has been plotted as a function of the number of points N for different spacing of the precision points for the case where $\phi_s = 60°$ and $\Delta\phi = 120°$ of Example 5.5. This shows that equally spaced points lead to larger errors when the end points are excluded than when they are included. The latter leads to a smaller maximum error than with Chebyshev spacings for $6 < N < 9$. It is interesting to note that for $N > 4$ the error remains constant at 1.65° with Chebyshev spacings but increases towards 2.2° for $N > 9$ with equally spaced points including the end points. The smallest value of the maximum error in this case is 1.5° for $N = 9$. The maximum error has been reduced to 1.50°, as compared with 2.41° obtained

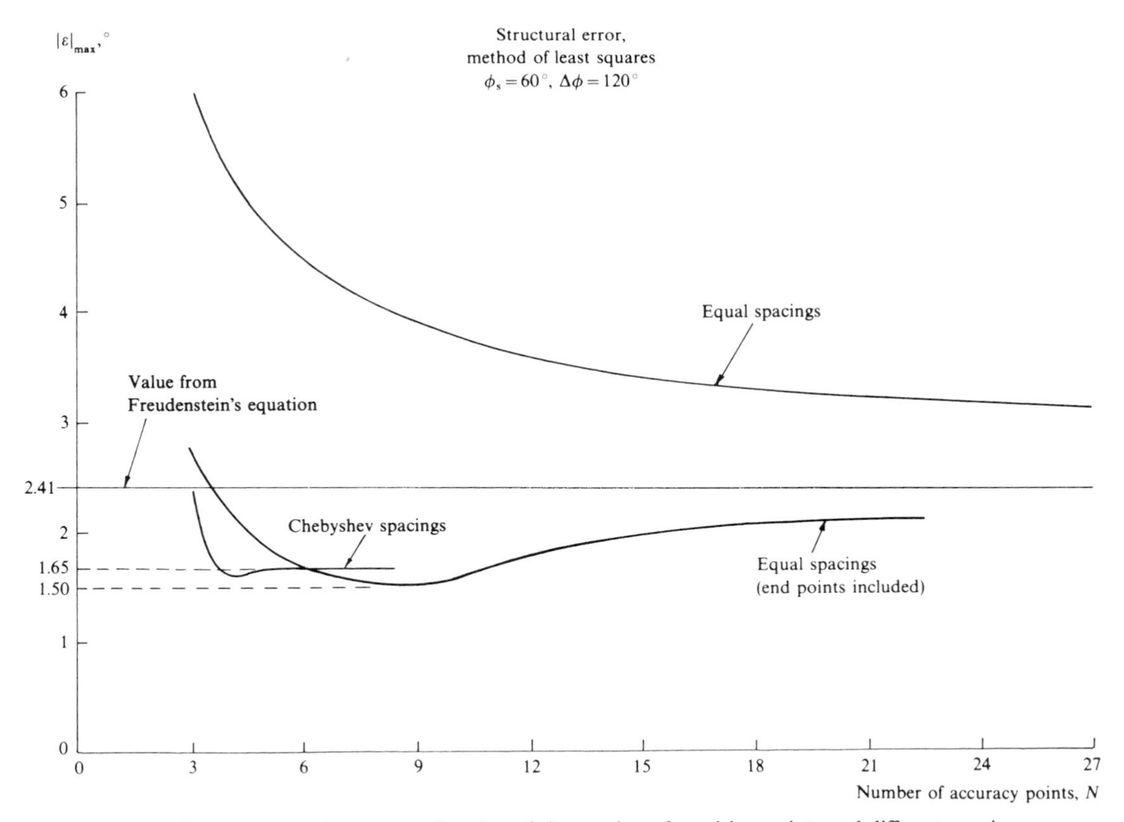

Figure 5.16 Maximum structural error as a function of the number of precision points and different spacings.

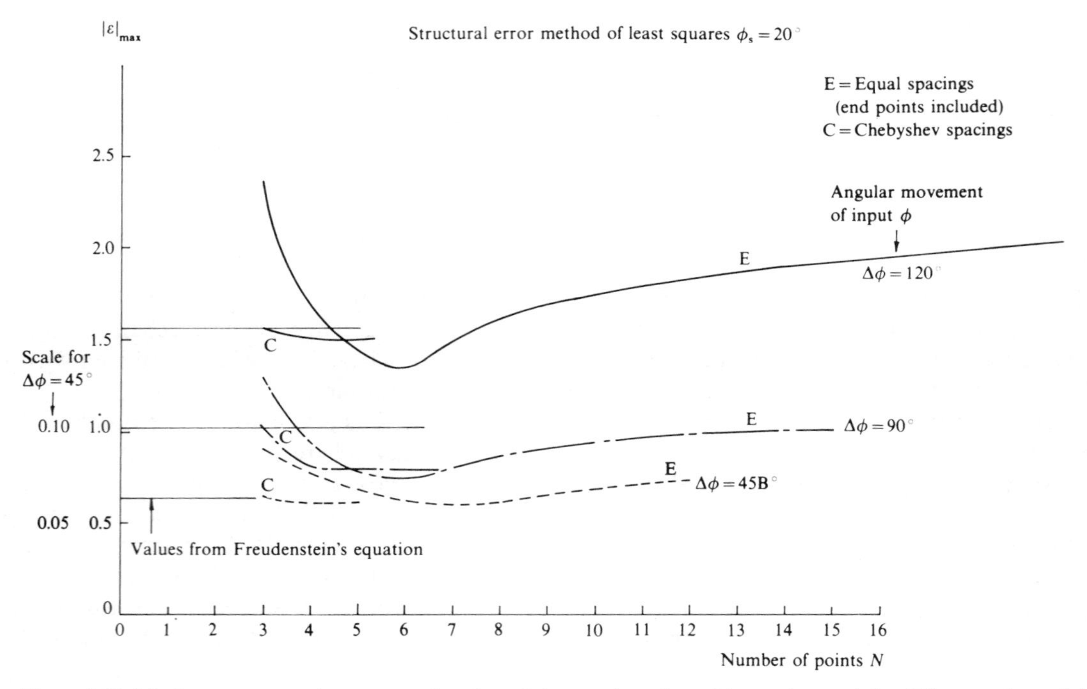

Figure 5.17 Maximum structural error as a function of the number of precision points and for different values of the total angular movement $\Delta\phi$ of the input.

when solving Freudenstein's equation using Chebyshev spacings (see Table 5.5), i.e., a reduction of nearly 38 per cent.

In Fig. 5.17 the influence of the number of points N on the value of the maximum structural error for a give initial input ϕ_s of 20° and varying anglar movement $\Delta\phi$ is shown. The results demonstrate that there is a definite improvement in the structural error compared with the values obtained by solving Freudenstein's equation with three precision points; see Table 5.5. There is a further improvement with equal spacings including the end points; in this case a minimum value is reached, whereas there is no further improvement with $N > 4$ using Chevyshev spacings.

These results also show that the error is sensitive to the angular movement of the input whatever spacings are employed; the greater the movement the greater the structural error.

In Fig. 5.18 the root mean square value of the structural error (ε_{rms}) has been plotted as a function of the number of precision points N. It clearly shows that for $N > 4$, ε_{rms} remains constant with Chebyshev spacings and that for $N > 7$, ε_{rms} drops very slightly with equal spacings (including the end points) and remains very close to that obtained using Chebyshev spacings.

The results obtained above are valid only for the function

$$\psi = 240 + 0.095\phi^{1.5}$$

which we wanted to generate, and the reader must not assume that other functions will behave in the same way.

The method of least squares does lead to an improvement in the value of the structural

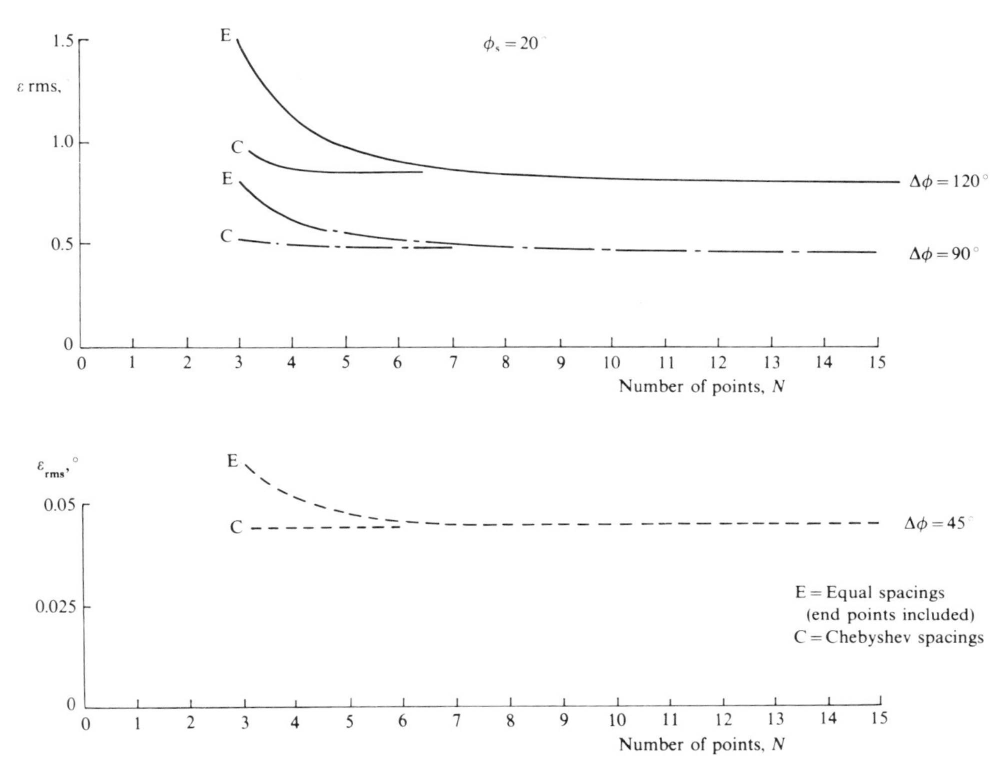

Figure 5.18 Root mean square value of the structural error as a function of the number of precision points for given values of the total movement $\Delta\phi$ of the input.

error and is more stable if the precision points are selected in accordance with Chebyshev spacings.

The design of a linkage mechanism to generate a particular function, i.e. $\psi = f(\phi)$ should be thoroughly investigated, as in the above case, from which an optimum solution will emerge, and the reader is advised to consult the Bibliography.

5.9 COORDINATION OF N POINTS

If there is a requirement for coordinating N discrete positions ($N > 3$) of the input with N positions of the output the method of least squares is the simplest to use, since it leads to three simultaneous linear equations in the three unknown parameters K_1, K_2, and K_3 in Freudenstein's equation.

Example 5.6 The striking mechanism of a weaving machine consists of a four-bar linkage. In a particular case such a linkage is required to coordinate five positions of the rocker follower with five positions of the input crank. Using the data give in the table, calculate the dimensions of the links to the nearest mm. Investigate also the quality of transmission of the resulting linkage.

Input angle ϕ, °	Output angle (usual conventions) ψ, °
40	−110
45	−104
50	−97
55	−89
60	−80

Length of the fixed link = 180 mm.

SOLUTION In Eqs (5.38) we need to calculate the following expressions:

$$\sum_{i=1}^{5} \cos^2 \phi_i = \cos^2 40 + \cos^2 45 + \cos^2 50 + \cos^2 55 + \cos^2 60 = \underline{2.078\,99}$$

$$\sum_{i=1}^{5} \cos \phi_i \cos \psi_i = \cos 40 \cos 110 + \cos 45 \cos 104 + \cos 50 \cos 97 + \cos 55 \cos 89 + \cos 60 \cos 80 = \underline{-0.414\,57}$$

since $\cos(-\theta) = \cos \theta$

$$\sum_{i=1}^{5} \cos \phi_i = \cos 40 + \cos 45 + \cos 50 + \cos 55 + \cos 60 = \underline{3.189\,52}$$

$$\sum_{i=1}^{5} \cos \phi_i \cos(\phi_i - \psi_i) = \cos 40 \cos(40+110) + \cos 45 \cos(45+104) + \cos 50 \cos(50+97) + \cos 55 \cos(55+89) + \cos 60 \cos(60+80) = \underline{-2.655\,67}$$

$$\sum_{i=1}^{5} \cos^2 \psi_i = \cos^2 110 + \cos^2 104 + \cos^2 97 + \cos^2 89 + \cos^2 80 = \underline{0.220\,81}$$

$$\sum_{i=1}^{5} \cos \psi_i = \cos 110 + \cos 104 + \cos 97 + \cos 89 + \cos 80 = \underline{-0.514\,71}$$

$$\sum_{i=1}^{5} \cos \psi_i \cos(\phi_i - \psi_i) = \cos 110 + \cos 150 + \cos 104 \cos 149 + \cos 97 \cos 147 + \cos 89 \cos 144 + \cos 80 \cos 140 = \underline{0.458\,63}$$

$$\sum_{i=1}^{5} \cos(\phi_i - \psi_i) = \cos 150 + \cos 149 + \cos 147 + \cos 144 + \cos 140 = \underline{-4.136\,92}$$

Substituting in Eqs (5.38) yields

$$2.078\,99K_1 - 0.414\,57K_2 - 3.189\,52K_3 = -2.655\,67$$
$$-0.414\,57K_1 + 0.220\,81K_2 + 0.514\,71K_3 = 0.458\,63$$
$$3.189\,52K_1 - 0.514\,71K_2 - 5K_3 = -4.136\,92$$

Solving for K_1, K_2, and K_3 we obtain

$$K_1 = 1.605\,27$$
$$K_2 = 1.020\,06$$
$$K_3 = 1.746\,38$$

With $d = 180$ mm it follows that

$$a = d/K_2 = 180/1.020\,06 = 176.46 \text{ mm}$$
$$c = d/K_1 = 180/1.605\,27 = 112.13 \text{ mm}$$
$$b = \sqrt{a^2 + c^2 + d^2 - 2acK_3}$$
$$= \sqrt{176.46^2 + 112.13^2 + 180^2 - 2 \times 176.46 \times 112.13 \times 1.746\,38}$$

Hence $b = 83.68$ mm.

The lengths of the links to the nearest millimetre are

$$a = 176 \text{ mm}$$
$$b = 84 \text{ mm}$$
$$c = 112 \text{ mm}$$
$$d = 180 \text{ mm}$$

The transmission angles are calculated by substituting in Eq. (5.13). When $\phi = 40°$ we have

$$\cos \mu = (b^2 + c^2 - a^2 - d^2 + 2ad \cos \phi)/2bc$$
$$= (84^2 + 112^2 - 176^2 - 180^2 + 2 \times 176 \times 180 \cos 40)/2 \times 84 \times 112$$

Hence $\mu = 75.3°$

Proceeding in a similar way for the other values of ϕ we find the following:

ϕ, °	μ, °
40	75.3
45	86.9
50	99.3
55	113.3
60	130.0

Since the values of the transmission angle μ lie inside the recommended range, $40° \leqslant \mu \leqslant 140°$, we conclude that the resulting linkage will have a good quality of transmission, i.e., it will operate smoothly, without jerky movements.

The above calculations were performed on a pocket scientific calculator working to five decimal places. The problem was also solved on a microcomputer using the author's program; the results are shown in Table 5.7. The table shows that there are no significant differences in the values obtained, which is to be expected, but the obvious advantage in using a computer is the speed with which new solutions are obtained when changing the

Table 5.7

```
*****************************
 DESIGN OF 4-BAR LINKAGES
       USING
THE METHOD OF LEAST SQUARES
  FOR THE COORDINATION OF
       N POINTS
*****************************

N MUST BE EQUAL TO OR GRATER THAN 3

NUMBER OF POINTS N= 5

PHI= 40              PSI=-110
PHI= 45              PSI=-104
PHI= 50              PSI=-97
PHI= 55              PSI=-89
PHI= 60              PSI=-80

THE LINK RATIOS ARE:

K1= 1.60527042
K2= 1.02006425
K3= 1.74638423

THE LINK LENGTHS ARE:

INPUT CRANK A0A= 176.459473
   COUPLER  AB = 83.6750573
  FOLLOWER  BB0= 112.13064
FIXED LINK A0B0= 180

TRANSMISSION ANGLE MU DEGREES

PHI= 40              MU= 75.49
PHI= 45              MU= 87.07
PHI= 50              MU= 99.6
PHI= 55              MU= 113.64
PHI= 60              MU= 130.55
```

data. For example, suppose we wanted to see if we could improve the transmission angle throughout the operating range; we could try altering the values of the input angle ϕ and the output angle ψ while maintaining the same increments $\Delta\phi$ and $\Delta\psi$. This was in fact carried but and the results are shown in Table 5.8, indicating quite clearly that an improvement in the transmission angle was achieved; however, the lengths of the links are very different. The two linkages drawn to the same scale in the extreme positions are shown in Fig. 5.19.

We notice from Fig. 5.19 that the second linkage is more compact than the first and will therefore occupy less space, which is an added advantage.

The reader should realize once more that there are a number of solutions to a given problem, and that the one selected must be that which gives the best performance.

Table 5.8

```
****************************
 DESIGN OF 4-BAR LINKAGES
        USING
THE METHOD OF LEAST SQUARES
  FOR THE COORDINATION OF
        N POINTS
****************************

N MUST BE EQUAL TO OR GRATER THAN 3

NUMBER OF POINTS N= 5

PHI= 35                 PSI=-105
PHI= 40                 PSI=-99
PHI= 45                 PSI=-92
PHI= 50                 PSI=-84
PHI= 55                 PSI=-75

THE LINK RATIOS ARE:

K1= 3.23441567
K2= 1.769605
K3= 2.95688616

THE LINK LENGTHS ARE:

INPUT CRANK A0A= 101.717615
    COUPLER  AB  = 111.208186
   FOLLOWER  BB0= 55.6514741
FIXED LINK A0B0= 180

TRANSMISSION ANGLE MU DEGREES

PHI= 35                 MU= 77.33
PHI= 40                 MU= 86.43
PHI= 45                 MU= 96.44
PHI= 50                 MU= 107.6
PHI= 55                 MU= 120.48
```

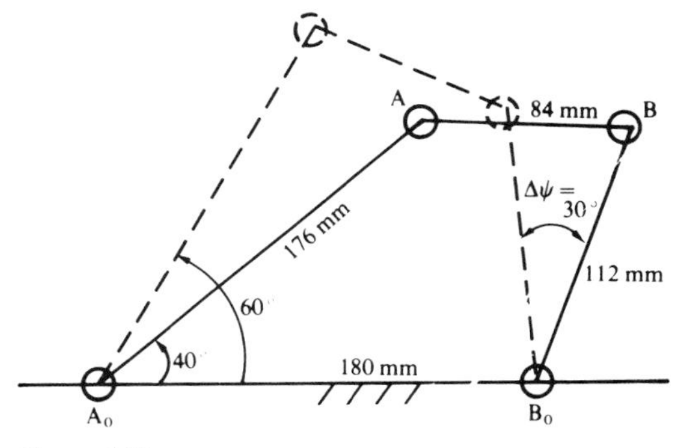

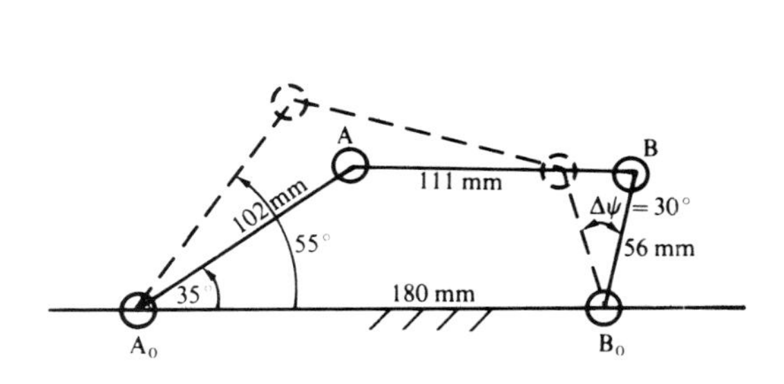

Figure 5.19

5.10 COORDINATIONS OF DERIVATIVES (VELOCITIES, ACCELERATIONS AND JERKS)

5.10.1 Introduction

The following expressions and Fig. 5.20 are for reference:

$$K_1 = \frac{d}{c} \qquad K_2 = \frac{d}{a} \qquad K_3 = \frac{a^2 - b^2 + c^2 + d^2}{2ac}$$

It is often convenient or desirable to work with velocities, accelerations, and jerks referred to the input ϕ, since $\psi = f(\phi)$, instead of the time t, as discussed in Sec. (3.4). We recall the expressions for the characteristic derivatives namely:

$$n_1 = \frac{d\psi}{d\phi}: \quad \text{characteristic velocity}$$

$$n_2 = \frac{d^2\psi}{d\phi^2}: \quad \text{characteristic acceleration}$$

$$n_3 = \frac{d^3\psi}{d\phi^3}: \quad \text{characteristic jerk}$$

The actual velocities, accelerations, and jerks (time derivatives) are then obtained from the following relationships:

$$\frac{d\psi}{dt} = \frac{d\psi}{d\phi} \cdot \frac{d\phi}{dt} = n_1 \dot{\phi}$$

$$\frac{d^2\psi}{dt^2} = \frac{d}{dt}(n_1\dot{\phi}) = n_1\ddot{\phi} + \frac{d}{dt}(n_1)\dot{\phi} = n_1\ddot{\phi} + \frac{d}{dt}\left(\frac{d\psi}{d\phi}\right)\dot{\phi} = n_1\ddot{\phi} + \frac{d}{d\phi}\left(\frac{d\psi}{d\phi}\right)\frac{d\phi}{dt}\dot{\phi}$$

$$= n_1\ddot{\phi} + n_2\dot{\phi}^2$$

Differentiating this last expression yields:

$$\frac{d^3\psi}{dt^3} = n_1\dddot{\phi} + 3n_2\dot{\phi}\ddot{\phi} + n_3\dot{\phi}^3$$

These time derivatives become characteristic derivatives when $\dot{\phi} = 1$. If $\dot{\phi} =$ a constant, ω say, which is often the case at an input then $\dot{\psi} = \omega n_1$, $\ddot{\psi} = n_2\omega^2$, and $\dddot{\psi} = n_3\omega^3$. Recalling Freudenstein's equation, we note that

$$K_1 \cos\phi + K_2 \cos\psi - K_3 = \cos(\phi - \psi) \tag{5.49}$$

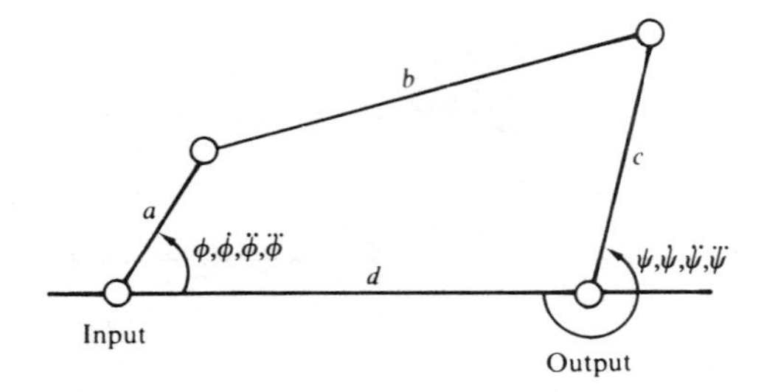

Figure 5.20

Then by successive differentiation we obtain:

$$K_1 \sin\phi + K_2 a_1 \sin\psi = c_1 \sin(\phi - \psi) \tag{5.50}$$

$$K_1 \cos\phi + K_2(a_2 \sin\psi + b_2 \cos\psi) = c_2 \sin(\phi - \psi) + d_2 \cos(\phi - \psi) \tag{5.51}$$

$$K_1 \sin\phi + K_2(a_3 \sin\psi + b_3 \cos\psi) = c_3 \sin(\phi - \psi) + d_3 \cos(\phi - \psi) \tag{5.52}$$

The coefficients a_i, b_i, c_i, and d_i, which are functions of the characteristic derivatives $n_i (i = 1, 2, 3)$, are given in the table.

i	a_i	b_i	c_i	d_i
1	n_1	0	$1 - n_1$	0
2	n_2	n_1^2	$-n_2$	$(1 - n_1)^2$
3	$n_1^3 - n_3$	$-3n_1n_2$	$(1 - n_1)^3 + n_3$	$3(1 - n_1)n_2$

5.10.2 Generation of Functions Around a Specified Position

In a given situation if we fix values for n_1, n_2, and n_3 for a particular value for ψ we have a system of four equations with four unknowns (K_1, K_2, K_3, and ϕ) to solve. Since Eqs (5.50), (5.51), and (5.52) contain K_1, K_2, and ϕ, we can eliminate K_1 and K_2 to obtain a quadratic equation in $\tan\phi$, thus:

$$(f + g \tan^2\psi) \tan^2\phi + (h \tan\psi) \tan\phi + k \tan^2\psi = 0 \tag{5.53}$$

Alternatively, given n_1, n_2, and n_3 and a particular value for ϕ, instead of ψ, Eq. (5.53) becomes a quadratic in $\tan\psi$, thus:

$$(k + g \tan^2\phi) \tan^2\psi + h \tan\phi \tan\psi + f \tan^2\phi = 0 \tag{5.54}$$

where

$$\left.\begin{aligned} f &= c_2b_3 - c_3b_2 + c_1b_2 \\ g &= d_2a_3 - d_3a_2 - a_1d_2 \\ k &= c_1a_3 - c_3a_1 \\ h &= g - f - k \end{aligned}\right\} \tag{5.55}$$

Note: This is true only if $\tan\psi$ or $\tan\phi$ is defined; i.e., provided that ψ (or ϕ) is not a right angle. This case would have to be examined separately.

Using Eq. (5.53), there are two values for ϕ, namely ϕ^+ and ϕ^-. Substituting the value obtained in Eqs (5.49), (5.50), and (5.51), we have a system of three linear equations which we can solve for K_1, K_2, and K_3, and hence the size of the required linkage to generate locally the desired function.

These equations are:

$$K_1 \cos\phi + K_2 \cos\psi - K_3 = \cos(\phi - \psi) \tag{5.56}$$

$$K_1 \sin\phi + K_2 n_1 \sin\psi = (1 - n_1) \sin(\phi - \psi) \tag{5.57}$$

$$K_1 \cos\phi + K_2(n_2 \sin\psi + n_1^2 \cos\psi) = -n_2 \sin(\phi - \psi) + (1 - n_1)^2 \cos(\phi - \psi) \tag{5.58}$$

Example 5.7 A four-bar linkage to replace geared segments is required such that the crank angular velocity ratio should be locally constant and equal to 0.7 when $\psi = 120°$.

Note: This is Example 3.5, which we solved graphically.

SOLUTION

$$n_1 = 0.7 \qquad n_2 = n_3 = 0 \qquad \psi_0 = 120^\circ$$

From the table of derivatives we find:

$$\begin{array}{llll} a_1 = 0.7 & b_1 = 0 & c_1 = 0.3 & d_1 = 0 \\ a_2 = 0 & b_2 = 0.4900 & c_2 = 0 & d_2 = 0.0900 \\ a_3 = 0.3430 & b_3 = 0 & c_3 = 0.0270 & d_3 = 0 \end{array}$$

Substituting in Eqs (5.55) we find

$$\begin{aligned} f &= -0.027 \times 0.49 + 0.3 \times 0.49 = 0.133\,77 \\ g &= 0.09 \times 0.343 - 0.7 \times 0.09 = -0.032\,13 \\ k &= 0.3 \times 0.343 - 0.027 \times 0.7 = 0.084\,00 \\ h &= g - f - k = -0.249\,90 \end{aligned}$$

Substituting in Eq. (5.53) and solving for ϕ yields

$$\phi^- = -84.79^\circ \text{ and } \phi^+ = -31.585^\circ$$

Let us examine these two solutions:

1. *With* $\phi^+ = -31.585^\circ$ *and* $\psi = 120^\circ$ the coefficients K_1, K_2, and K_3 from Eqs (5.56), (5.57), and (5.58) have the following values:

$$K_1 = -0.2138 \qquad K_2 = -0.4202 \qquad K_3 = 0.9076$$

Taking $d = 10$, say (units of length), the other link lengths are:

$$a = -23.79 \qquad \text{b} = 28.80 \qquad c = -46.77$$

The required linkage is shown in Fig. 5.21(a). It can be seen that the transmission angle μ of 43.8° is acceptable.

The values in brackets are those obtained using a graphical solution i.e., the Carter–Hall construction in reverse discussed in Sec. 3.4.8, Example 3.5, with $d = 10$. The accuracy should be noted showing that with care it is possible to obtain an acceptable degree of accuracy.

2. *With* $\phi^- = -84.79^\circ$ *and* $\psi = 120^\circ$, solving the equations as above yields

$$K = 0.099\,063 \qquad K_2 = 0.370\,222 \qquad K_3 = 0.731\,745$$

With $d = 10$ we have for the link lengths

$$a = 27.01 \qquad b = 83.84 \qquad c = 100.95$$

The resulting linkage is shown in Fig. 5.21(b). The transmission angle of 13.8° is quite unacceptable.

5.10.3 Structural Error

Let us now investigate the structural error in the neighbourhood of $\psi = 120^\circ$. To simplify the calculations we find it easier to measure the angles as shown in Fig. 5.22 so that $\psi = 120^\circ$ is

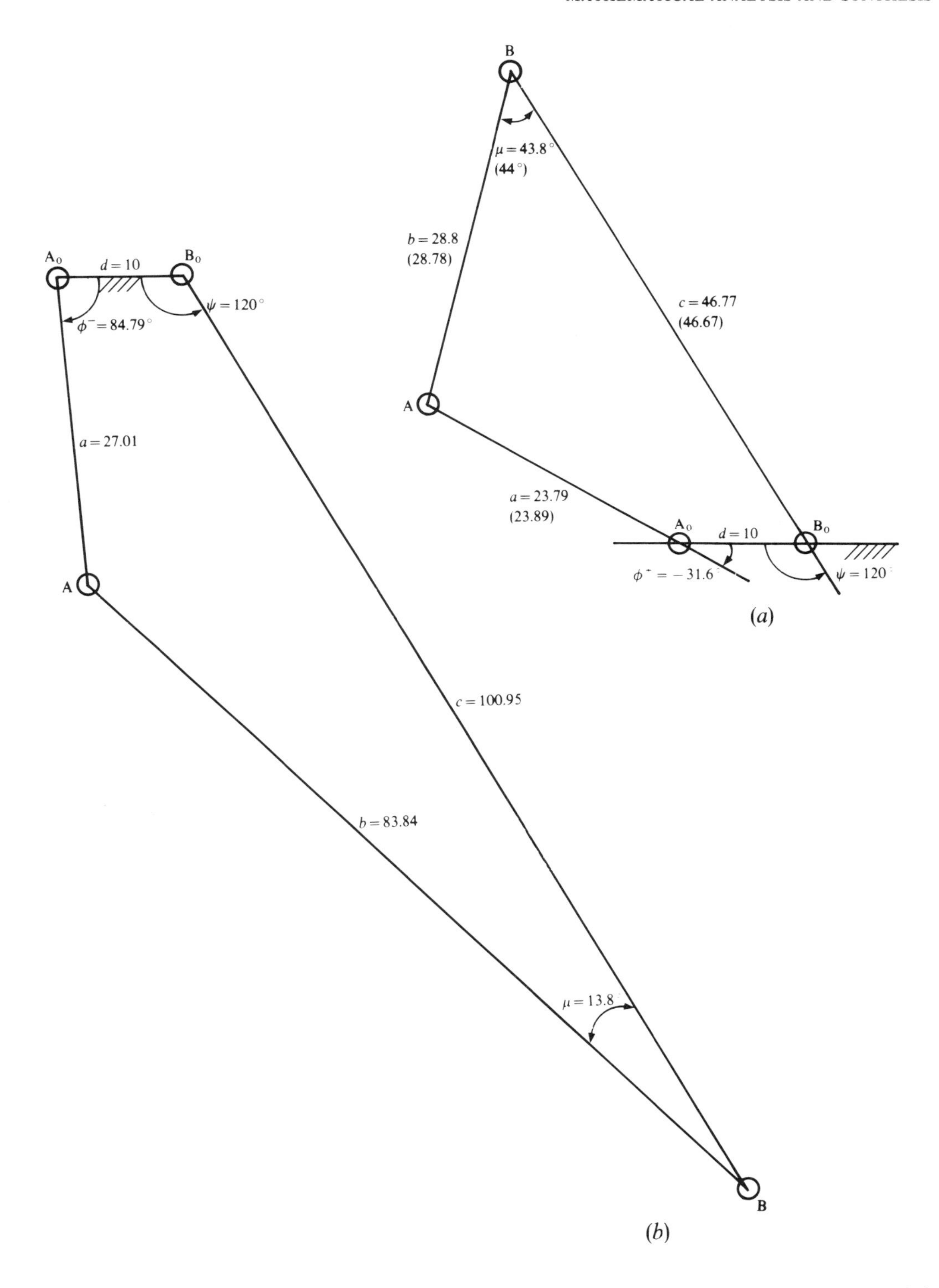

Figure 5.21 The two solutions for the size of the linkage: (*a*) practical linkage with good quality of transmission; (*b*) impractical solution—very poor quality of transmission due to the large link ratios.

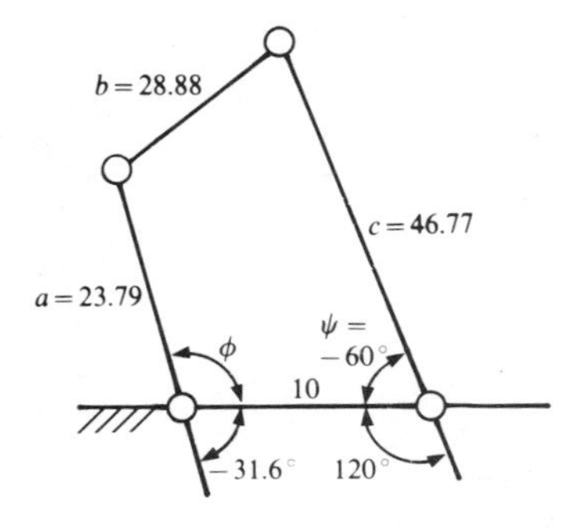

Figure 5.22

equivalent to $\psi = -60°$ and $\phi = -31.6°$ is equivalent to $\phi = 148.4$. The values of the link lengths a and c are now positive.

The relationship $d\psi/d\phi = 0.7$ is equivalent to generating the function $\psi = 0.7\phi + \text{constant}$. With $\psi = -60$ when $\phi = 148.4$, constant $= -163.88$.

Hence $\psi_d = 0.7\phi - 163.88$.

Where ψ_d is the function we wish to generate, i.e., it is the desired function.

To calculate the value of the function ψ_g generated by the linkage using the link lengths given above we require

$$\psi_g^+ = 2 \arctan \left\{ \frac{A + \sqrt{A^2 + B^2 - C^2}}{(B + C)} \right\} \qquad \text{(see Eq. (5.11))}$$

where

$$A = \sin \phi,$$
$$B = \cos \phi - K_2',$$
$$C = K_1' \cos \phi - K_3'$$

with

$$K_2' = -K_2, \ K_1' = -K_1, \text{ and } K_3' = K_3$$

owing to the change in the measurement of the angles ϕ and ψ.

Hence $A = \sin \phi$, $B = \cos \phi - 0.4202$, $C = 0.2138 \cos \phi - 0.9076$.

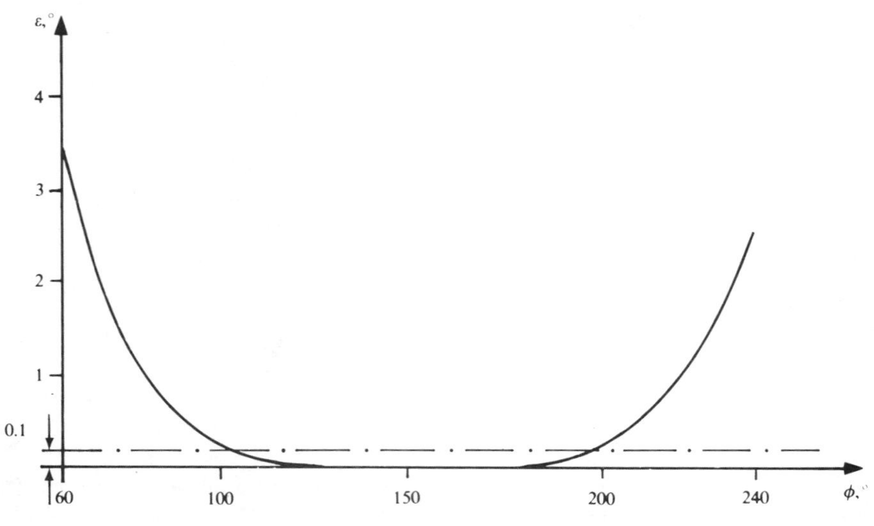

Figure 5.23 Structural error as a function of the input angle ϕ.

The structural error given by $\varepsilon=\psi_d-\psi_g$ has been calculated and is plotted (Fig. 5.23) for $60\leqslant\phi\leqslant240°$, showing clearly that the error is less than $0.1°$ for $110°\leqslant\phi\leqslant190°$. This means that the linkage can operate over quite a large angle around $\psi=120$ (or $-60)°$ with very little error; thus local generation of functions using the first three derivatives is, in general, quite accurate. It is equivalent to generating a function using the first three derivatives in a Taylor expansion.

5.11 ANALYSIS OF THE SIX-BAR LINKAGE

5.11 Introduction

The four-bar linkage we have discussed, analysed, and synthetized so far is the 'building brick' for many useful mechanisms, but by adding more links we can increase the number of possible mechanisms with many practical applications.

An example of an experimental low-cost mechanism using a six-bar linkage to enable a disabled person to drink from a cup is shown in Fig. 5.24(*a*). At (*a*) the cup is being filled and at (*b*) it has been brought into a position close enough to enable the patient to drink from the cup. The mechanism is powered by a small d.c. motor operated by pneumatic switches, one for raising and one for lowering the cup, which can be easily activated by the patient. Trials with various patients have been very succssful and further development of this mechanism is progressing.

Thus let us consider the four-bar linkage shown in Fig. 5.25(*a*) in which A_0A is the input. We know that either the point C on the coupler ABC or the point D on the follower B_0BD, or both, can be an output. For example, the crane shown in Fig. 1.7 is a four-bar linkage in which C is the output. By adding two links CD and DE, with E fixed, in Fig. 5.25(*b*) we obtain a six-bar linkage which is, in fact, the Stephenson type 3 mechanism of Fig. 1.3.

If on the other hand we add two links DE and EF, with F fixed, we obtain another six-bar linkage (Fig. 5.25*c*) which is the Watt type II mechanism of Fig. 1.3. All these mechanisms are derived from kinematic chains such as the ones shown in Fig. 1.2. A number of possible mechanisms derived from six-bar linkages are shown in Fig. 1.3. In some cases double joints are used.

To analyse such mechanisms we can employ some of the graphical techniques discussed previously or adopt a mathematical approach. As we have said before, graphical methods are quick and visual but many cases are required to cover a complete cycle of operation of the mechanism whether the input makes complete revolutions or oscillates through an angle of less than $360°$.

The mathematical approach enables us to derive algebraic expressions that are easily handled by pocket or microcomputers. The complete performance of the output of the mechanism, i.e., position, velocity, acceleration, and forces, can be tabulated and/or displayed graphically and the effect of a change in one of the parameters, e.g., a length or an angle, can be seen at a glance.

To illustrate the mathematical approach we propose to investigate the six-bar linkage shown in Fig. 5.25(*b*).

5.11.2 Displacement of the Output

Figure 5.26 shows the linkage in which the input link $A_0A=a$ makes an angle ϕ with the horizontal axis A_0X of the frame of reference XA_0Y. B_0 is a fixed pivot at a distance d from A_0.

(*a*)

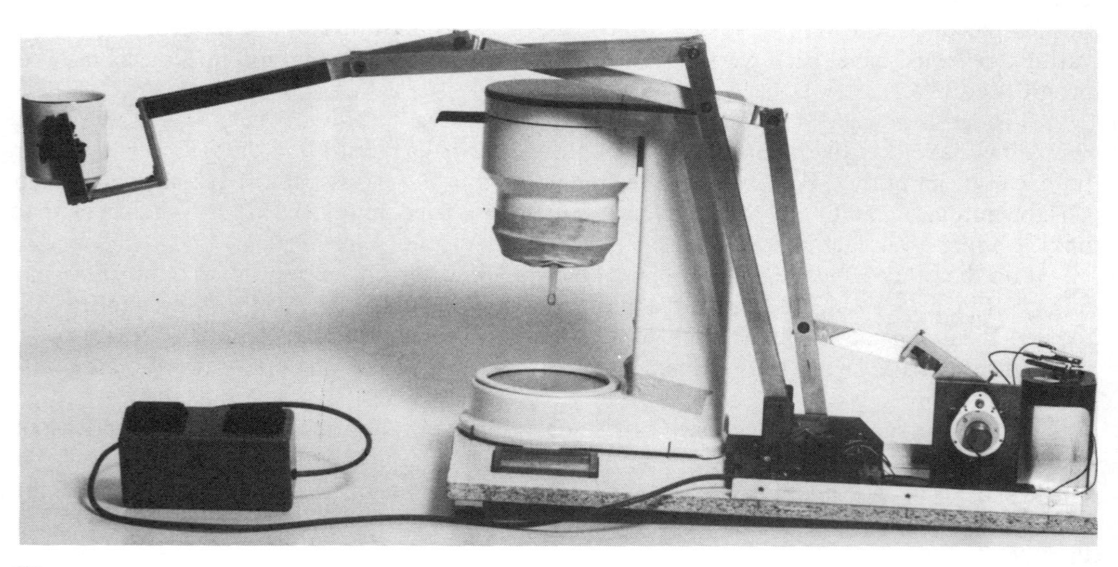

(*b*)

Figure 5.24 Experimental six-bar linkage to enable a disabled person to drink from a cup: (*a*) cup being filled from a standard coffee/tea machine; (*b*) cup brought up to a suitable position for drinking. (*Courtesy of the University of Bath.*)

The length of the coupler AB $=b$ and that of the follower $B_0B=c$. The position of the coupler point C is defined by the length $l=AC$ and the angle α relative to AB. The connecting link CD has a length e and makes an angle β with a line through C parallel to A_0X and in the direction shown. The output link DE has a length f at an angle γ with a line through E parallel to A_0X and in the direction indicated. The position of the coupler point C is defined by the instantaneous coordinates x_c and y_c; the fixed point E has coordinates x_e and y_e.

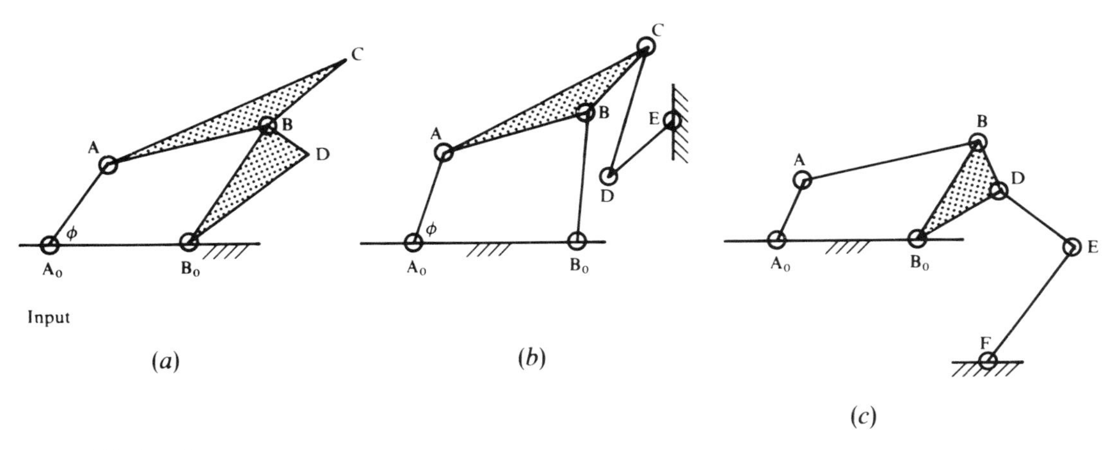

Figure 5.25 Possible six-bar linkages.

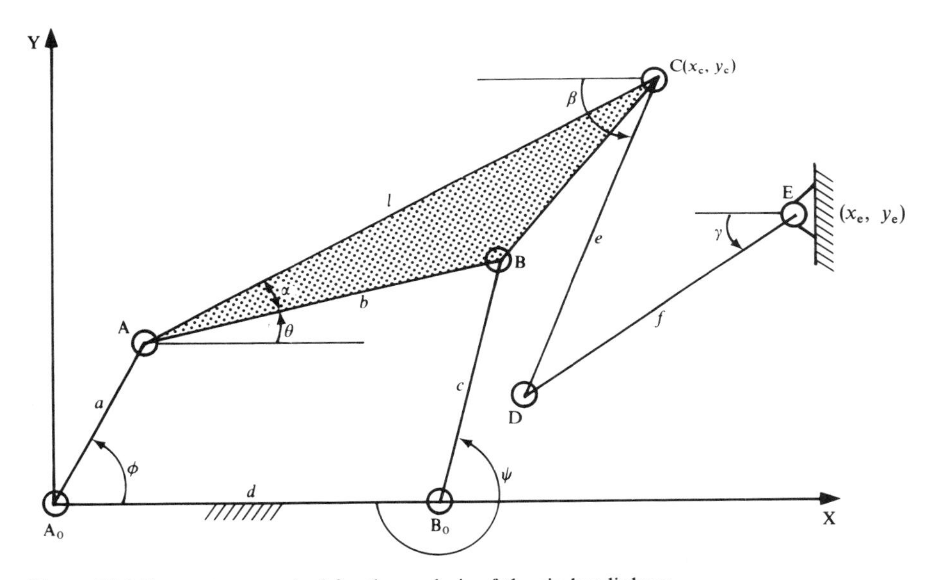

Figure 5.26 Parameters required for the analysis of the six-bar linkage.

For the links CD and DE, resolving along A_0X and A_0Y, we have

$$e \sin \beta = (y_c - y_e) + f \sin \gamma \tag{5.59}$$

$$e \cos \beta = f \cos \gamma - (x_e - x_c) \tag{5.60}$$

Squaring, adding, and collecting terms yields

$$A \sin \gamma + B \cos \gamma = C \tag{5.61}$$

where

$$A = 2f(y_c - y_e)$$
$$B = -2f(x_e - x_c)$$
$$C = e^2 - f^2 - (y_c - y_e)^2 - (x_e - x_c)^2$$

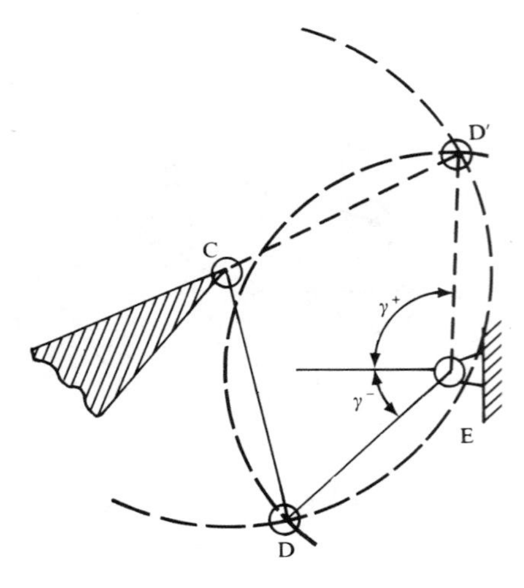

Figure 5.27 The two possible arrangements of the last two links of the six-bar linkage.

To solve the trigometric Eq. (5.61) we use the half-angle method as shown on page 90 and obtain the following solution for γ, the position of the output link DE,

$$\gamma = 2 \arctan\left(\frac{A \pm D}{B + C}\right) \tag{5.62}$$

where $D = \sqrt{A^2 + B^2 - C^2}$.

There are two values for γ corresponding to the two possible positions D and D′ of joint D as shown in Fig. 5.27 for a given value of ϕ and hence a given position for the point C on the coupler ABC.

Position D corresponds to the minus sign while position D′ corresponds to the positive sign in Eq. (5.62).

5.11.3 Velocity of the Output

The angular velocity of the link DE, i.e., $\dot{\gamma}$, is obtained by differentiating Eq. (5.61) rather than Eq. (5.62).

Hence differentiating Eq. (5.61) with respect to time yields

$$\dot{A} \sin \gamma + A \cos \gamma \dot{\gamma} + \dot{B} \cos \gamma - B \sin \gamma \dot{\gamma} = \dot{C}$$

Solving for $\dot{\gamma}$ we obtain the following expression for the angular velocity of the output link DE:

$$\dot{\gamma} = \frac{\dot{C} - \dot{B} \cos \gamma - \dot{A} \sin \gamma}{A \cos \gamma - B \sin \gamma} \tag{5.63}$$

where

$$\left.\begin{aligned} \dot{A} &= 2f \dot{y}_c \\ \dot{B} &= 2f \dot{x}_c \\ \dot{C} &= 2(x_e - x_c)\dot{x}_c - (y_c - y_e)\dot{y}_c \end{aligned}\right\} \tag{5.64}$$

We notice that in Eq. (5.64) the velocity of the point C on the coupler ABC is required, C being the input to the links CD and DE.

Referring to Fig. 5.26, the coordinates at C are

$$\left.\begin{aligned} x_c &= a\cos\phi + l\cos(\theta+\alpha) \\ y_c &= a\sin\phi + l\sin(\theta+\alpha) \end{aligned}\right\} \tag{5.65}$$

and

Differentiating these equations with respect to time yields

$$\dot{x}_c = -a\sin\phi\,\dot{\phi} - l\sin(\theta+\alpha)\dot{\theta} \tag{5.66}$$

for the velocity of C in a direction parallel to A_0X

and

$$\dot{y}_c = a\cos\phi\,\dot{\phi} + l\cos(\theta+\alpha)\dot{\theta} \tag{5.67}$$

for the velocity of C in a direction parallel to A_0Y.

Furthermore, in Eqs (5.65), (5.66), and (5.67) we require the value of θ, the position of the coupler line AB, and the angular velocity $\dot{\theta}$ of the coupler ABC as a function of the input angle ϕ. These are calculated from Sec. 5.5, Eqs (5.22) and (5.23), namely

$$\theta = 2\arctan\left(\frac{D \pm \sqrt{D^2+E^2-F^2}}{E+F}\right) \tag{5.68}$$

where

$$D = \sin\phi$$

$$E = \cos\phi - d/a$$

$$F = \frac{c^2-(a^2+b^2+d^2)}{2ab} + \frac{d}{b}\cos\phi$$

We recall that there are two values of θ, corresponding to the fact that a four-bar linkage of given link lengths can be assembled in two ways as shown in Fig. 5.2, an 'OPEN' configuration A_0ABB_0 and a 'CROSSED' configuration $A_0AB'B_0$.

The angular velocity $\dot{\theta}$ is then given by

$$\dot{\theta} = \left(\frac{\sin(\phi-\theta) - \mathrm{d}\sin\phi/\mathrm{b}}{\sin(\phi-\theta) + \mathrm{d}\sin\theta/\mathrm{a}}\right)\dot{\phi} \tag{5.69}$$

Thus it follows that in order to compute the variation in the value of the angular velocity $\dot{\gamma}$ of the output link DE as a function of the input angle ϕ the following sequence of operation should be adopted:

Is the mechanism A_0ABB_0 of the 'OPEN' or the 'CROSSED' type?

↓

'CROSSED' θ^+ or θ^{-1} 'OPEN' — from Eq. (5.68)

↓

Required θ

↓

Angular velocity of coupler ABC $\dot{\theta}$ — from Eq. (5.69)

↓

Coordinates of point C $x_c y_c$ — from Eq. (5.65)

↓

Coefficients ABC

↓

Configuration of the output of the mechanism
(refer to Fig. 5.27)
↓
'OPEN' γ^+ or γ^- 'CROSSED' — from Eq. (5.62)
↓
required γ
↓
Velocities of point C $\dot{x}_c, \dot{y}_c$ — from Eqs (5.66) and (5.67)
↓
Time derivatives of the coefficients A, B, C: $\dot{A}$, $\dot{B}$, $\dot{C}$ — from Eq. (5.64)
↓
Required angular velocity of output link DE: $\dot{\gamma}$ — from Eq. (5.63)

To illustrate these steps let us consider the following example for one particular position of a given mechanism.

Example 5.8 For the mechanism shown in Fig. 5.28 calculate the position and angular velocity of the output link DE when the input makes an angle $\phi = 160°$ with the fixed link A_0B_0 and is rotating at a constant angular velocity $\omega = 150$ rad/s in the direction indicated. Dimensions: $a = 40$ mm; $b = 150$ mm; $c = 90$ mm; $d = 131$ mm; $l = 136$ mm; $\alpha = 36°$; $e = 84$ mm; $f = 213$ mm; $x_e = 276$ mm; $y_e = 125$ mm.
Required: γ and $\dot{\gamma}$ for $\phi = 160°$.

SOLUTION

1. We first require the value of θ, i.e., the position of the coupler ABC for the given position of the input ϕ. Hence the following values of D, E and F are needed:
$D = \sin\phi = \sin 160 = 0.3420$
$E = \cos\phi - d/a = \cos 160 - 131/40 = -4.2147$

$$F = \frac{c^2 - (a^2 + b^2 + d^2)}{2ab} + \frac{d}{b}\cos\phi$$

$$= \frac{90^2 - (40^2 + 150^2 + 131^2)}{2 \times 40 \times 150} + \frac{131}{150}\cos 160 = -3.5841$$

Let $G = \sqrt{D^2 + E^2 - F^2} = 2.2439$

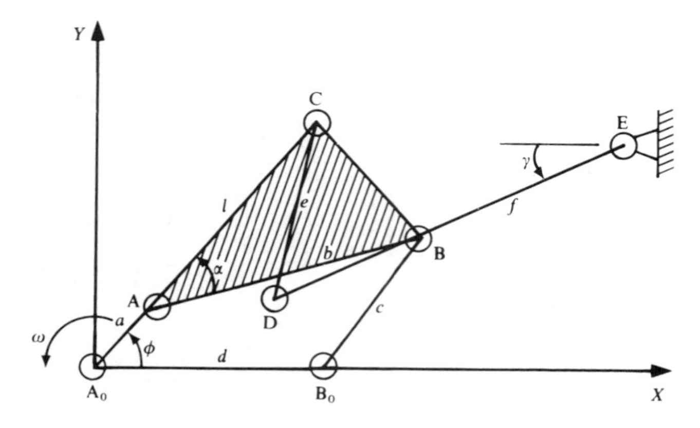

Figure 5.28

In Eq (5.68) we require the minus sign, i.e., θ^-, since this mechanism is of the 'OPEN' type.

Substituting in (5.68) yields

$$\theta = 2\arctan(D-G)/(E+F) = 27.41^\circ$$

2. To calculate the angular velocity of the coupler ABC we substitute in Eq. (5.69) and obtain

$$\dot{\theta} = \frac{\sin(160-27.41) - 131 \times \sin 160/150}{\sin(160-27.41) + 131 \times \sin 27.41/40} \times 150 = +29.25 \text{ rad/s, same direction as } \phi$$

3. The position of the coupler point C is given by Eq. (5.65), i.e.,

$$x_c = 40\cos 160 + 136\cos(27.41+36) = 23.286$$
$$y_c = 40\sin 160 + 136\sin(27.41+36) = 135.296$$

4. Before we can calculate the value of γ, the position of the output link DE, we need the values of the parameters A, B, and C in Eq. (5.61):

$$A = 2\times 213(135.296-125) = 4386$$
$$B = -2\times 213(276-23.286) = -107\,656$$
$$C = 84^2 - 213^2 - (135.296-125)^2 - (276-23.286)^2 = -102\,283$$

also $D = \sqrt{4386^2 + 107\,656^2 - 10\,228^2} = 33\,870$

To obtain the value of γ we note that the configuration of this mechanism is that shown by CDE in Fig. 5.22. We therefore require the value of γ^-, i.e., corresponding to the minus sign in Eq. (5.62). Hence substituting in (5.62) yields

$$\gamma = 2\arctan\left(\frac{4386-33\,870}{-107\,656-102\,285}\right) = 15.99^\circ$$

5. The velocities of the coupler point C in the x- and y-directions are given by Eqs (5.66) and (5.67) respectively. Hence

$$\dot{x}_c = -40\sin 160\times 150 - 136\sin(27.41+36)\times 29.25 = -5609 \text{ mm/s}$$
$$\dot{y}_c = 40\cos 160\times 150 + 136\sin(27.41+36)\times 29.25 = -3858 \text{ mm/s}$$

6. We require the derivatives $\dot{A}$, $\dot{B}$, and $\dot{C}$ of the three parameters A, B, and C; their values are given by Eq. (5.64), i.e.,

$$\dot{A} = 2\times 213\times -3858 = -1\,643\,508$$
$$\dot{B} = 2\times 213\times -5609 = -2\,389\,434$$
$$\dot{C} = 2(276-23.286)(-5609) - 2(135.296-125)(-3858)$$
$$= -2\,755\,502$$

7. We are now in a position to calculate the value of $\dot{\gamma}$, the angular velocity of the output link DE. Substituting in Eq. (5.63) yields

$$\dot{\gamma} = \frac{(-2\,755\,502) - (-2\,389\,434\cos 15.99) - (-1\,643\,508\sin 15.99)}{(4386\cos 15.99) - (-107\,656\sin 15.99)}$$

Hence $\underline{\dot{\gamma} = -0.17 \text{ rad/s}}$

Table 5.9

```
Analysis of the 6-bar linkage
****************************
    by J. GROSJEAN
    ~~~~~~~~~~~~~~

link lengths & coordinates are in mm
~~~~~~~~~~~~~~~~~~~~~~~~~~~~~~~~~~~~

a= 40          b= 150          c= 90          d= 131
coupler length:l= 136          alpha= 36
coord. of output shaft Xe & Ye= 276  125
lengths of the links CD=e & DE=f= 84  213

input ang. vel.,rad/sec= 1  input ang. accel.,rad/sec^2= 0
~~~~~~~~~~~~~~~~~~~~~~~~~~~~~~~~~~~~~~~~~~~~~~~~~~~~~~~~~~~~
```

Input:	Output:	Output vel. & accel.:	
phi degrees	gamma degrees	Ang. velocity rad/sec	Ang. acceleration rad/sec^2
+0.0000	+22.3444	-0.1304	-0.1663
+10.0000	+20.9293	-0.1489	-0.0402
+20.0000	+19.4443	-0.1444	+0.0869
+30.0000	+18.1037	-0.1213	+0.1672
+40.0000	+17.0459	-0.0897	+0.1863
+50.0000	+16.3071	-0.0588	+0.1631
+60.0000	+15.8506	-0.0337	+0.1239
+70.0000	+15.6107	-0.0154	+0.0858
+80.0000	+15.5217	-0.0033	+0.0551
+90.0000	+15.5300	+0.0043	+0.0321
+100.0000	+15.5953	+0.0083	+0.0155
+110.0000	+15.6883	+0.0099	+0.0033
+120.0000	+15.7875	+0.0097	-0.0057
+130.0000	+15.8771	+0.0081	-0.0123
+140.0000	+15.9455	+0.0055	-0.0169
+150.0000	+15.9847	+0.0023	-0.0194
+160.0000	+15.9904	-0.0011	-0.0193
+170.0000	+15.9629	-0.0043	-0.0157
+180.0000	+15.9087	-0.0064	-0.0076
+190.0000	+15.8421	-0.0065	+0.0066
+200.0000	+15.7881	-0.0036	+0.0280
+210.0000	+15.7840	+0.0037	+0.0568
+220.0000	+15.8797	+0.0165	+0.0904
+230.0000	+16.1330	+0.0351	+0.1217
+240.0000	+16.5966	+0.0581	+0.1389
+250.0000	+17.2989	+0.0821	+0.1319
+260.0000	+18.2278	+0.1027	+0.0999
+270.0000	+19.3285	+0.1160	+0.0510
+280.0000	+20.5173	+0.1201	-0.0047
+290.0000	+21.6978	+0.1143	-0.0612
+300.0000	+22.7717	+0.0988	-0.1160
+310.0000	+23.6436	+0.0740	-0.1683
+320.0000	+24.2223	+0.0403	-0.2165
+330.0000	+24.4245	-0.0010	-0.2552
+340.0000	+24.1844	-0.0475	-0.2718
+350.0000	+23.4760	-0.0935	-0.2472
+360.0000	+22.3444	-0.1304	-0.1663

The minus sign means that the angular velocity of the output link DE is, at that instant, clockwise in accordance with our sign convention.

In practice we would most likely require the variation in the output angular velocity as a function of the input angle. In this case it would be advisable to program a pocket or microcomputer.

A suitable program entitled 'Analysis of the 6-bar linkage' has been devised by the author on a pocket computer (Sharp PC 1360) as well as on an OPUS PC microcomputer. The printout from the latter for one complete revolution of the crank A_0A and a 1-rad/s input angular velocity is shown in Table 5.9.

Our calculation above gave $\dot{\gamma} = -0.17$ rad/s, hence for a 1 rad/s input velocity this would have been $\dot{\gamma} = -0.17/150 = -0.0011$ rad/s which is the value shown in Table 5.9 corresponding to $\phi = 160°$.

The program also calculates the angular acceleration of the output link DE, and the reader will observe that the input angular acceleration of the crank has been assumed to be zero. This would be the case when the input crank is driven by a constant-speed motor; however, if the linkage is driven by the output of another linkage then the input angular acceleration would not be zero, e.g., as for the mechanisms shown in Fig. 1.9 and in Fig. 1.13. The reader will recall that in the case of a four-bar linkage driven at constant speed its output, whether taken from a coupler point or from the follower, will possess an acceleration, as well as a jerk, which varies in value throughout its cycle of operation. Furthermore, the output acceleration or jerk can reach very high values in many cases in practice, particularly in high-speed mechanisms; high values of the acceleration lead to high inertia torques (see Chapter 8 of this text), which affect the input torque requirement as well as the stresses in the links.

5.11.4 Acceleration of the Output

The angular acceleration of the output link DE is obtained by differentiating Eq. (5.63) with respect to time. In this equation

let $$\lambda_1 = \dot{C} - \dot{B}\cos\gamma - \dot{A}\sin\gamma$$

and $$\lambda_2 = A\cos\gamma - B\sin\gamma$$

so that Eq. (5.63) may be expressed thus:

$$\dot{\gamma} = \lambda_1/\lambda_2 \tag{5.70}$$

Differentiating (5.70) yields

$$\ddot{\gamma} = \frac{\lambda_2\dot{\lambda}_1 - \lambda_1\dot{\lambda}_2}{\lambda_2^2} \tag{5.71}$$

where $$\dot{\lambda}_1 = \dot{\gamma}(\dot{B}\sin\gamma - \dot{A}\cos\gamma) + \ddot{C} - \ddot{B}\cos\gamma - \ddot{A}\sin\gamma \tag{5.72}$$

and $$\dot{\lambda}_2 = -\dot{\gamma}(A\sin\gamma + B\cos\gamma) + \dot{A}\cos\gamma - \dot{B}\sin\gamma \tag{5.73}$$

It is left as an exercise for the reader to show that for a constant 1 rad/s input the angular acceleration of the output link DE is -0.019 rad/s^2 as shown in Table 5.9.

Case (1): Four-bar linkage

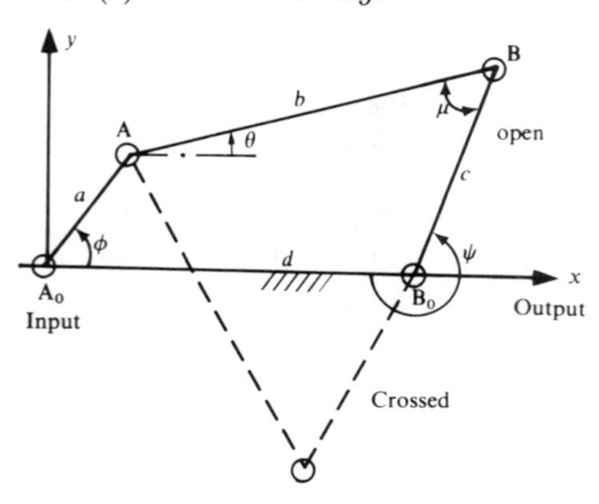

For a given input:
Crank A_0A:

$\phi \; \dot{\phi} \; \ddot{\phi} \; \dddot{\phi}$

the output is:

$\psi \; \dot{\psi} \; \ddot{\psi} \; \dddot{\psi}$	follower
$\theta \; \dot{\theta} \; \ddot{\theta}$	coupler
μ	transmission angle

Case (2): Point C
(a) on the coupler

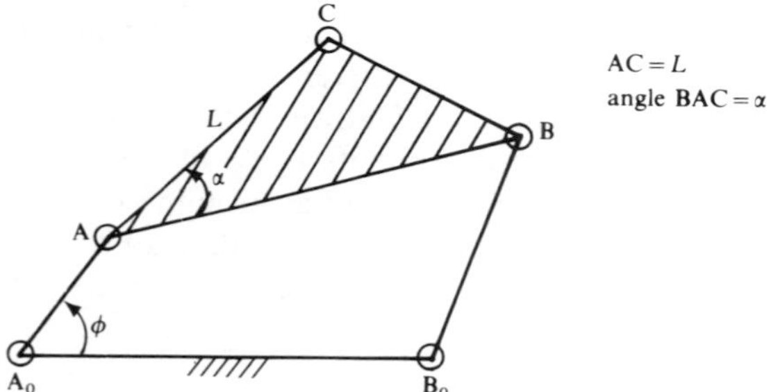

output:

$x_c \;\; y_c$	coordinates of C
$\dot{x}_c \;\; \dot{y}_c$	velocities of C
$\ddot{x}_c \;\; \ddot{y}_c$	accelerations of C

(b) on the follower

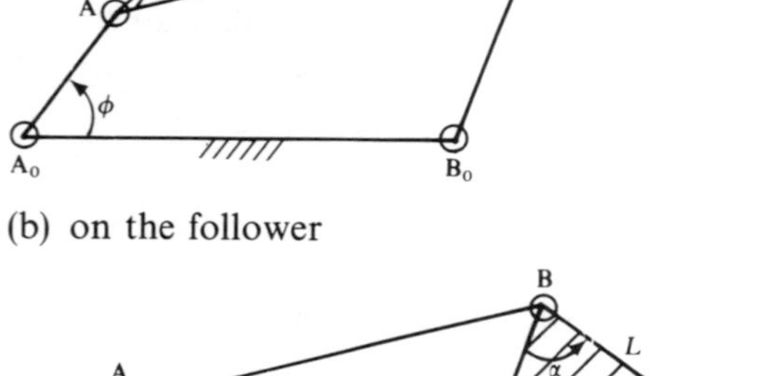

Case (3): Six-bar linkage by the addition of two links CD and DE

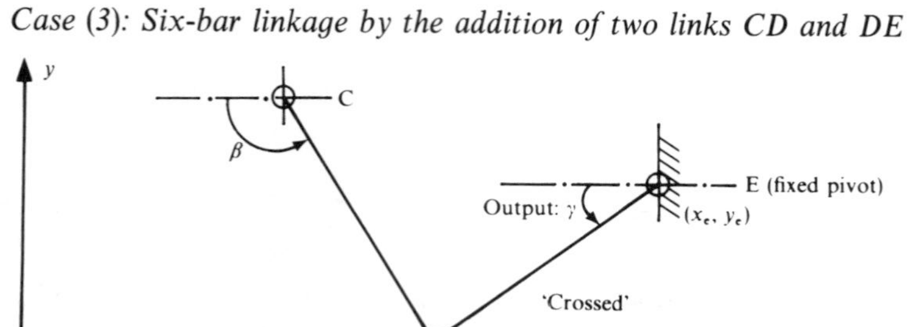

output:

$\beta \;\; \gamma$	positions
$\dot{\beta} \;\; \dot{\gamma}$	velocities
$\ddot{\beta} \;\; \ddot{\gamma}$	accelerations

Figure 5.29 Diagrammatical representation of the four-bar and six-bar linkages showing the parameters used in the computer programms A1 and A2.

5.11.5 Computer Programs

The author's program in Appendix A, entitled 'Analysis of the Four-Bar and Six-Bar Linkages', enables all parameters of a given linkage to be calculated using either a pocket computer which uses BASIC or a desk computer programmable in FORTRAN or BASIC.

This program is intended for use by a reader (student or practising engineer) who does not have access to some of the well-known ones such as KAPCA, ADAMS, and IMP.

Figure 5.29 shows the four cases covered by the program. The meaning of the question in the program, 'is the linkage "OPEN" or "CROSSED"?', is illustrated in the figure. It should be pointed out that since the input angular velocity of a linkage may not be constant, particularly if it is derived from a previous linkage or other device, the program asks for the values of the input angular acceleration and jerk (if known) as well as the input position and velocity.

5.11.6 Models of Mechanisms

In the experience of the author as a teacher and designer, models of linkage mechanisms play a very important role in the understanding of their construction and kinematic behaviour. A few examples of working precision models and a teaching kit developed by the author for teaching and design purposes are shown in Fig 5.30 to 5.34.

Figure 5.30 shows a variety of models (part of a display for use by students) from the basic four-bar linkage and its inversions to the double four-bar linkage used on the space shuttle for the opening and closing of the payload doors (see Exercise 3.14) and the multi-linkage of an early shoe-stitching mechanism. Some of the models illustrate the use of coupler curves; one

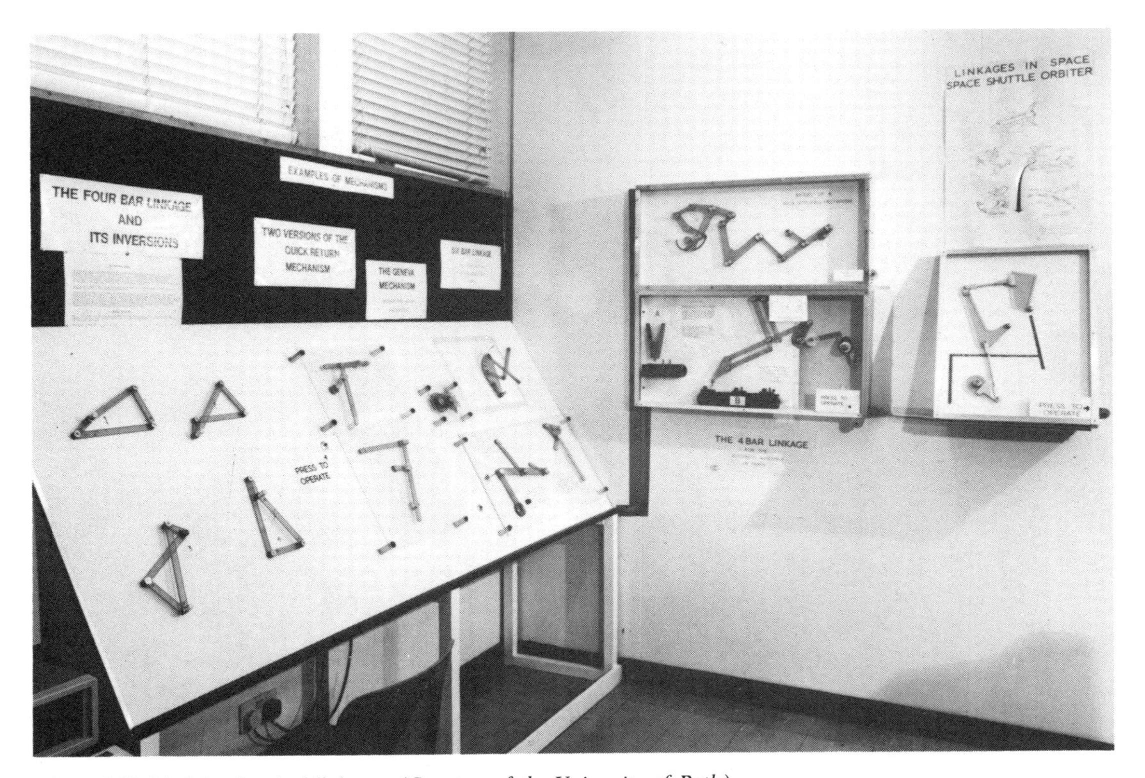

Figure 5.30 Models of typical linkages. (*Courtesy of the University of Bath.*)

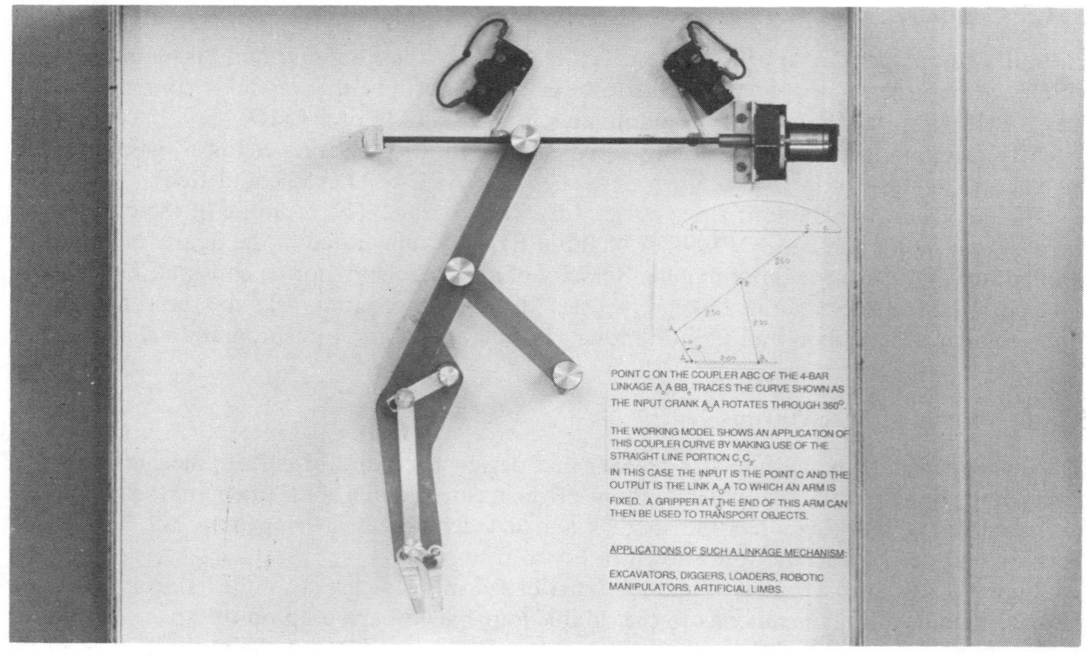

(*a*)

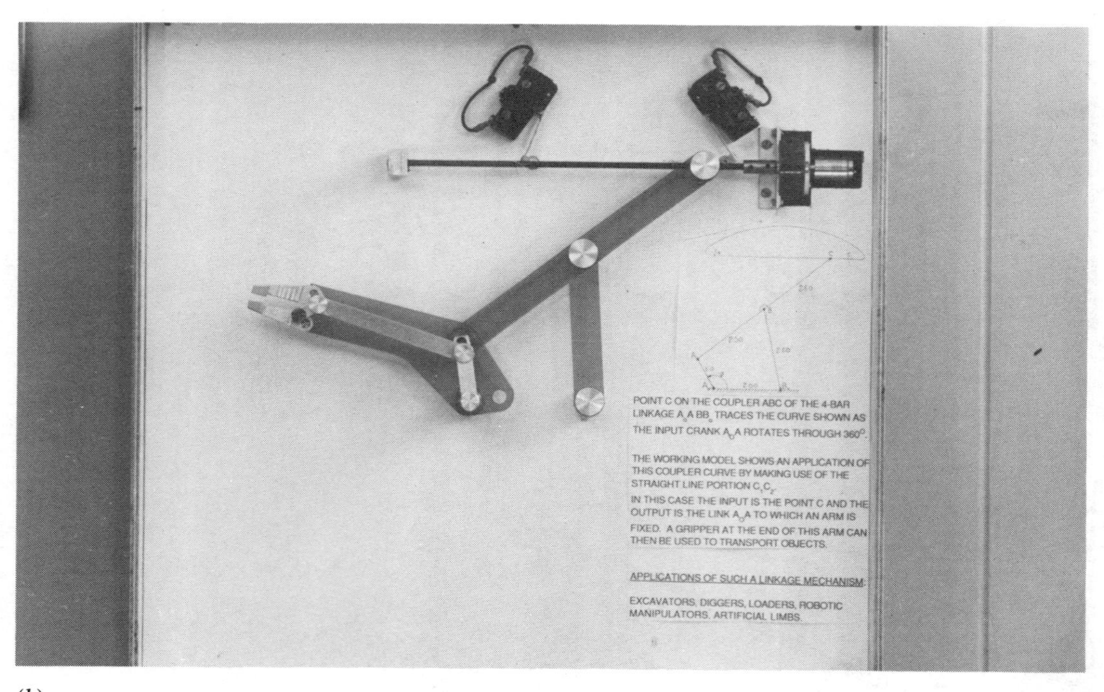

(*b*)

Figure 5.31 Four-bar linkage: model illustrating a possible practical design of a manipulator for industrial application. (*Courtesy University of Bath.*)

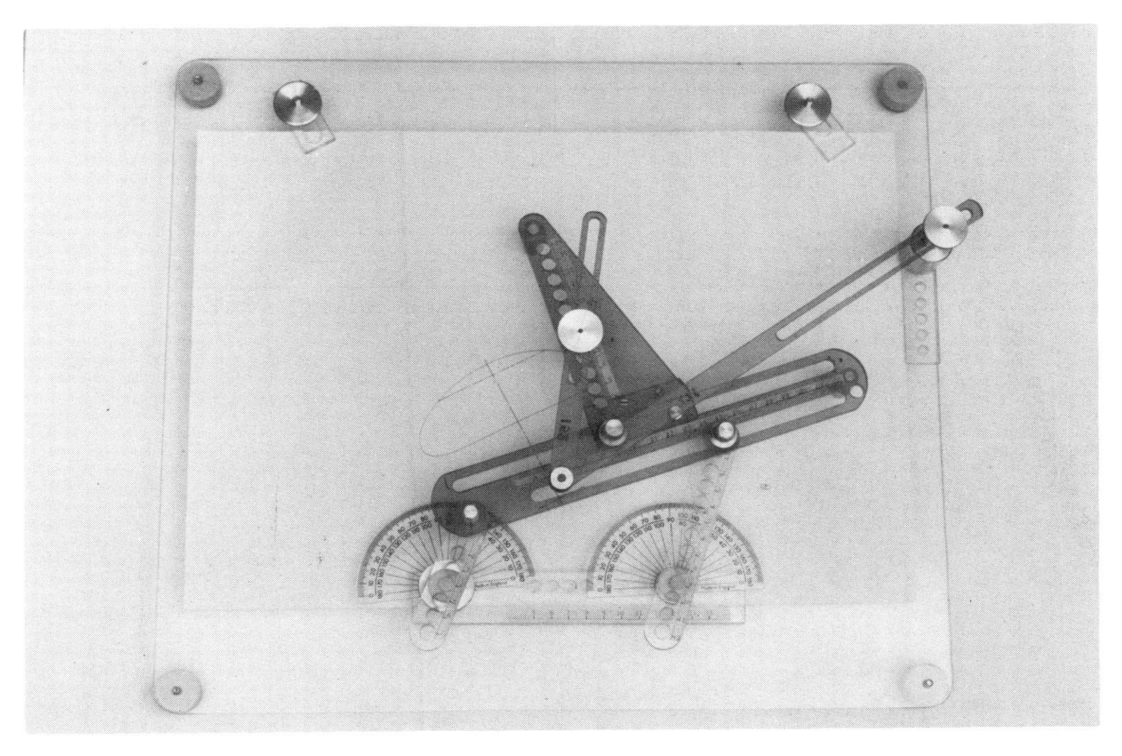

Figure 5.32 Multi-position four-bar and six-bar linkage teaching kit. (*Courtesy of the University of Bath.*)

such mechanism in its extreme positions is shown in Figure 5.31(*a*) and (*b*) (see also Example 8.1). It is a four-bar linkage where the input is a point on part of the coupler curve operated by a long screw driven by an electric motor via a large reduction gearbox to give the desired speed; the function of the two switches is to reverse the direction of motion. The output is an arm which could in practice be as long as necessary when the linkage is used as a manipulator performing repetitive tasks; the gripper is itself a four-bar linkage of the 'crossed' type remotely operated by means of other linkages. The picture shows quite clearly the large movement of the arm.

Figure 5.32 shows a teaching kit which when assembled makes up a four- or six-bar linkage and is capable of a very large range of movements; it is an improved version of an earlier one used by first-year engineering students who have to assemble the mechanism before undertaking a number of tasks (Fig. 5.33).

Other models are shown in Figs 1.13, 7.3, 7.4, and 7.22, the latter is a particular case of that shown in Fig. 5.34.

Figure 5.34 shows a spatial mechanism capable of many movements for demonstration purposes and for use by students. The angle between the input and output shafts can be altered by repositioning the angled pillar; the upright pillar, when selected, can also be moved to any of the positions shown on the base of the model but in this case the angle between input and output shafts is fixed at 90°.

In order to understand the behaviour of a particular mechanism fairly quickly, much simpler and cheaper materials and tools can be used such as stiff cardboard, balsa wood, drawing pin, paper fasteners, glue, etc. Making the model takes less time than drawing the many

Figure 5.33 First-year engineering students learning about linkages with the use of the kit. (*Courtesy of the University of Bath.*)

diagrams needed to analyse the motion. Furthermore, when looking for a coupler curve to design a six-bar linkage to produce a desired motion, it is possible to arrive at a solution fairly quickly using cruder models. It is always possible to refine the mechanism subsequently.

Although the computer is a very powerful tool it does not give the student or the practising engineer the 'feel' of the mechanism they are studying or designing. For example, when the transmission angle is poor during part of the cycle one can 'feel' that the motion is not smooth. For students a model is preferable to a line diagram on a blackboard because in handling the model they are being active. In the author's experience such an activity has overcome some of the difficulties encountered by students when confronted by the subject for the first time. Another important aspect which models help to appreciate is the spacing between links, i.e., how the linkage is to be assembled, something that the line diagram does not show.

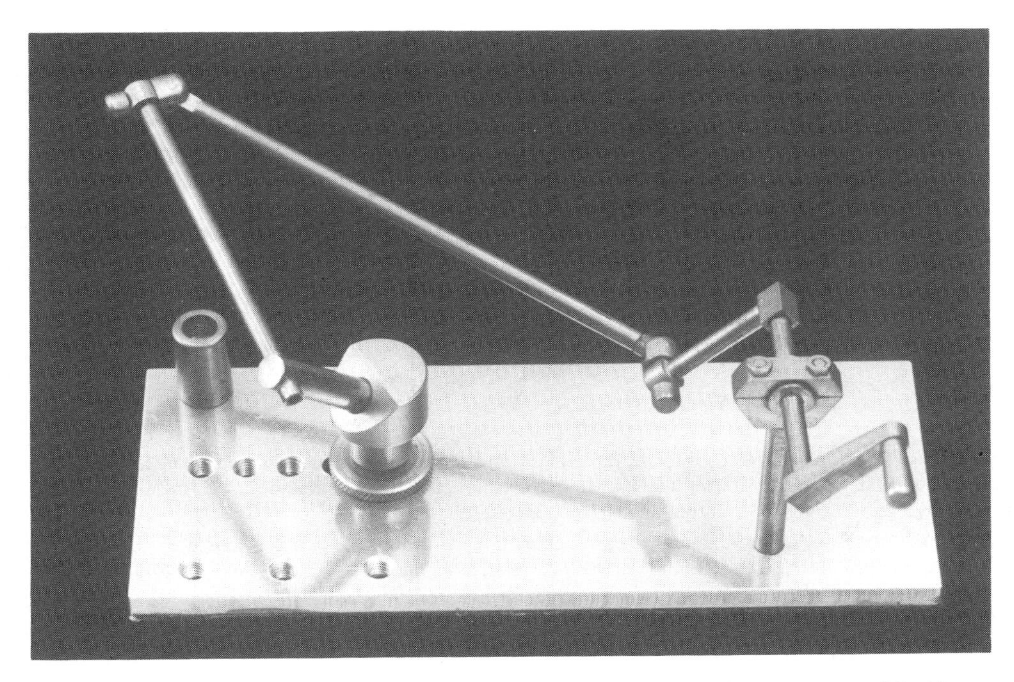

Figure 5.34 Model of a four-bar multi-position spatial linkage. (*Courtesy of the University of Bath.*)

EXERCISES

5.1 Figure (5.35) shows a four-bar linkage A_0ABB_0. Calculate (*a*) the angular velocity of the output link BB_0, (*b*) the velocity of the point B, (*c*) the angular acceleration of the coupler, (*d*) the angular acceleration of the output link. Dimensions are in metres.

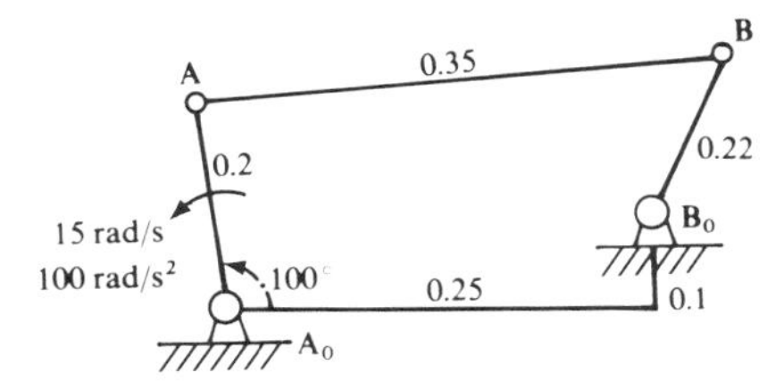

Figure 5.35

5.2 Calculate the angular acceleration of the links CP, PA, and AQ as well as the acceleration of the slider P, shown in Fig. 5.36. The crank OC rotates at 250 rev/min. Dimensions are in millimetres.

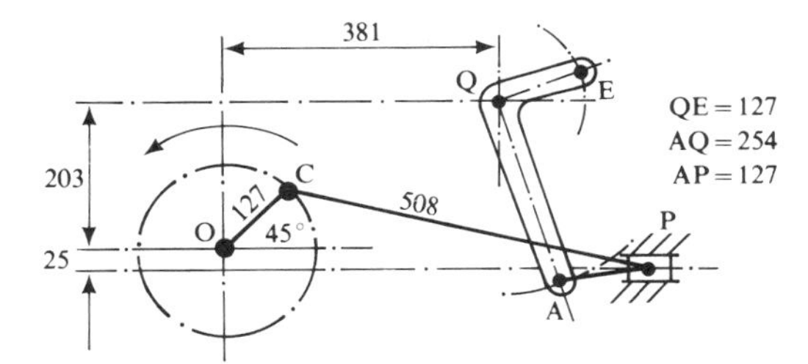

Figure 5.36

5.3 Figure 5.37 a four-bar linkage in which the output is taken from the coupler at the point C.

Using the following data calculate the magnitude of the acceleration of the point C when $\phi=0$ and the input link AA_0 rotates at 120 rev/min anticlockwise. $a=A_0A=50$ mm; $b=AB=100$ mm; $c=BB_0=100$ mm; $d=A_0B_0=150$ mm; $l=AC=197$ mm; $s=BC=100$ mm.

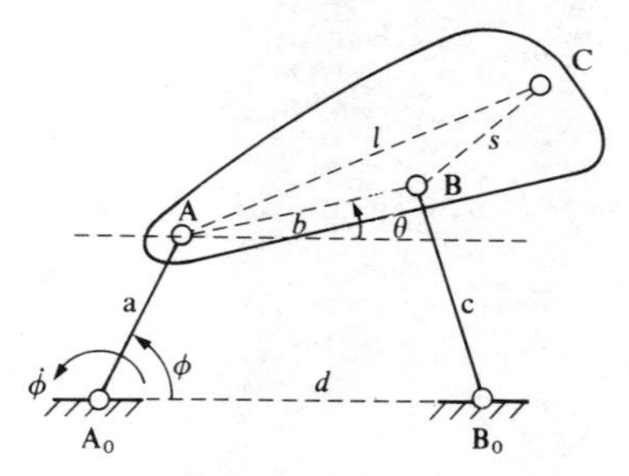

Figure 5.37

5.4 A four-bar linkage is to be used to coordinate three input and three output positions and to fit in a space of 450 mm × 350 mm. Calculate the dimensions of the links to one decimal place. Calculate also the transmission angle for each position. Input angle, 30°, 45°, 60°; output angle, 200°, 235°, 270°.

5.5 Figure 5.38 shows the shovel of a mechanical digger such as the one in Fig. 1.9(*b*); it is actuated by the hydraulic cylinder via the four-bar linkage ABCD. AB is the 'fixed link' relative to the outer-arm assembly ABE of the digger. The shovel is required to move through an angle of 200° while the link BC moves through an angle of 110°.

If the link BC is not to be less than 425 mm calculate the lengths of the other links, given that $\phi=20°$ and $\psi=120°$ in the initial position of the shovel.

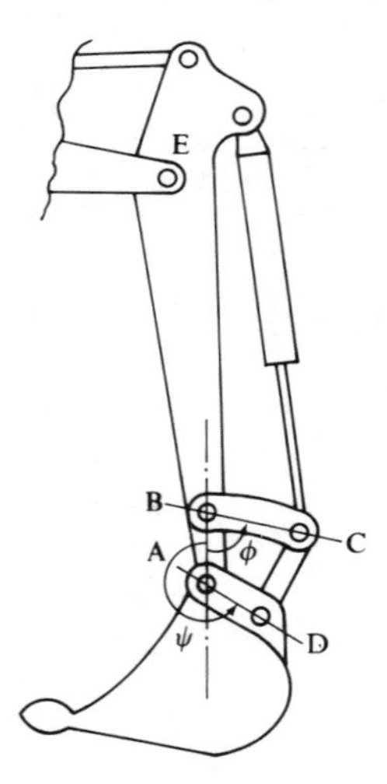

Figure 5.38

5.6 A four-bar linkage is required to generate the function $\psi=245+0.35\phi$ degrees for $20° \leqslant \phi \leqslant 150°$ and to fit in a space of 450 × 150 mm. Calculate the size of each link such that the structural error is minimized.

Check the transmission angle in the worst position and comment on its value.

Figure 5.39 is for reference.

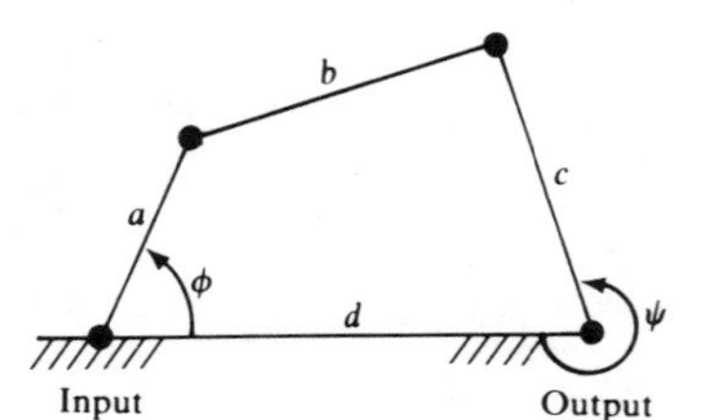

Figure 5.39

5.7 A four-bar linkage designed to generate the function $\psi = 0.045\phi^2$ degrees in the range $35 \leqslant \Phi \leqslant 40$ has the following link lengths: $a = 96.62$ mm, $b = 48.99$ mm, $c = -29.29$ mm, and $d = 100$ mm. d is the fixed link, a the input link, b the coupler, and c the output link. ϕ and ψ are the input and output angles respectively, measured in accordance with the usual convention. Chebyshev spacings with three precision points were used in the design calculations.

Show that the maximum structural error of the above linkage is less than 0.02 degrees.

5.8 The angular velocities of two oscillating, parallel shafts A and B are to be coordinated using a four-bar linkage as shown in Fig. 5.40. Using the data given below design a suitable linkage and obtain an expression for the angular acceleration of the output shaft.

Data: when $\phi_1 = 30°$, angular velocity $\dot{\phi}_1 = 1$ rad/s
$\psi_1 = 195°$, angular velocity $\dot{\psi}_1 = 1.71$ rad/s
and when $\phi_2 = 45°$, $\dot{\phi}_2 = 2$ rad/s
$\psi_2 = 220°$ $\dot{\psi}_2 = 3.28$ rad/s

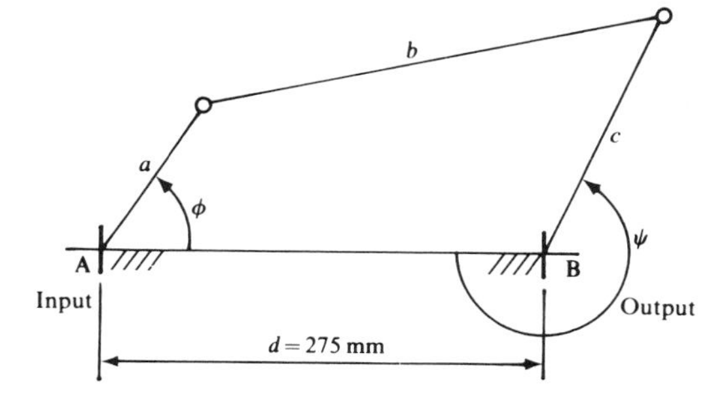

Figure 5.40

5.9 A four-bar linkage is to be designed for an output-to-input velocity ratio of 0.25 and for $n_2 = n_3 = 0$, in the neighbourhood of $\phi = 60°$; the input has a constant angular velocity.

Calculate the lengths of the links a, b, c, and d to two decimal places so that the mechanism will fit within the space shown in Fig. 5.41.

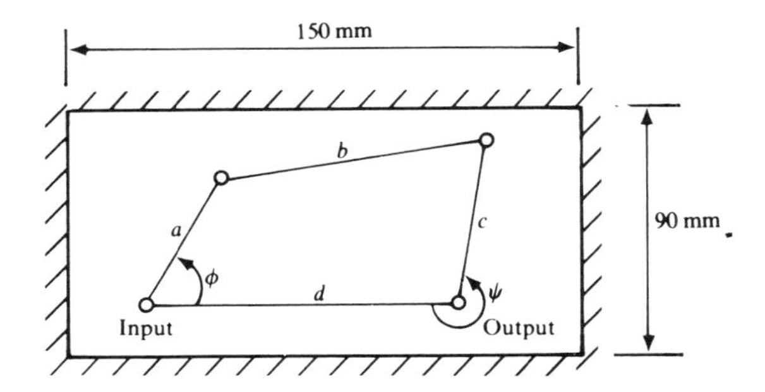

Figure 5.41

5.10 A slider–crank mechanism is to be designed to coordinate three positions of the slider S with three positions of a knob K as shown in Fig. 5.42. The knob is connected by a shaft to gear G1 which meshes with the geared segment G2 rigidly attached to the crank A_0A; the gear ratio is 7.2 to 1.

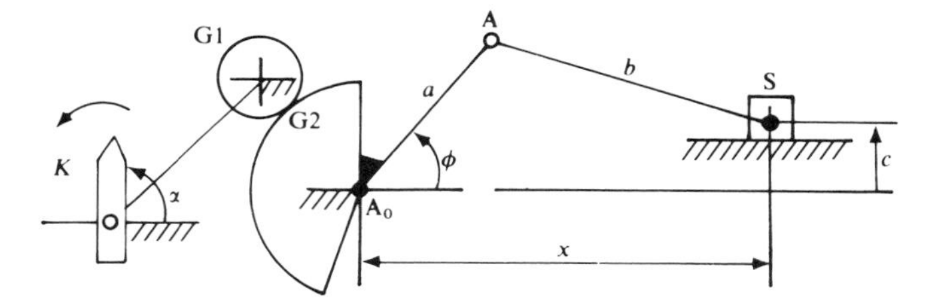

Figure 5.42

The required relationship between the position α of the knob and slider position x is as follows:

α, °	x, mm
90	64
180	81
270	96

and with $\phi = 60°$ as a first trial.

Show that the relationship between ϕ and x may be expressed by

$$K_1 \sin\phi + K_2 x \cos\phi + K_3 = x^2$$

where $K_1 = 2ac$

$K_2 = 2a$

$K_3 = b^2 - a^2 - c^2$

Hence calculate the lengths of a, b, and c to two decimal places and sketch the resulting mechanism.

Calculate also the transmission angle for each position and explain briefly what could be done to improve the quality of transmission.

5.11 For the six-bar linkage shown in Fig. 5.43 calculate the position, angular velocity, and angular acceleration of links CD and DE when $\phi = 90°$ and $\dot{\phi} = 315$ rad/s. $AA_0 = 50$ mm; $AB = 150$ mm; $BC = 30$ mm; $CB_0 = 100$ mm; $CD = 110$ mm; $DE = 160$ mm; $A_0B_0 = 150$ mm; $A_0F = 60$ mm; $FE = 300$ mm.

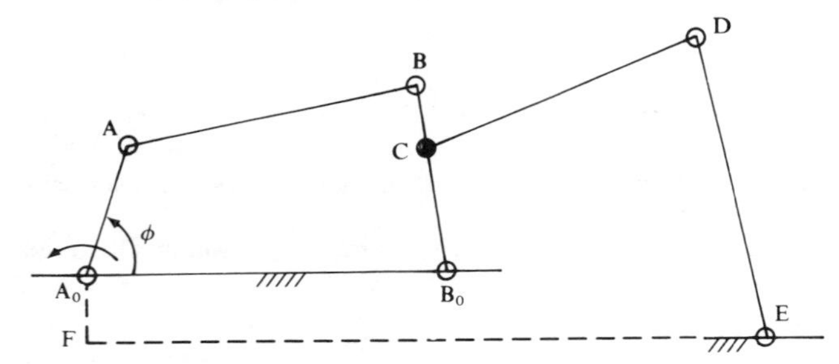

Figure 5.43

5.12 Referring to Fig. 5.26 in the text, derive an expression for the position β and the angular velocity $\dot{\beta}$ of the link CD which connects the coupler point C to the input link DE, and hence calculate β and $\dot{\beta}$ in Example 5.8 on page 140 when $\phi = 160°$. Check your solution graphically.

5.13 For the mechanism in Fig. 5.44 calculate the position, velocity, and acceleration of the coupler point C in the x- and y-directions when $\phi = 90°$. Hence calculate the angular velocity and acceleration of the output link DE. $a = 50$ mm; $b = 150$ mm; $c = 100$ mm; $d = 150$ mm; $l_1 = 175$ mm; $l_2 = 143$ mm; $e = 151$ mm; $f = 216$ mm; $x_e = -90$ mm; $y_e = -50$ mm; $\omega = 200$ rad/s.

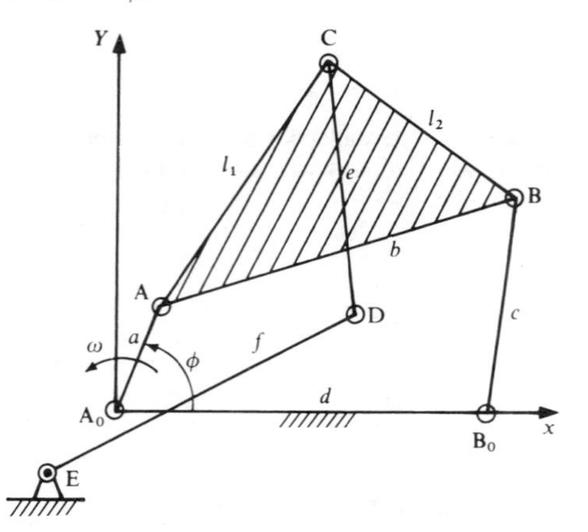

Figure 5.44

5.14 The input of the high-speed mechanism shown in Figure 5.45 rotates at 3000 rev/min. Calculate the output angular velocity and acceleration of a shaft at E located on the same axis as B_0 when $\phi=0°$. Dimensions: $a=25$ mm; $b=55$ mm; $c=50$ mm; $d=62.5$ mm; $e=76$ mm; $f=87.5$ mm; $l=65$ mm; $\alpha=110°$.

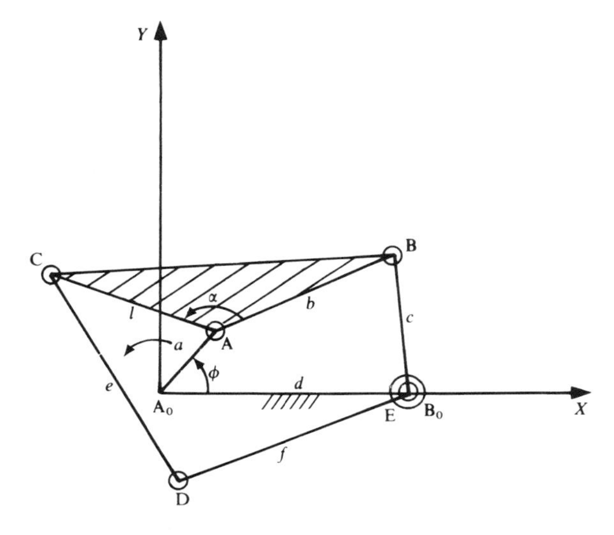

Figure 5.45

CHAPTER

SIX

GEAR TRAINS

6.1 INTRODUCTION

There are many situations in engineering where the transmission of rotary motion from one shaft to another is a fundamental requirement, the following being typical instances:

1. A car where the output rotation of the engine is transferred to the driving wheels by means of the gearbox and differential. The gearbox allows the driver to select different gear ratios depending upon the traffic situation, whereas the differential has a fixed value. The speed of the engine in this case is not constant since it is under the control of the driver.
2. A constant-speed electric motor driving a conveyor belt through a fixed-ratio gearbox to maintain a given forward speed.
3. A gas turbine rotating at constant speed and driving the propeller of a turboprop aircraft via a gearbox of fixed ratio, typically 10:1.

The car engine, the electric motor, and the gas turbine are examples of what are commonly referred to as *prime movers*. The car, the conveyor, and the aircraft are referred to as *loads*. Figure 6.1 shows diagrammatically the common situation of a prime mover driving a load via some kind of *converter*.

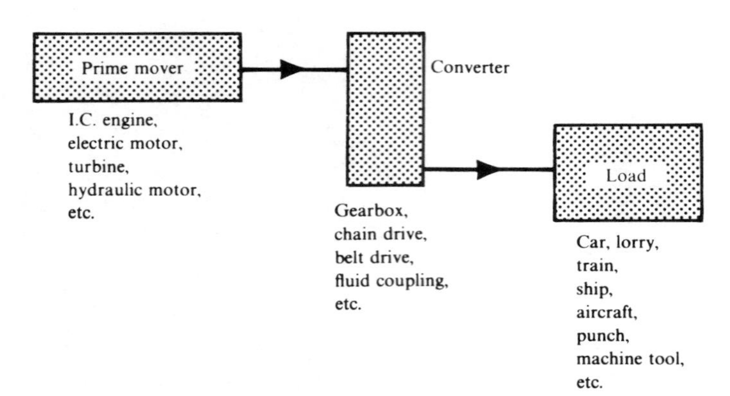

Figure 6.1 Diagram illustrating the common situation of a load driven by some kind of prime mover via a converter.

Each prime mover has its own characteristic such as those shown in Fig. 6.2. The torque and power output are usually functions of the speed; similarly the load also has its own characteristic, which is either constant or a function of speed. The resistance to motion of a train (the load), for example, is of the form $A+BV+CV^2$ where A, B, and C are constants and V is the speed of the train.

The purpose of the 'converter' is to transform the characteristic of the prime mover in order to match the load demand. To illustrate this point let us consider the case of a car having a mass of 950 kg accelerating from 30 km/h to 60 km/h in 3 seconds, the rolling radius of the

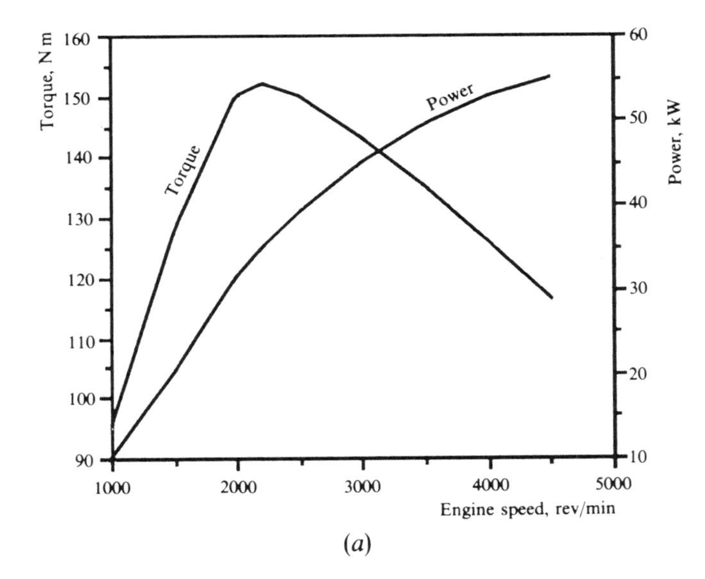

Figure 6.2(*a*) Torque and power curves of the Ford 1.8-litre turbo-diesel engine. (*Courtesy of the Ford Motor Company.*)

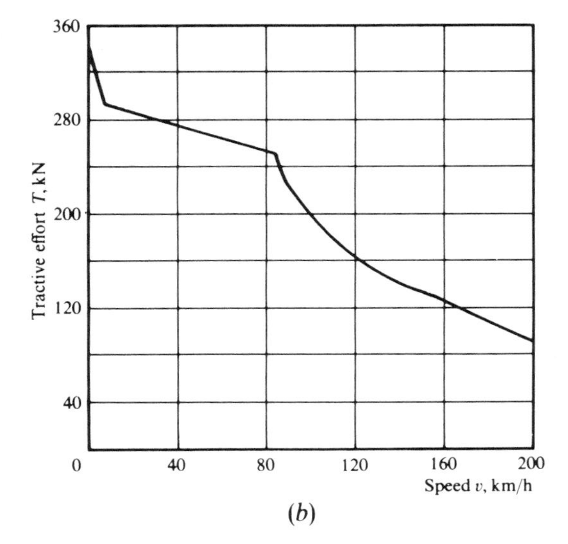

Figure 6.2(*b*) Tractive effort versus speed for a locomotive. (*Courtesy of Stanley Thornes, Cheltenham.*)

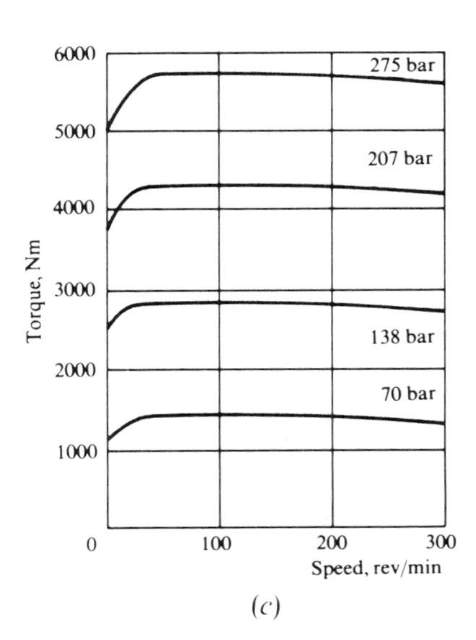

Figure 6.2(*c*) Characteristics of a hydraulic motor. (*Courtesy of Stanley Thornes, Cheltenham.*)

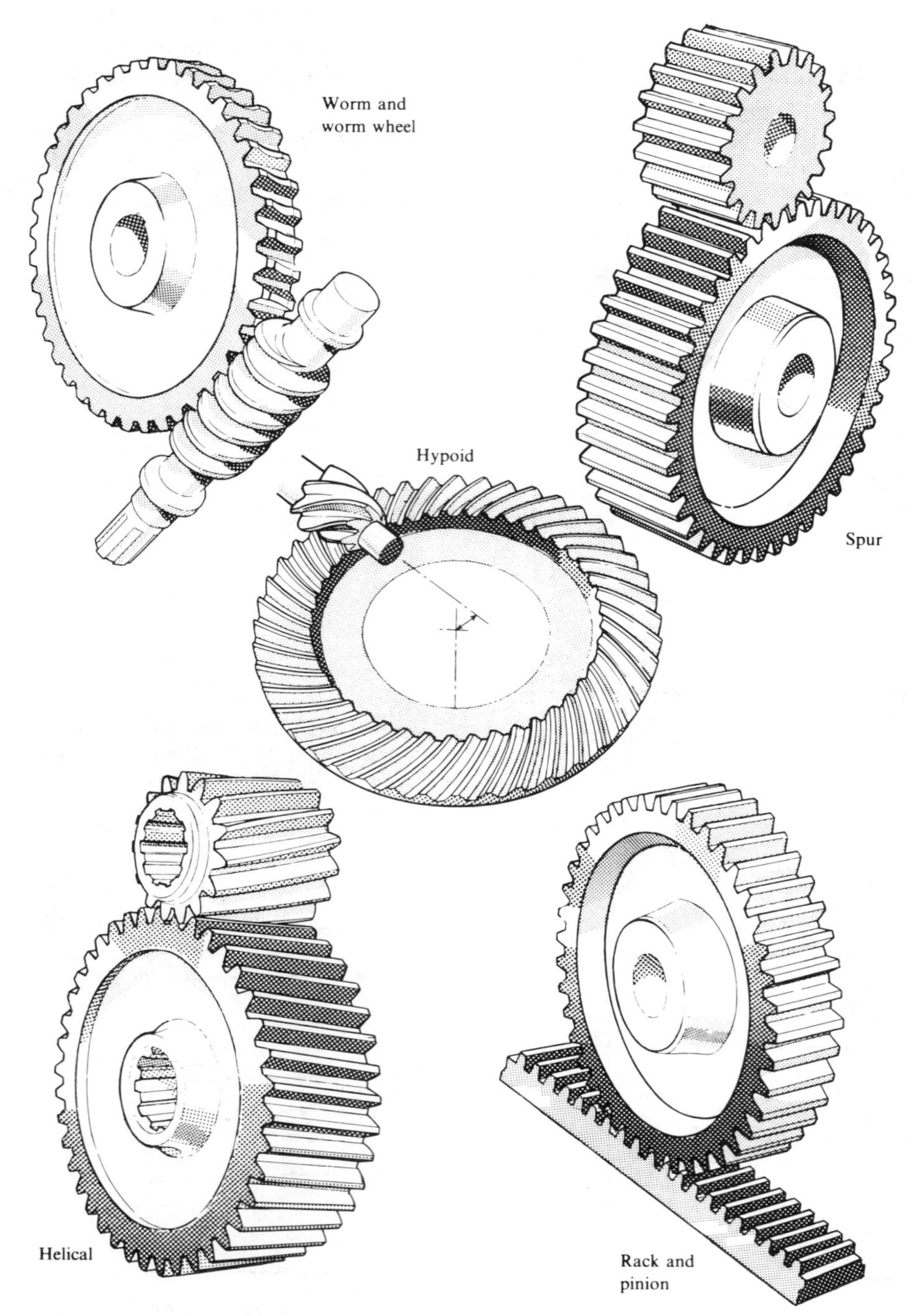

Figure 6.3(*a*) Types of gears. (*Courtesy of Brown Gears Industries.*)

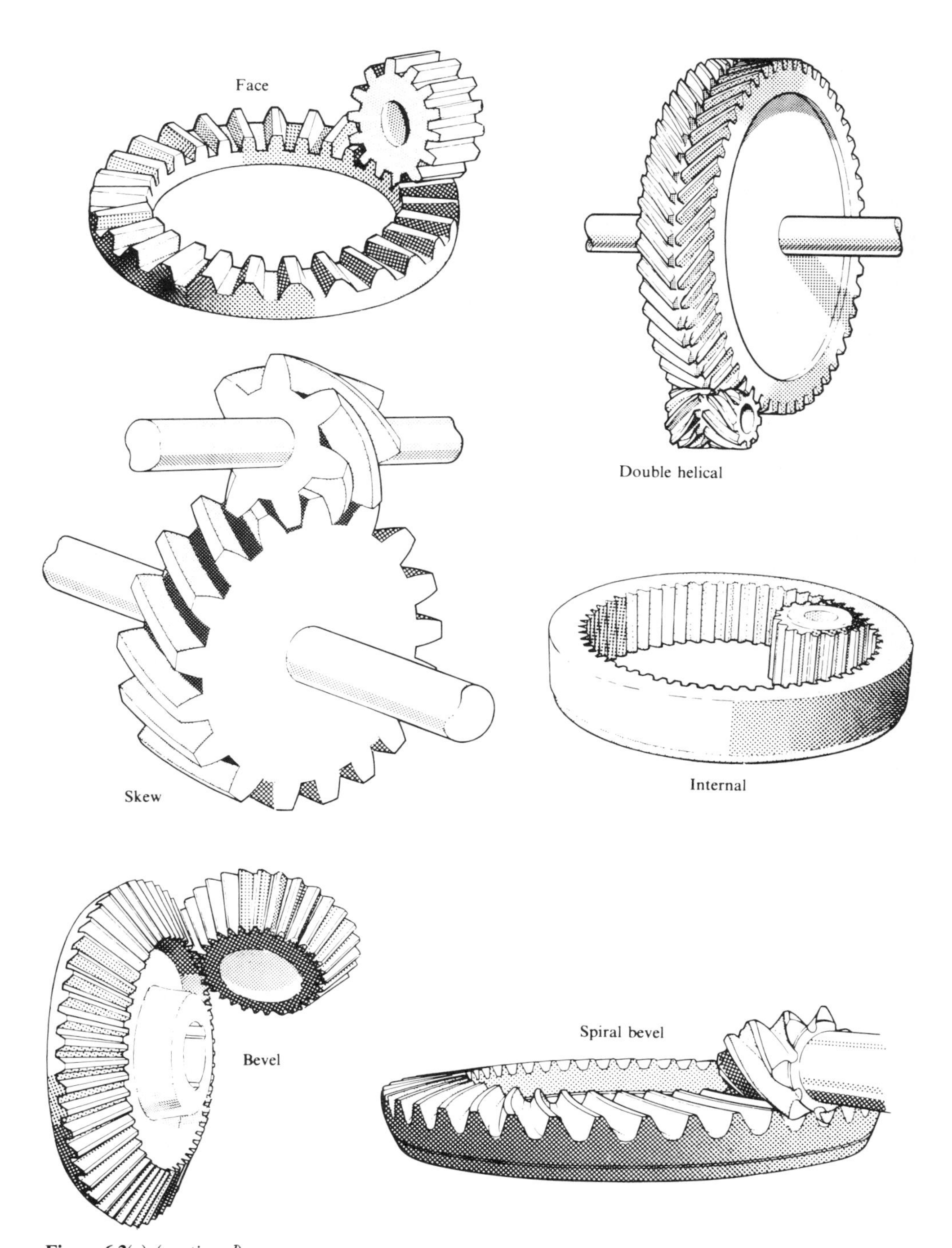

Figure 6.3(*a*) (*continued*)

Figure 6.3(*b*) Multispeed gear unit. (*Courtesy of Brown Gears Industries.*)

Figure 6.3(*c*) Examples of gears, belt drives, and Hooke's joints. (*Courtesy of Davall Gears.*)

driving wheels being 280 mm and the vehicle's resistance to motion being $450+0.5V^2$ N.

To simplify the calculations let us consider average values only; thus:

Acceleration of the car $a=(V_2-V_1)/t=[(60/3.6)-(30/3.6)]/3=2.78\ \text{m/s}^2$
Average resistance to motion $=450+0.5\{[(60/3.6)^2+(30/3.6)^2]/2\}=537$ N
Force to accelerate the car $=950\times 2.78=2641$ N
Total force required $=2641+537=3178$ N
Hence average torque required at the driving wheels $=3178\times 0.28=890$ N m
Average speed $=(30+60)/2/3.6=12.5$ m/s
Average angular velocity of the wheels $=12.5\times 60/0.28/2/\pi=426$ rev/min

Let us assume that this car is equipped with the engine whose characteristics are those shown in Fig. 6.2(*a*) and whose average output torque measured at the crank is 143 N m in the speed range 1500–3500 rev/min. Thus we see that there is an incompatibility between the load demand of 890 N m and the torque available at the crankshaft of the engine of 143 N m; hence without the intervention of some form of converter, in this case the gearbox followed by the differential, the car will not accelerate.

For the car to accelerate the gearbox plus differential needs to have an overall ratio of $890/143=6.22$; the corresponding speed of the engine is $6.22\times 426=2652$ rev/min. If the differential of the car has a typical ratio of 3.36:1, then the gearbox will have to have a ratio of $6.22/3.36=1.85$, in which case the driver will very likely select second gear.

In this chapter we consider the type of converter that employs gears only and analyse their different arrangements from a kinematic and kinetic point of view, starting with the simplest combination of a two-gear train and then going on to that of the planetary or epicyclic configuration. Such converters are better known as gearboxes and may include different kinds of gears: spur, helical, bevel, worm and wormwheel, hypoid, etc., as shown in Fig. 6.3(*a*); also illustrated is an internal gear arrangement. A multi-speed unit is shown in Fig. 6.3(*b*). Figure 6.3(*c*) shows a number of small gears, a belt drive and Hooke's joints.

In the analysis of gear trains the following dynamic relationships for rotational motion will be needed:

1. Torque = moment of inertia × angular acceleration, or in symbols $C=I\alpha$, where C is measured in N m, I in kg m^2, and α in rad/s^2. (6.1)
2. Kinetic energy $T=\frac{1}{2}I\omega^2$ where ω is measured in rad/s and T in joules (J). (6.2)
3. Power = torque × angular velocity, or in symbols $P=C\omega$ where P is measured in watts (W). (6.3)

6.2 GEOMETRY OF GEARS: DEFINITIONS

The geometry of gears and their manufacture are not discussed in this text but as a reminder to the reader formulae and definitions for spur and helical gears are to be found in Fig. 6.4.

6.3 SIMPLE GEAR TRAIN

The simplest trains consists of two gears as shown diagrammatically in Fig. 6.5(*a*). Gear 1 is a *pinion* and gear 2 is a *gearwheel*. If we assume the pinion is the *driver* or *input* then the gearwheel will be the *driven gear* or *output*.

Spur and Helical Gear - Formulae and Definitions

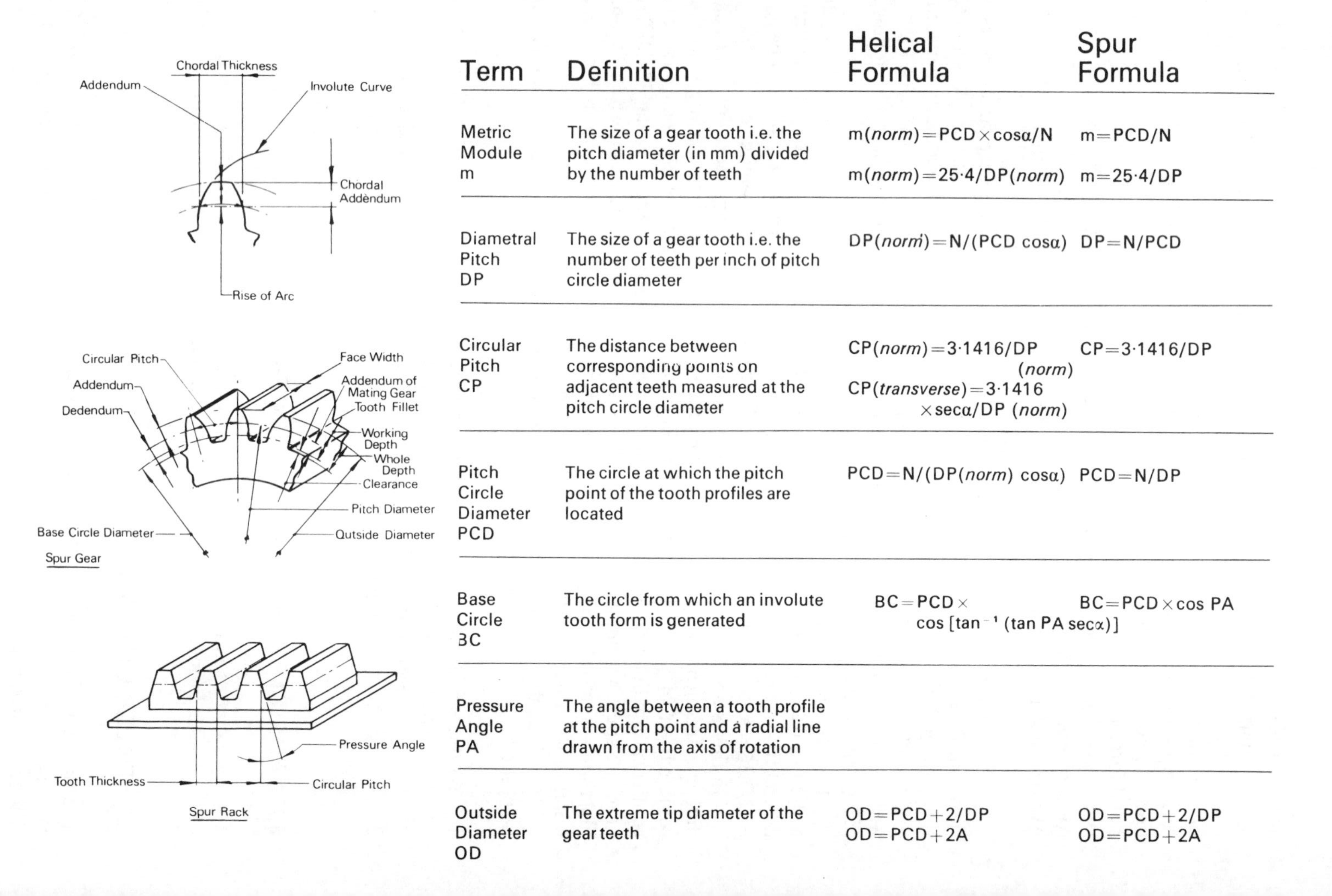

Term	Definition	Helical Formula	Spur Formula
Metric Module m	The size of a gear tooth i.e. the pitch diameter (in mm) divided by the number of teeth	m(*norm*)=PCD×cosα/N m(*norm*)=25·4/DP(*norm*)	m=PCD/N m=25·4/DP
Diametral Pitch DP	The size of a gear tooth i.e. the number of teeth per inch of pitch circle diameter	DP(*norm*)=N/(PCD cosα)	DP=N/PCD
Circular Pitch CP	The distance between corresponding points on adjacent teeth measured at the pitch circle diameter	CP(*norm*)=3·1416/DP (*norm*) CP(*transverse*)=3·1416 ×secα/DP (*norm*)	CP=3·1416/DP
Pitch Circle Diameter PCD	The circle at which the pitch point of the tooth profiles are located	PCD=N/(DP(*norm*) cosα)	PCD=N/DP
Base Circle BC	The circle from which an involute tooth form is generated	BC=PCD× cos [$\tan^{-1}$ (tan PA secα)]	BC=PCD×cos PA
Pressure Angle PA	The angle between a tooth profile at the pitch point and a radial line drawn from the axis of rotation		
Outside Diameter OD	The extreme tip diameter of the gear teeth	OD=PCD+2/DP OD=PCD+2A	OD=PCD+2/DP OD=PCD+2A

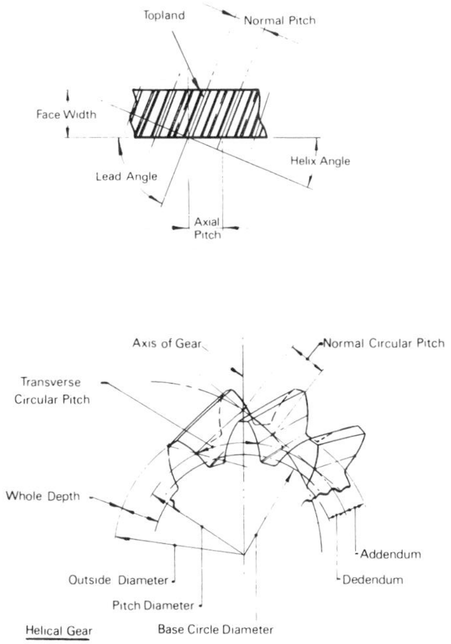

Term	Definition	Helical	Spur
Helix Angle α	The angle of inclination of tooth flank with respect to the transverse plane or plane of rotation	$\cos \alpha = N/(PCD \times DP)$ (*norm*)	
Addendum A	The radial distance between the PCD and the OD	A = 1/DP (*norm*)	A = 1/DP
Dedendum D	The radial distance between the PCD and the tooth root	D = ht − A	D = ht − A
Whole Depth ht	The whole tooth depth i.e. the addendum plus dedendum	20DP (1·25m) & coarser ht = 2·25/DP (norm) (BS436) Finer than 20DP (1·25m) ht − 2·4/DP (norm) (BS978)	20DP (1·25m) & coarser ht = 2·25/DP (BS 436) Finer than 20DP (1·25m) ht = 2·4/DP (BS978)
Working Depth hw	The depth of engagement of two mating gears	hw = 2/DP (*norm*)	hw = 2/DP
Backlash Bl	The amount at the pitch line by which the tooth space exceeds the mating tooth thickness	Bl = Cr × 2 tan PA	Bl = Cr × 2 tan PA
Normal Tooth Thickness Tn	The distance at the pitch cylinder between opposite and normal faces of the same tooth	Tn = 1·5708/DP (*norm*)	Tn = 1·5708/DP
Number of Teeth N	The number of teeth round the PCD of either the pinion or gear	N = DP(*norm*) × PCD × cosα	N = PCD × DP

Cr = Radial Clearance — The amount by which the running centre distance exceeds the actual gear centres when close in mesh

Figure 6.4 Definitions for spur and helical gears. (*Courtesy of Davall Gears.*)

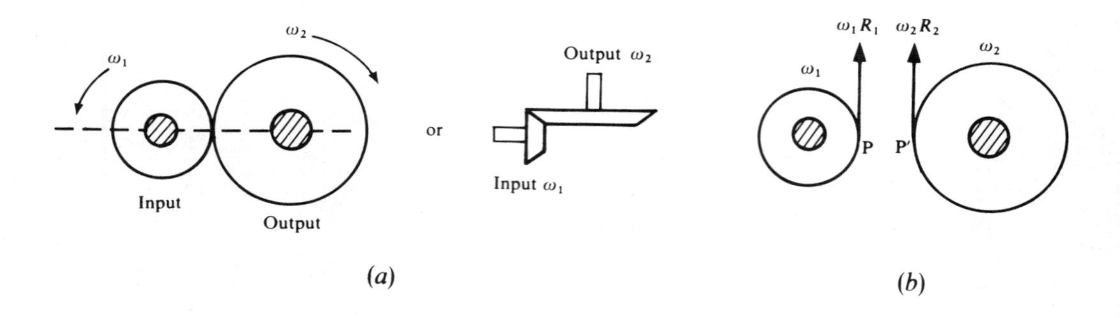

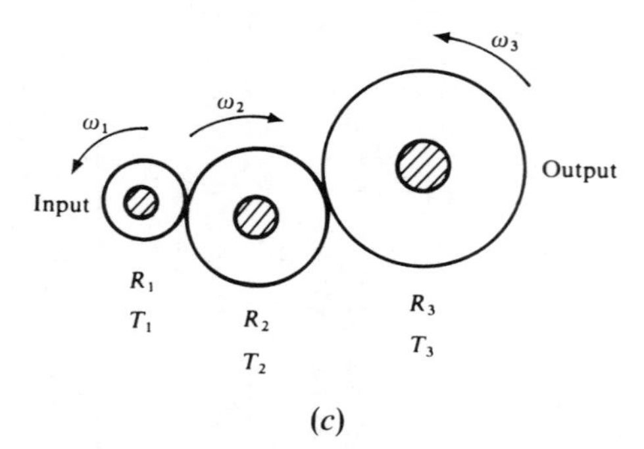

Figure 6.5 Simple gear train.

Let R_1 = the pitch radius of the pinion
T_1 = number of teeth
ω_1 = angular velocity,
R_2 = the pitch radius of the gearwheel,
T_2 = the number of teeth
ω_2 = the angular velocity

Consider the pinion and gear wheel separately (Fig. 6.5*b*); then at the pitch point we have:

Linear velocity at P, $V_p = \omega_1 R_1$

Linear velocity at P′, $V_{p'} = \omega_2 R_2$

For compatibility at the pitch point we must have $V_p = V_{p'}$, i.e., $\omega_1 R_1 = \omega_2 R_2$;

therefore
$$\frac{\omega_2}{\omega_1} = -\frac{R_1}{R_2} \tag{6.4}$$

The minus sign indicates a change in direction, i.e., if the pinion rotates in an anticlockwise direction then the gearwheel will rotate in a clockwise direction. However in the case of an internal gear arrangement as in Fig. 6.3(*a*), both the pinion and the gearwheel rotate in the same directions.

If p is the circular pitch, i.e., the distance between two teeth, then

$$2\pi R_1 = pT_1 \qquad \text{and} \qquad 2\pi R_2 = pT_2$$

therefore
$$\frac{R_1}{R_2} = \frac{T_1}{T_2}$$

Hence Eq. (6.4) becomes

$$\frac{\omega_2}{\omega_1}=\frac{T_1}{T_2} \tag{6.5}$$

Therefore the ratio of the speeds of gear 2 and gear 1 is inversely proportional to the number of teeth. Since $T_2 > T_1$ it follows that the gearwheel will have a slower rate of rotation; the *gear ratio* G in this case is $G=T_2/T_1$ down.

Let us interpose between the pinion and the gearwheel another gear of radius R_3 and having T_3 teeth as shown in Fig. 6.5(*c*). Then, applying Eq. (6.4) twice, we obtain

$$\frac{\omega_3}{\omega_1}=\frac{\omega_3}{\omega_2}\times\frac{\omega_2}{\omega_1}=\left(-\frac{R_2}{R_3}\right)\left(-\frac{R_1}{R_2}\right)=+\frac{R_1}{R_3}=\frac{T_1}{T_3} \tag{6.6}$$

We see that gear 3 will rotate in the same direction as the pinion; thus gear 3 is an *idler* whose purpose is to change the direction of rotation. Although it appears that radius or the number of teeth of the idler is immaterial from a kinematic point of view, it must be capable of transmitting the required torque and power.

Hence if a simple gear train consists of n gears then by Eq. (6.4) the speed ratio is given by

$$\frac{\text{Output}}{\text{Input}}=\frac{\omega_n}{\omega_1}=\pm\frac{T_1}{T_n}$$

where the positive sign is to be taken if n is odd, and the minus sign if n is even.

6.4 COMPOUND GEAR TRAIN

A *compound* gear train is one in which some shafts carry two gears locked together as illustrated in Fig. 6.6, gears 2 and 3, and as can be seen in Fig. 6.3(*b*). Thus gears 2 and 3 rotate at the same speed and in the same direction. The output is taken from gear 4, which meshes with gear 3. One important advantage of compound trains is that large speed ratios can be obtained.

By proceeding as for the simple train we find:

$$\frac{\omega_2}{\omega_1}=-\frac{\mathrm{T}_1}{\mathrm{T}_2} \quad \text{and} \quad \frac{\omega_4}{\omega_3}=-\frac{\mathrm{T}_3}{\mathrm{T}_4} \quad \text{also } \omega_2=\omega_3$$

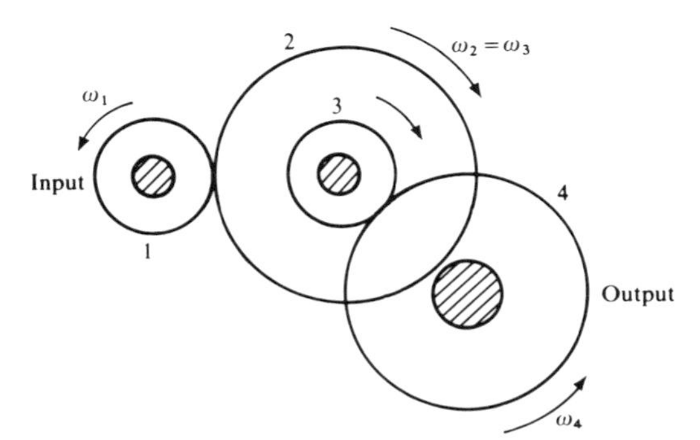

Figure 6.6 Compound gear train.

therefore
$$\frac{\omega_4}{\omega_1}=\frac{\omega_2}{\omega_1}\frac{\omega_4}{\omega_3}=\left(-\frac{T_1}{T_2}\right)\left(-\frac{T_3}{T_4}\right)=\frac{T_1}{T_2}\times\frac{T_3}{T_4} \tag{6.7}$$

In this case the output gear has the same direction of rotation as the input gear, and

$$\text{Velocity ratio}=\frac{\text{product of teeth in driving wheels}}{\text{product of teeth in driven wheels}}$$

Example 6.1 The hoist of a gantry crane is required to lift a load at a speed of 4.25 m/min. The drum on which the rope is wound has a diameter of 650 mm and the driving motor rotates at 720 rev/min. The arrangement of the rope is such that its speed is twice that of the speed of the load. If the gearbox consists of pairs of spur gears and assuming that the ratio of each pair should not exceed about 5 or 6 to 1 (typical values in practice), calculate the number of pairs required, and hence suggest suitable number of teeth on each gear.

SOLUTION Speed of the drum $=2\times 4.25/0.65\pi=4.16$ rev/min. Hence total reduction required $=720/4.16=173:1$. If we take 6 as the maximum ratio for each pair of gears, it looks as though three reductions will be required; each ratio G would, therefore, have to be

$$G=\sqrt[3]{173}=5.57$$

We could choose 20 teeth for each pinion as a minimum which will give $20\times 5.57=111$ teeth for each gear, in which case the total reduction will be

$$G=\left(\frac{111}{20}\right)^3=171:1$$

This value may not be considered close enough to that required. Let us increase the last two gears by one tooth, in which case

$$G=\left(\frac{111}{20}\right)\left(\frac{112}{20}\right)\left(\frac{112}{20}\right)=174$$

which may be more acceptable.

An advantage of the compound gear train is that it enables the input and output shafts to be in line as shown in Fig. 6.7, in which BC is a compound wheel. Such an arrangement is known as a *reverted train* and is used in the gearboxes of motor cars. It follows from the geometry that

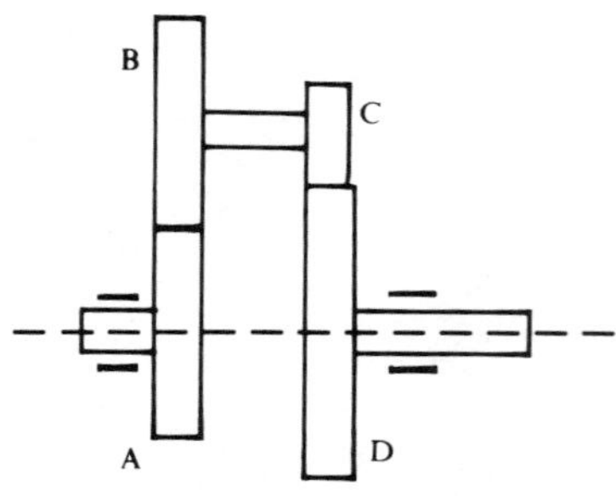

Figure 6.7 Reverted gear train.

the radii of the gears must be such that

$$R_A + R_B = R_C + R_D \tag{6.8}$$

There is another relationship which must be satisfied, namely

$$\frac{mT_A}{2} + \frac{mT_B}{2} = \frac{m'T_C}{2} + \frac{m'T_D}{2} \tag{6.9}$$

where the m's are the modules and the T's the number of teeth. If the modules are equal, then

$$T_A + T_B = T_C + T_D \tag{6.9a}$$

6.5 EPICYCLIC GEAR TRAINS

6.5.1 Introduction

An *epicyclic* gear train is one in which there is relative motion between some of the axes of the gears making up the train, i.e., the axes of rotation of some of the gears are not fixed. It is not always easy to visualize the motion of such trains; however, one thing we should bear in mind is the fact that the relative motion between a pair of mating gears is always the same, whether the shafts involved are stationary or moving.

Some important aspect of epicyclic gears are as follows:

1. the input and output shafts lie on a common axis;
2. high reductions are possible within very compact arrangements, by connecting two or more planetary trains in series;
3. loads can be spread between a number of gears;
4. in variable gearboxes for cars the gears can remain in mesh all the time while brakes are applied to various elements in turn to obtain the ratios required; this reduces wear on the gears.

Epicyclic gear trains are used a great deal in aircraft, helicopters, mechanisms, machine tools, automatic transmissions in cars and lorries, etc. Examples are shown in Fig. 6.8(*a*) and (*b*). (*a*) shows the differential assembly and planetary gearing for the operation of the payload doors of the space shuttle, while (*b*) shows the 11:1 gear reduction of a turboprop aircraft.

Figure 6.9(*a*) is a diagrammatical representation of the simple planetary or epicyclic train. It consists of four elements: a *sun* S which meshes with *planetary gears* P (usually three or more) free to rotate on shafts fixed to an *arm* A; the last element is the *ring gear* R having internal teeth which mesh with those of the planetary gears. It will be noticed that all the shafts are co-axial on XX. Any one of the shafts can be used as an input or output. Figure 6.9(*b*) shows a possible arrangement of a series combination of two simple epicyclic gear trains.

Many types of epicyclic gear trains are in common use and Fig. 6.10 illustrates diagrammatically three basic arrangements which we propose to investigate in the next section. There are a number of methods for analysing epicyclic gear trains, the following being most common:

1. an analytical approach based on the relative motion between the elements;
2. a tabular approach.

In this text we propose to concentrate on method (1) and indicate the steps involved using method (2) in Example 6.3.

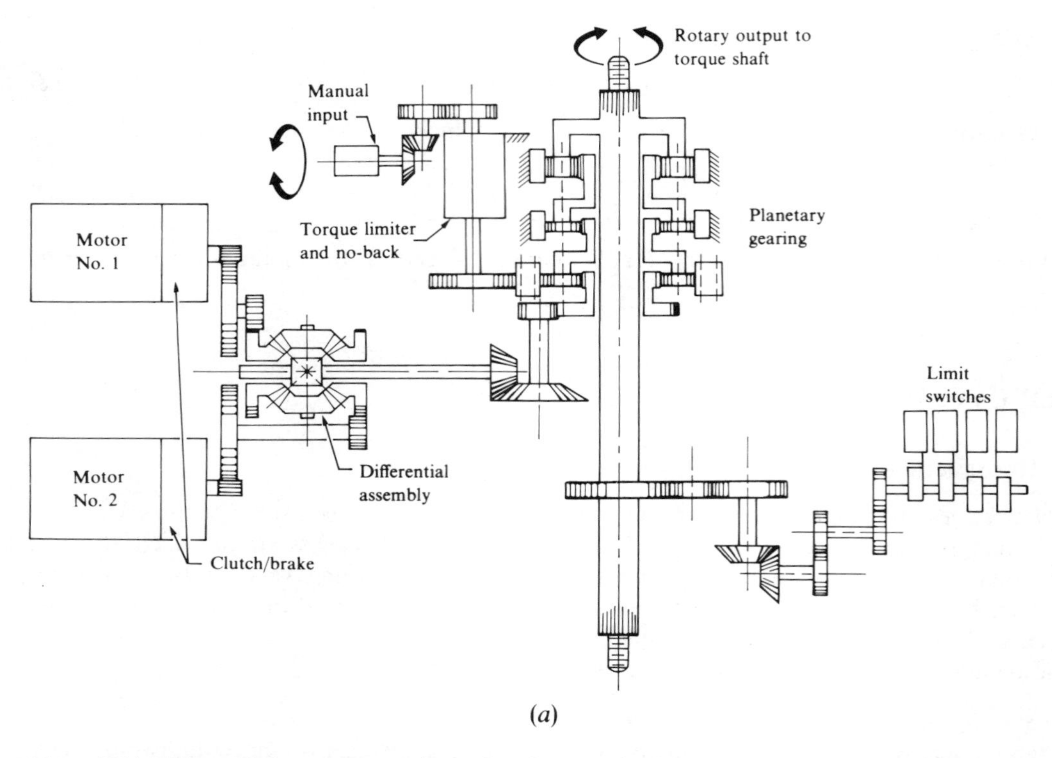

Figure 6.8(*a*) Differential planetary gear trains used on the space shuttle. (Reproduced with the permission of the Council of the Institution of Mechanical Engineers.)

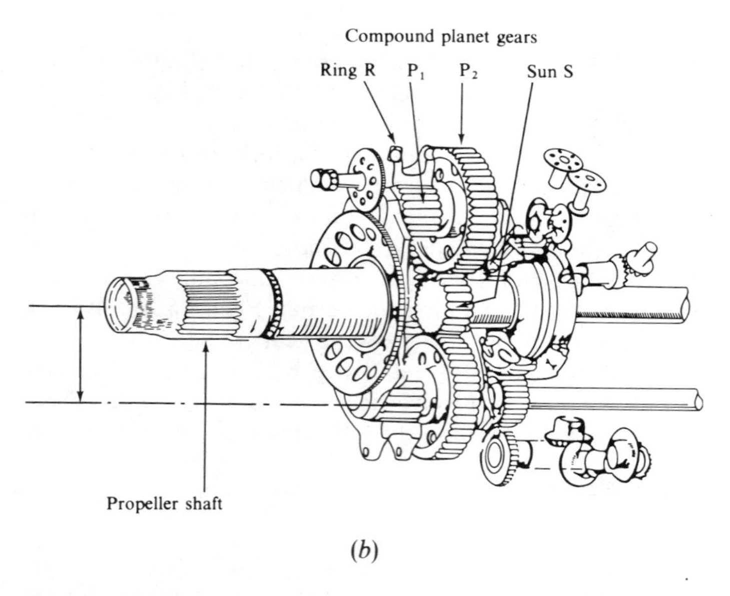

Figure 6.8(*b*) Simple planetary gear reduction for a propeller-driven aircraft. (*Courtesy of Rolls-Royce plc.*)

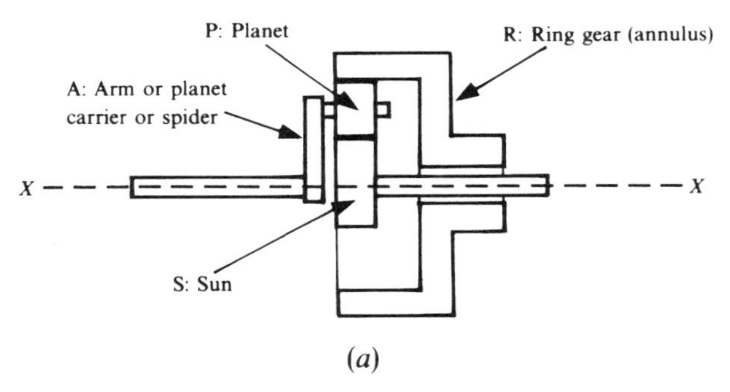

Figure 6.9(*a*) Simple planetary gear train.

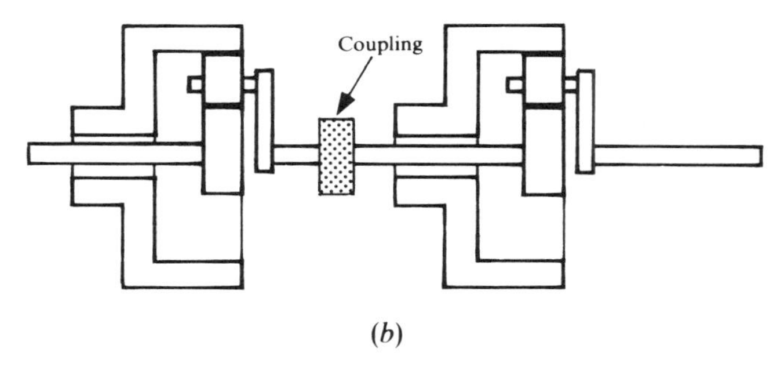

Figure 6.9(*b*) Planetary gear trains in series.

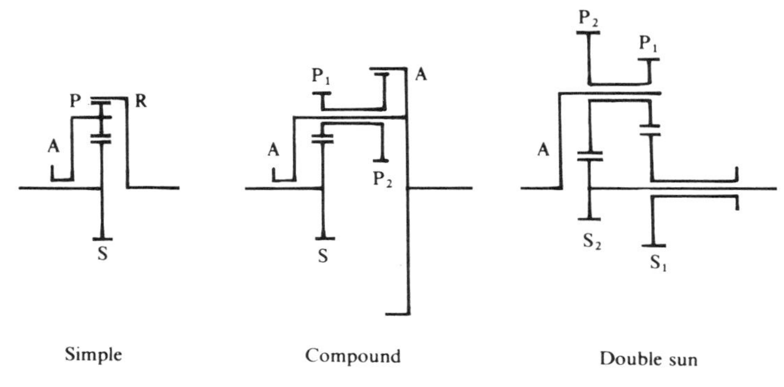

Figure 6.10 Types of epicyclic gear trains.

6.5.2 Analysis of Planetary Gear Trains

(*a*) **Simple epicyclic** Referring to Fig. 6.10,

let ω_S = absolute angular velocity of the sun S

ω_R = absolute angular velocity of the ring R

ω_A = absolute angular velocity of the arm A

T_S, T_R, and T_P = number of teeth in the sun, the ring, and the planet respectively.

Consider motion relative to the arm: an observer sitting on the arm would see the following relative velocities:

Velocity of the sun relative to the arm: $\omega_{SA} = \omega_S - \omega_A$

Velocity of the ring relative to the arm: $\omega_{RA} = \omega_R - \omega_A$

If we divide these two equations we get

$$\frac{\omega_{SA}}{\omega_{RA}} = \frac{\omega_S - \omega_A}{\omega_R - \omega_A} \tag{6.10}$$

This equation expresses the ratio of the relative velocity of the sun to that of the ring; this ratio is constant and is often referred to as the *train value*. Since we are looking at motion relative to the arm, we have effectively reduced the system to a simple train as discussed in Sec. 6.3; hence

$$\frac{\omega_S - \omega_A}{\omega_R - \omega_A} = \left(-\frac{T_P}{T_S}\right)\left(+\frac{T_R}{T_P}\right) = -\frac{T_R}{T_S} \tag{6.11}$$

Note the positive sign in the second gear ratio; this is because the ring gear is an internal gear.

Three ratios are possible when one element at a time is fixed, as shown in Fig. 6.11.

Case 1 The sun is the input and the ring gear is fixed: $\omega_R = 0$.
Substituting in Eq. (6.11) we obtain

$$-\frac{\omega_S}{\omega_A} + 1 = -\frac{T_R}{T_S} \quad \Rightarrow \quad \frac{\omega_S}{\omega_A} = 1 + \frac{T_R}{T_S}$$

so that the speed of the arm is given by

$$\omega_A = \frac{\omega_S}{\left(1 + \frac{T_R}{T_S}\right)}$$

in the same direction of rotation as the input.

Case 2 The arm is the input and the sun is fixed: $\omega_S = 0$.
Substituting in Eq. (6.11) we obtain

$$\frac{-\omega_A}{\omega_R - \omega_A} = -\frac{T_R}{T_S}$$

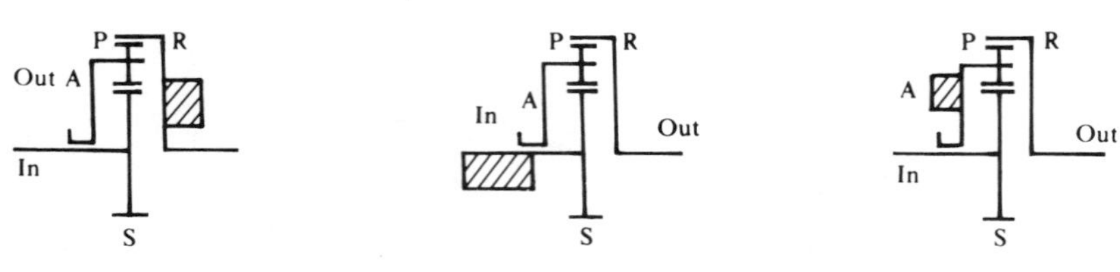

Figure 6.11 Fixing different elements of a simple planetary gear train results in different output/input ratios.

hence the speed of the ring is given by

$$\omega_R = \left(\frac{T_S + T_R}{T_R}\right)$$

also in the same direction as the rotation of the input.

Case 3 The sun is the input and the arm is fixed, $\omega_A = 0$.
Substitution once more in Eq. (6.11) yields

$$\omega_R = -\frac{T_S}{T_R}\omega_S$$

In this case the rotation of the ring is in the opposite direction to the input.

Example 6.2 A simple epicyclic gear train (Fig. 6.9*a*) has the following gear teeth: $T_S = 35$, $T_P = 25$, $T_R = 85$. Calculate (*a*) the speed and direction of rotation of the ring gear if the sun is fixed and the arm rotates at 255 rev/min clockwise, (*b*) the speed and direction of rotation of the arm if the ring gear rotates at 110 rev/min anticlockwise and the sun rotates at 255 rev/min clockwise.

SOLUTION For motion relative to the arm we have

$$\frac{\omega_R - \omega_A}{\omega_S - \omega_A} = -\frac{T_S}{T_R} \qquad \text{by Eq. (6.11)}$$

(*a*) With $\omega_S = 0$ and substituting numerical values we get

$$\frac{\omega_R - 255}{0 - 255} = -\frac{35}{85}$$

Solving for ω_R yields

$$\omega_R = 255 + 255 \times \frac{35}{85} = 360 \text{ rev/min clockwise}$$

(*b*) In this case no shaft is fixed and we have two inputs. It is important to assign the correct sign for the rotations of the inputs: these are $\omega_R = +255$ and $\omega_R = -110$.
Hence substituting in the equation yields

$$\frac{255 - \omega_A}{-110 - \omega_A} = -\frac{85}{35} \qquad \Rightarrow \qquad \omega_S\left(1 + \frac{85}{35}\right) = 255 - \frac{85}{35} \times 110$$

Hence we get $\omega_A = -3.54$ rev/min, i.e., the arm rotates in an anticlockwise direction at 3.54 rev/min.

(*b*) Compound epicyclic train Referring to Fig. 6.10 and considering motion relative to the arm, we have

$$\frac{\omega_R - \omega_A}{\omega_S - \omega_A} = \left(-\frac{T_S}{T_1}\right)\left(+\frac{T_2}{T_R}\right) = -\frac{T_S}{T_1}\frac{T_2}{T_R}$$

where T_1 and T_2 are the number of the compound planet gears P_1 and P_2 respectively.
By fixing any one of the elements of the train a number of ratios can be obtained.

Example 6.3 In the epicyclic train shown in Fig. 6.12 the arm A carries compound wheels P_1 and P_2 having 36 and 20 teeth respectively. Planet P_2 meshes with the ring gear R and planet P_1 meshes with the sun S having 22 teeth. Given that the modules of P_1 and S are 4 and that those of P_2 and R are 6, calculate (*a*) the number of teeth on the ring gear; (*b*) the speed ratio between the sun and the arm.

SOLUTION (*a*) From Eq. (6.8) the pitch diameters are connected by the following relationships:

$$\tfrac{1}{2}D_R = \tfrac{1}{2}D_2 + \tfrac{1}{2}D_1 + \tfrac{1}{2}D_S$$

Referring to Fig. 6.4, we have:

for P_2 with a module of 6, $D_2 = m_1 T_2 = 6 \times 20 = 120$ mm
for P_1 with a module of 4, $D_1 = m_2 T_1 = 4 \times 36 = 144$ mm
for S with a module of 4, $D_S = m_2 T_S = 4 \times 22 = 88$ mm

Hence $D_R = 120 + 144 + 88 = 352$ mm.
Since the module of the ring gear is 4, the number of teeth $T_R = D_R/m_2 = 352/4 = 88$.
(*b*) For this part of the question we propose to illustrate the tabular method of solution. The steps are as follows:

1. Imagine the whole train locked; rotate it through one complete revolution clockwise.
2. Give the ring gear an anticlockwise rotation with the arm fixed.
3. Add all columns.
4. Calculate the relative speeds.

	R	P_1 and P_2	S	A
Step 1	+1	+1	+1	+1
Step 2	−1	$-\frac{88}{20} = -4.4$	$\frac{88}{20}\frac{36}{22} = 7.2$	0
Step 3	0	1−4.4	1+7.2	+1
Relative speeds	0	−3.4	8.2	1

Hence $\qquad$ Speed ratio, $\dfrac{\text{Speed of the arm}}{\text{Speed of the sun}} = \dfrac{1}{8.2}$

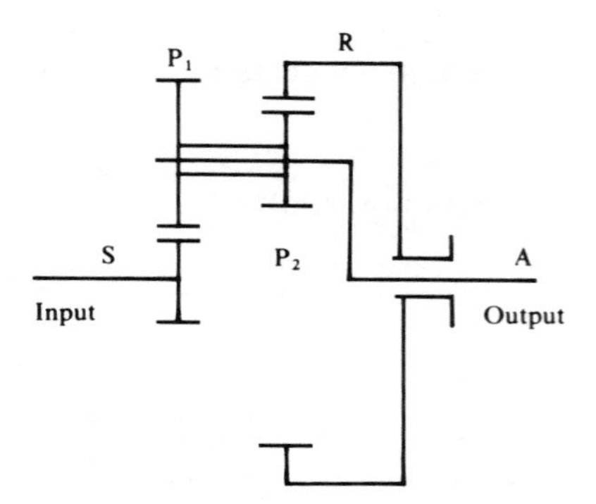

Figure 6.12

The reader is advised to check this result using the relative velocity method. We see from this example that large speed ratios are possible with this kind of arrangement; thus in terms of the number of teeth we have:

$$\frac{\text{Output motion}}{\text{Input motion}}=\frac{1}{1+\dfrac{T_R}{T_1}\dfrac{T_2}{T_S}}$$

The ratio will large if $T_R \gg T_1$ and $T_2 > T_S$.

(*c*) **Double sun epicyclic train** Referring to Fig. 6.10, and applying the relative velocity approach, we obtain:

$$\frac{\omega_{S2}-\omega_A}{\omega_{S1}-\omega_A}=\left(-\frac{T_{S1}}{T_1}\right)\left(-\frac{T_2}{T_{S2}}\right)=\frac{T_{S1}}{T_1}\frac{T_2}{T_{S2}}$$

In this case if the arm is fixed then the arrangement becomes the simple compound train of Sec. 6.4.

6.6 DYNAMICS OF GEARED SYSTEMS

6.6.1 Torque Required to Accelerate Geared Systems: Equivalent Inertia

Consider a typical system consisting of a prime mover driving a load through a gearbox of ratio $G:1$ down (the load is slower than the prime mover) as shown in Fig. 6.13. For the prime mover,

let C_m = torque developed which may or may not be constant; in many cases it is a function of speed (see Fig. 6.2).
α_m = angular acceleration
I_m = moment of inertia

and for the load,

let L = torque demand or resisting torque; not necessarily constant
α_L = angular acceleration
I_L = moment of inertia

If we neglect the inertia of the gears and apply Eq. (6.1) we have:

Torque required at A to accelerate the prime mover, $C_P = I_m \alpha_m$

Torque required at B to accelerate the load, $C_L = I_L \alpha_L$ but

Torque at A to accelerate the load, $C_{AL} = \dfrac{C_L}{G}$ hence

Total torque at A to accelerate the system, $C_m = C_P + C_{AL}$

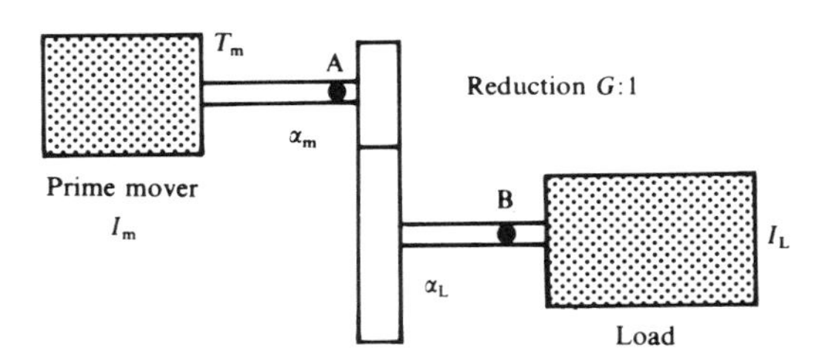

Figure 6.13 Diagram of a simple speed reduction system between prime mover and load.

Therefore
$$C_m = I_m\alpha_m + \frac{I_L\alpha_L}{G}$$

From the kinematics of the gears, we have

$$\alpha_L = \frac{\alpha_m}{G}$$

Substituting yields
$$C_m = \alpha_m\left(I_m + \frac{I_L}{G^2}\right) \tag{6.12}$$

or
$$C_m = I_e\alpha_m$$

where I_e = equivalent moment of inertia of the system as seen by the prime mover.

Therefore
$$I_e = I_m + \frac{I_L}{G^2} \tag{6.13}$$

If I_1 is the moment of inertia of gears when referred to the prime-mover shaft and I_2 the moment of inertia of gears when referred to the load shaft, then

$$I_e = I_m + I_1 + \frac{(I_L + I_2)}{G^2}$$

The concept of equivalent moment of inertia is very useful when solving dynamical problems; we can always refer the motion of the whole system to the prime-mover output shaft.

Example 6.4 A simple type of actuator as shown in Fig. 6.14 consisting of a shunt motor is used to open and close a butterfly valve through a reduction gearbox of value G:1. If I_m is the moment of inertia on the motor shaft, I_V the moment of inertia on the valve shaft, C_m the constant torque delivered by the motor, L a constant resistive torque offered by the valve and θ the angle turned through by the valve in a time t_0, show that the gearbox ratio is given by

$$aG^2 + bG + c = 0$$

where a, b, and c are functions of the above parameters.

If $I_m = 3 \times 10^{-5}$ kg m^2, $I_V = 0.02$ kg m^2, $C_m = 0.35$ N m, $L = 12$ N m, $t_0 = 4$ seconds, and $\theta = 90°$ calculate the value of the gear ratio required.

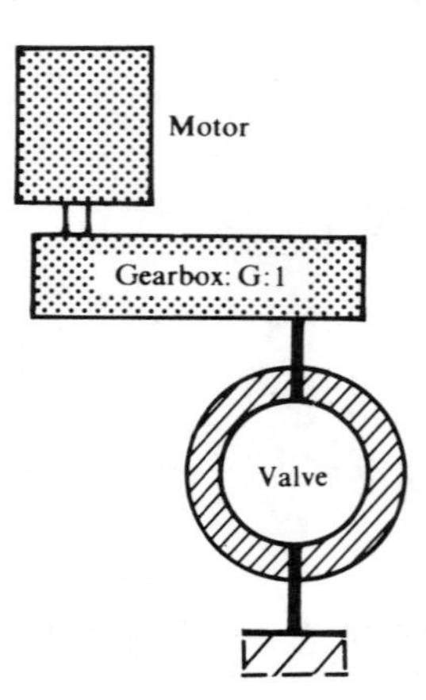

Figure 6.14

SOLUTION In this example it will be more convenient to refer the motion to the load side instead of the prime-mover side.

The equivalent moment of inertia of the system referred to the valve side is given by

$$I_e = I_m G^2 + I_v$$

and the torque delivered by the motor as seen by the load is equal to $C_m G$.

The equation of motion is

$$C_m G - L = I_e \alpha$$

where α is the angular acceleration of the valve. But

$$\alpha = \frac{d\omega}{dt} \qquad \text{or} \qquad \alpha = \omega \frac{d\omega}{d\theta}$$

where ω is the angular velocity of the valve. The equation of motion becomes

$$C_m G - L = I_e \frac{d\omega}{dt} \qquad \text{or} \qquad C_m G - L = I_e \omega \frac{d\omega}{d\theta}$$

Integrating both equations yields

$$\int_0^{t_0} (C_m G - L)\, dt = I_e \int_0^{\omega} d\omega \qquad \text{and} \qquad \int_0^{\theta} (C_m G - L)\, d\theta = I_e \int_0^{\omega} \omega\, d\omega$$

Hence

$$(C_m G - L) t_0 = I_e \omega \qquad \text{and} \qquad (C_m G - L)\theta = \tfrac{1}{2} I_e \omega^2$$

Eliminating ω from both equations, we get

$$(C_m G - L)\theta = \tfrac{1}{2} I_e \frac{(C_m G - L)^2 t_0^2}{I_e^2}$$

Substituting for I_e yields

$$(I_m G^2 + I_v)\theta = \tfrac{1}{2}(C_m G - L) t_0^2$$

Expanding and collecting terms, we obtain

$$I_m \theta G^2 - \tfrac{1}{2} C_m t_0^2 G + I_v \theta + \tfrac{1}{2} L t_0^2 = 0 \qquad \Leftrightarrow \qquad aG^2 + bG + c = 0$$

where $a = I_m \theta$

$b = -\tfrac{1}{2} C_m t_0^2$

$c = I_v \theta + \tfrac{1}{2} L t_0^2$

Substituting numerical values yields

$$a = 3 \times 10^{-5} \times \frac{\pi}{2} = 4.712 \times 10^{-5} \qquad b = -\tfrac{1}{2} \times 0.35 \times 4^2 = -2.8 \qquad c = 0.02 \times \frac{\pi}{2} \times 12 \times 4^2 = 96.03$$

Substituting and solving the resulting quadratic equation, we get the value of the gear ratio required, namely $G = 34.3$. The other solution giving $G = 59\,383$ is not practical.

Such an actuator is typical of many to be found in practice; Figs. 6.15 and 6.16 illustrate two types used in aircraft and helicopters for operating a variety of environmental equipment,

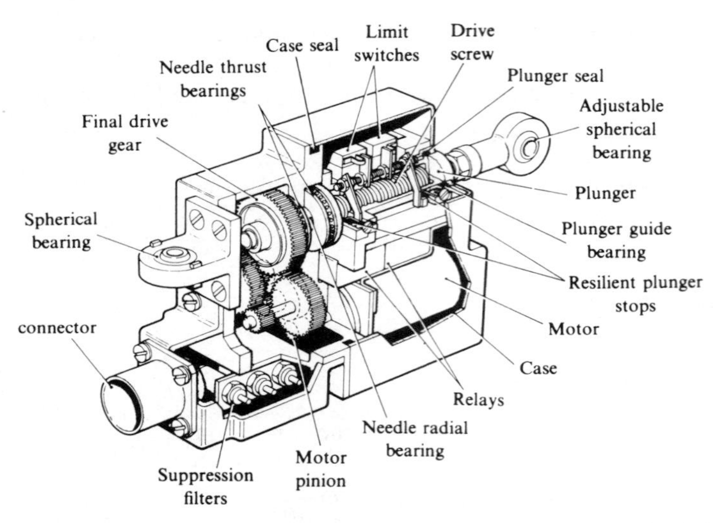

Figure 6.15 Typical arrangement for a small compact electromechanical actuator. (*Courtesy of the Institution of Mechanical Engineers.*)

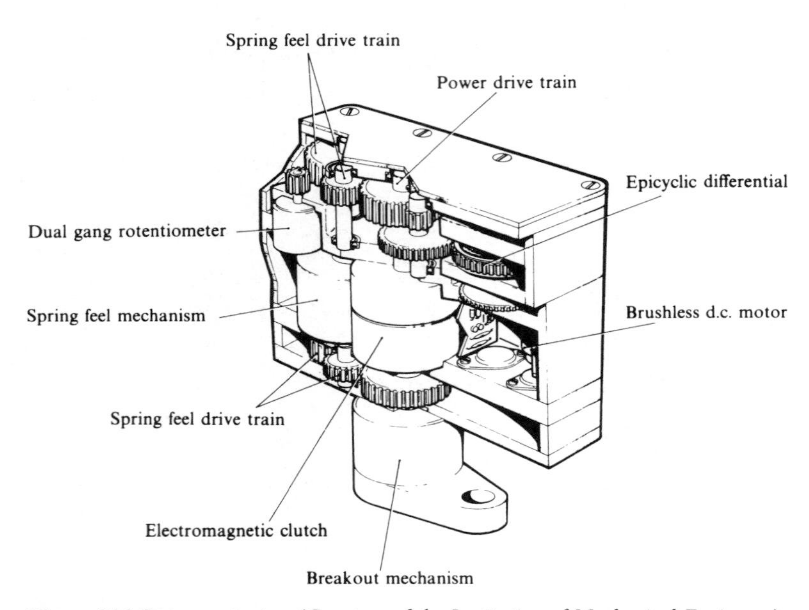

Figure 6.16 Rotary actuator. (*Courtesy of the Institution of Mechanical Engineers.*)

release mechanisms, fuel valves, etc. Figure 6.15 shows a typical arrangement of a small, compact electromechanical actuator. Figure 6.16 shows a rotary type where the rotor of the brushless d.c. motor is connected to a spur gear stage of 10.17:1 reduction followed by an epicyclic differential stage having a 61.7:1 reduction. The output of the epicyclic forms the input to a further spur reduction of 5.57:1 ratio.

Example 6.5 Figure 6.17 shows a simple lifting device consisting of a motor M driving a drum D through a reduction gearbox of ratio G:1. The load is lifted by means of a cable

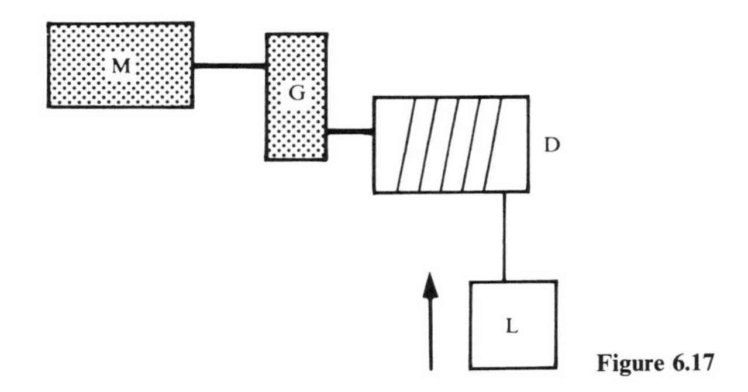

Figure 6.17

wrapped on the drum. Using the data given below, calculate the vertical acceleration of the load neglecting the mass of the cable. Data:

Moment of inertia of the motor and gears on the motor shaft = 0.04 kg m^2.
Moment of inertia of the drum and gears on tne drum shaft = 25 kg m^2.
Gear reduction = 30:1. Drum diameter = 800 mm. Mass of load = 500 kg.
Motor torque = 104 N m, assumed constant.
Motor mechanical efficiency = 85 per cent.

SOLUTION Let a be the upward acceleration of the load, and T the tension in the cable (Fig. 6.18). We can solve this problem in two ways as follows:

1. *Motion referred to the drum shaft, i.e., the load shaft*

$$T - mg = ma \qquad \text{therefore} \qquad T = mg + ma$$

Load torque on the drum axis $= TR = mgR + maR$
$= mgR + mR^2\alpha_d$, since $a = R\alpha_d$.

Equivalent moment of inertia referred to the drum axis, $I_e = I_1G^2 + I_2 + mR^2$.

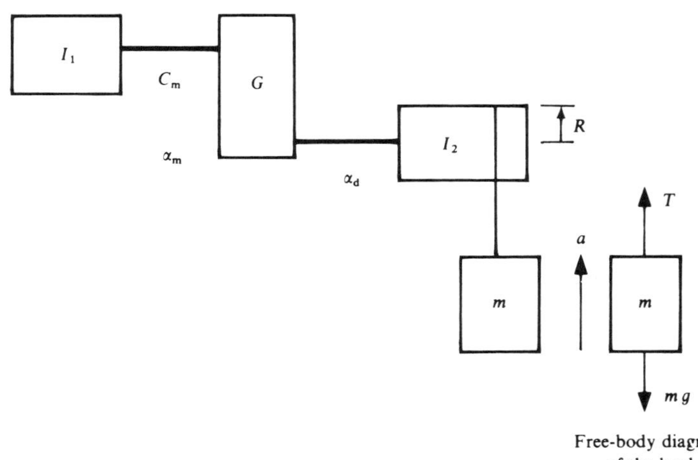

Figure 6.18

Applying the torque equation, we get

$$\eta C_m G - mgR = (I_1 G^2 + I_2 + mR^2)\alpha_d$$

where η is the efficiency.

Hence $$\alpha_d = \frac{\eta C_m G - mgR}{I_1 G^2 + I_2 + mR^2} = \frac{0.85 \times 104 \times 30 - 500 \times 9.81 \times 0.4}{0.04 \times 30^2 + 25 + 500 \times 0.4^2} = 4.89 \text{ rad/s}^2$$

but $$a = \alpha_d R = 4.89 \times 0.4 = 1.96 \text{ m/s}^2$$

2. *Motion referred to the motor shaft*

Equivalent moment of inertia,

$$I_{em} = I_1 + \frac{I_2 + mR^2}{G^2} = 0.04 + \frac{25 + 500 \times 0.4^2}{30^2} = 0.157 \text{ kg}^2$$

Load torque, $$C_L = \frac{mgR}{G} = \frac{500 \times 9.81 \times 0.4}{30} = 65.4 \text{ N m}$$

Applying the torque equation, we have

$$\eta C_m - C_L = I_{em}\alpha_m \qquad \text{but} \qquad \alpha_m = \frac{a}{R} G$$

Hence $$a = \frac{(\eta C_m - C_L)R}{I_{em} G} = \frac{(0.85 \times 104 - 65.4) \times 0.4}{0.157 \times 30} = 1.95 \text{ m/s}^2$$

6.6.2 Kinetic Energy in Epicyclic Gear Trains: Inertia Effects

Consider the arm A of an epicyclic train rotating about an axis through O and carrying three planets (typical number) as shown in Fig. 6.19.

Let R = length OG of the arm

I_A = moment of inertia of the arm about O

I_P = moment of inertia of a planet about its centre of mass G

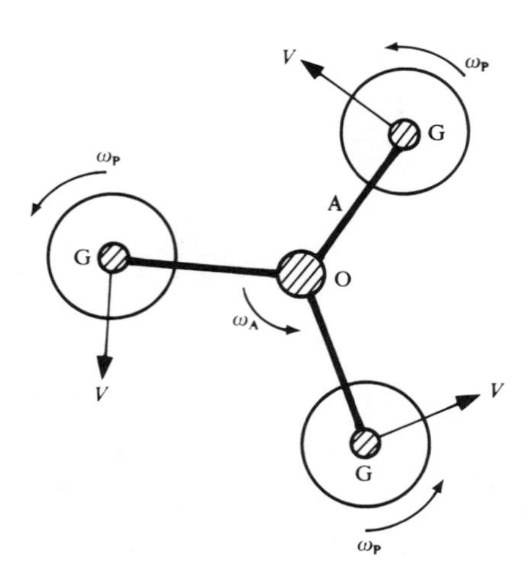

Figure 6.19 Diagram of the arm of an epicyclic gear train carrying three planets.

m_P = mass of a planet
ω_A = absolute angular velocity of the arm
ω_P = absolute angular velocity of a planet
V = absolute linear velocity of the centre of mass $G = \omega_A R$

Hence the total kinetic energy T of the system is given by

$$T = \tfrac{1}{2} I_A \omega_A^2 + 3.\tfrac{1}{2} m_P R^2 \omega_A^2 + 3.\tfrac{1}{2} I_P \omega_P^2$$

Let I_e be the equivalent moment of inertia of the system referred to the axis through O;

then
$$T = \tfrac{1}{2} I_e \omega_A^2 = \frac{1}{2}\left(I_A + 3m_P R^2 + 3I_P \frac{\omega_P^2}{\omega_A^2}\right)\omega_A^2 \tag{6.14}$$

and therefore
$$I_e = I_A + 3m_P R^2 + 3I_P\left(\frac{\omega_P}{\omega_A}\right)^2 \tag{6.15}$$

Thus the system may be considered as having a moment of inertia I_e about O and an angular velocity ω_A at a particular instant.

If the prime-mover torque is C_m and the load demand C_L referred to the axis O, the acceleration $\dot{\omega}_A = \alpha$ will be obtained from the following dynamic equation:

$$C_m - C_L = \left[I_A + 3m_P R^2 + 3I_P\left(\frac{\omega_P}{\omega_A}\right)^2\right]\alpha \tag{6.16}$$

If each planet meshes with another gear, e.g., such as a sun gear, of moment of inertia I_1, then the equivalent moment of inertia of the system referred to the arm axis through O will be:

$$I_e = I_A + 3m_P R^2 + 3I_P\left(\frac{\omega_P}{\omega_A}\right)^2 + I_1\left(\frac{\omega_1}{\omega_A}\right)^2 \tag{6.17}$$

where ω_1 is the absolute angular velocity of the gear.

This equation can be extended to include all the elements of a train.

Example 6.6 If in Example 6.3 the arm carries three sets of compound planets, calculate the torque required to accelerate the system from rest to 2000 rev/min in 4 seconds, assuming that the arm is the input and the sun the output. Data:

Moment of inertia of each compound planet $= 7.5 \times 10^{-4}$ kg m^2.
Mass of each compound planet $= 2$ kg.
Moment of inertia of the arm assembly $= 3.25 \times 10^{-3}$ kg m^2.
Radius OG $= 120$ mm.
Moment of inertia of the sun including the shaft $= 4.25 \times 10^{-4}$ kg m^2.
Load torque $= 0.85$ N m.

Solution From the solution to Example 6.3 the speed ratios are:

$$\frac{\text{Speed of the arm}}{\text{Speed of the sun}} = \frac{1}{8.2}$$

and for the planets $= 3.4$.
Applying Eq. (6.17), we have

$$I_e = 3.25 \times 10^{-3} + 3 \times 2 \times 0.12^2 + 3 \times 7.5 \times 10^{-4} \times 3.4^2 + 4.25 \times 10^{-4} \times 8.2^2 = 0.144 \text{ kg m}^2$$

Acceleration, $$\alpha = \frac{2\pi 2000}{60}\frac{1}{4}\frac{1}{8.2} = 6.39\ \text{rad/s}^2$$

Applying Eq. (6.16) yields the torque required to accelerate the system

$$C_m = I_e\alpha + C_L = 0.144 \times 6.39 + 0.85 \times 8.2 = 7.89\ \text{N m}$$

6.6.3 Torques on Gear Trains

Figure 6.20 shows a typical gearbox in diagrammatic form; it could represent a simple train, an epicyclic, or any other train.

Let ω_1 and ω_2 be the input and output velocities respectively; the gear ratio will be given by $\omega_2 = k\omega_1$ where k depends on the number of teeth on the gears that make up the train as we saw earlier.

If M_1, M_2, and M_C are, respectively, the input torque, the output torque and the torque required to hold the casing stationary; and, if the system is not accelerating, we must have

$$M_1 + M_2 + M_C = 0 \tag{6.18}$$

This relationship applies even if there is some friction.

Furthermore, the kinetic energy remains constant, hence

$$M_1\omega_1 + M_2\omega_2 = 0 \tag{6.19a}$$

The torque ratio is therefore given by

$$M_1/M_2 = -\omega_2/\omega_1 = -k$$

We should note that k may be positive or negative.

If on the other hand the casing is not stationary, it follows that

$$M_1\omega_1 + M_2\omega_2 + M_c\omega_c = 0 \tag{6.19b}$$

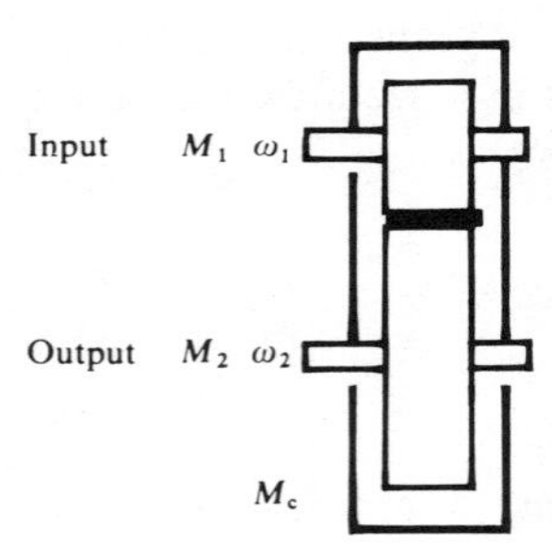

Figure 6.20 Diagrammatical repesentation of a typical gearbox.

EXERCISES

6.1 Figure 6.21 shows a typical three-speed car gearbox (with reverse omitted). The gears are in mesh all the time; for first and second gears (the intermediate gears), A and B respectively are locked to the output shaft. In third (top) gear the input and output shafts are locked together. For each of the intermediate gears calculate the output speed and torque, and the torque required to hold the casing, when the engine delivers 64 kW at 5000 rev/min.

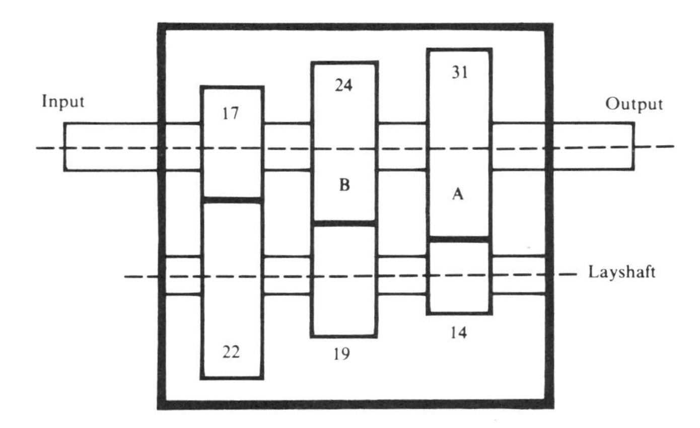

Figure 6.21

6.2 Figure 6.22 shows an early version of the epicyclic reduction gear between the turbine and the propeller shaft of the Proteus engine. The sun wheel S is on the turbine shaft; the planet gears P_1 and P_2 (of which there are four pairs) are fixed together, the planet carrier driving the propeller shaft. The internal ring gear R is fixed. Calculate the reduction ratio.

If the torque delivered to the turbine is 2400 N m, calculate the propeller torque, the torque required to hold R and the total tangential load carried by the bearings of one of the planet assemblies. What *total* torque about its axis does the engine exert on the airframe? (Neglect the small torque from the intake air and jet exhaust.)

Numbers of teeth: sun (S) = 24, planet (P_1) = 21, planet (P_2) = 49, ring (R) = 104.

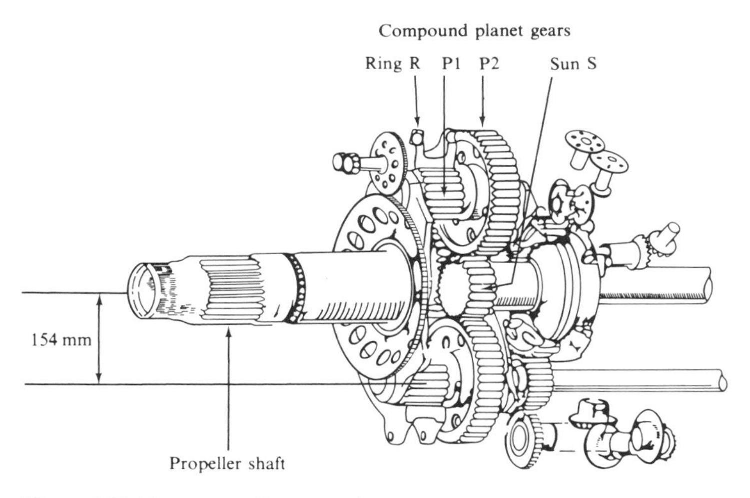

Figure 6.22 (*Courtesy Rolls-Royce plc.*)

6.3 An automatic radar tracker has a motor with moment of inertia 0.014 kg m^2 turning an aerial of moment of inertia 28.8 kg m^2 through a reduction gearbox.

(*a*) For a given motor torque, what should the reduction ratio be for maximum acceleration of the aerial?

(*b*) The motor characteristics are such that the maximum torque available falls as the motor speeds up, at the rate of 7 per cent of the standstill torque for each 20 rad/s. Calculate the reduction ratio for maximum acceleration of the aerial when it turns at 0.8 rad/s.

6.4 In the vehicle differential gear shown in Fig. 6.23 the gearbox drives the pinion P which meshes with the crown wheel C. C carries the bearings for the epicyclic bevels B, which mesh with bevels on the two half-shafts driving the wheels. Show that $\omega_L + \omega_R = 2\omega_p(N_p/N_c)$ and also that equal torque is delivered to each half-shaft under all conditions.

The gearbox of Exercise 6.1 drives a differential in which the gears all have 70 mm effective radius. P and C have 8 and 33 teeth respectively. Calculate the loads on the bearings of the epicyclic gears B when the vehicle is being driven at full torque in first gear. There are four epicyclic gears.

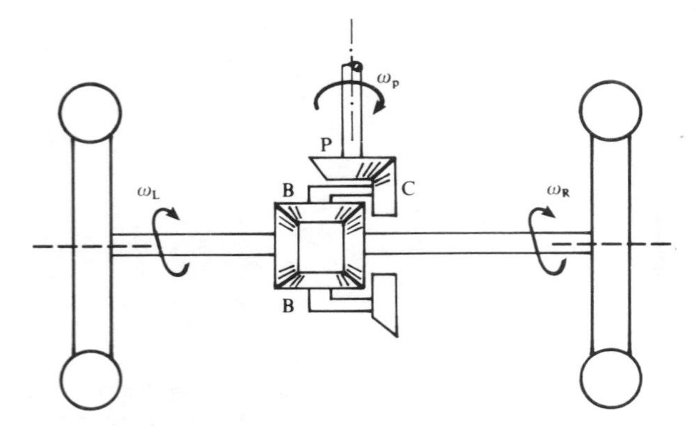

Figure 6.23

6.5 Figure 6.24 shows diagrammatically a proprietary variable-speed drive. The cone E is held by an arm D, which is driven at constant speed. There is a friction drive between the cone E and a rotationally fixed member C, and it may be assumed that there is no slip. This fixed member can be moved along the axis of the gear to change the gear ratio. If the fixed member C has an inner diameter of 150 mm and the diameter of the cone varies from 60 mm to 45 mm, find the range in output speed which can be obtained from an input speed of 5000 rev/min. A has 90 teeth and B 30.

Find also the torque transmitted to member C when it is in the extreme right-hand position, when the input power is 0.4 kW.

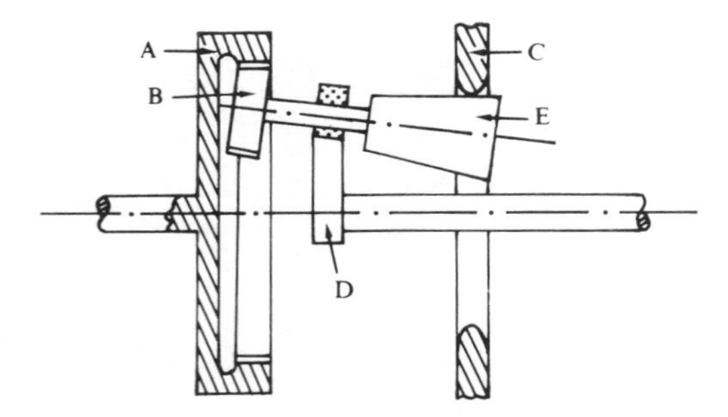

Figure 6.24

6.6 An electric motor is used to open or close a valve in a pipeline through a double-reduction gearbox: the first set of gears provides a reduction $G1{:}1$ and the second set a reduction $G2{:}1$. If I_m is the moment of inertia of the rotor of the motor, I_v that of the valve, and $k\theta$ the resistance offered by the valve, where θ is the angle turned through by the valve, derive an expression for the angular velocity ω of the valve after it has been rotated through an angle α from rest. The motor delivers a constant torque T.

Using the following data, calculate the value of ω when $\alpha = 90°$. $T = 10$ N m; $I_m = 0.025$ kg m^2; $I_v = 12.5$ kg m^2; $G1 = 4.5$; $G2 = 3.25$; $k = 25$(N m)/rad.

6.7 An actuator consisting of an electric motor driving a reduction gearbox is used to operate one of the flaps of an aircraft. Using the data given below, calculate the speed of rotation of the driving motor in rev/min when the flap has turned through an angle of 15° from its zero position. Data:

Constant torque delivered by the driving motor = 0.0125 N m.
Gear ratio = 100:1.
Moment of inertia of motor = 2.75×10^{-5} kg m^2.
Moment of inertia of the flap referred to the output shaft of the gearbox = 0.35 kg m^2.
Moment required to operate the flap referred to output shaft of the gearbox = 2.8θ N m, where θ is the angle turned through by the flap measured in radians.

6.8 A hoist used on a building site consists of a trolley running on inclined rails as shown in Fig. 6.25; its platform AB carries a maximum load of 125 kg. The trolley is designed in such a way that the wheels are always in contact with the rails. The trolley has a mass of 15 kg and its platform a mass of 7 kg; AB = 1 m. The trolley and its load are lifted by a motor driving a drum D through a double reduction. The moments of inertia are: 0.075 kg m^2 for the motor, 2 kg m^2. for C and 12 kg m^2 for the drum D.

If the motor delivers a constant torque of 25 N m calculate (*a*) the acceleration of the loaded trolley, and (*b*) the bending moment at the root A of the platform, assuming that the weight is concentrated half-way along the platform.

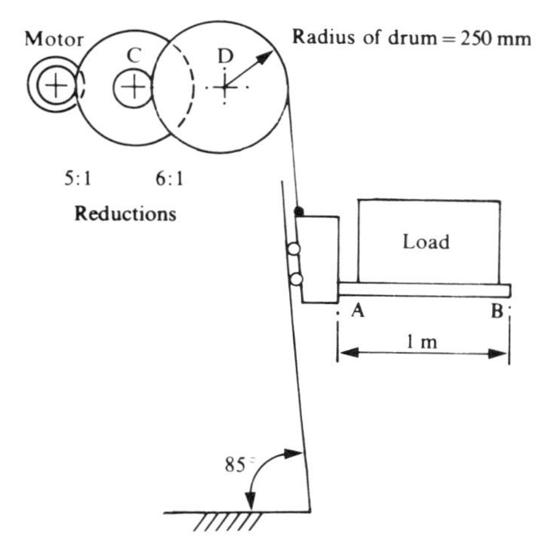

Figure 6.25

CHAPTER

SEVEN

INTRODUCTION TO SPATIAL MECHANISMS

7.1 INTRODUCTION

In the preceding chapters we have been concerned with the analysis and synthesis of plane linkages or mechanisms. It is not difficult to realize that many of them could be given three-dimensional motion by a simple rotation about an axis parallel to their plan of operation; e.g., the crane shown in Fig. 1.7 can rotate about a vertical axis at the same time as the input link BB_0 is rotating, thus giving the load three-dimensional motion. Another example is the mechanism shown in Fig. 1.13 for the automatic assembly of parts. If the castings were positioned on a carousel, for example, instead of a conveyor system then it is likely that the casting which is waiting for the part to be fitted to it would be out of the plane of the mechanism. In such a situation the mechanism could be made to rotate about a vertical axis simultaneously with the motion of the linkage, thus giving the part a three-dimensional motion.

It is, however, possible to obtain the motion of two shafts inclined at some angle γ to each other by means of a spatial linkage as illustrated in Fig. 7.1; such a linkage is in fact a four-bar

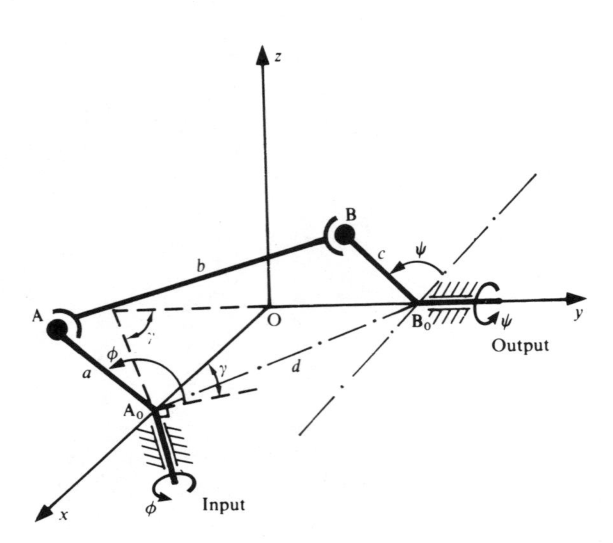

Figure 7.1 Typical four-bar spatial linkage showing two spherical joints and two revolute joints.

spatial linkage; it is relatively simple and is often found in practice. It consists of a crank A_0A, the input, and a crank B_0B, the output. The two are linked together by means of the coupler AB using spherical joints, and is thus referred to as an RSSR linkage since it consists of two revolutes (R) and two spherical (S) joints. Such a linkage has one degree of freedom and the input–output relationship (ϕ–ψ) for displacement, velocity, acceleration, and jerk is not affected by the fact that the coupler AB can spin about its own axis. If on the other hand the path traced by any point on the coupler is taken as an output then this possible motion would have to be taken into account.

The mobility criterion is given by the following relationship due to Grübler:

$$\text{Number of degrees of freedom, } F = 6(n - j - 1) + \sum f$$

where n = number of links
j = number of joints
f = number of freedoms at each joint

Thus for the space linkage shown in Fig. 7.1, $n = 4$, $j = 4$, and $\sum f = 1 \times 2 + 3 \times 2 = 8$.

Therefore $F = 6(4 - 4 - 1) + 8 = 2$; one degree of freedom is due to the free spin of the coupler AB about its own axis.

In Fig. 7.1 the lengths of the four links are $A_0A = a$, $AB = b$, $B_0B = c$, and $A_0B_0 = d$ as for the plane four-bar linkage.

The spatial four-bar linkage can be a crank–rocker or a double crank or double rocker.

For a given value of ϕ, measured as shown, there are two possible values for ψ, since the coupler AB with centre A can intersect the circle traced by B, radius C, in two points as for the planar four-bar linkage.

If in Fig. 7.1 we replace B_0B by a slider then we obtain an RSSP (P ≡ prismatic) spatial linkage as shown in Fig. 7.2, where B is the slider on the rod B_0 and whose position along the y-axis is y corresponding to the input position ϕ of link A_0A.

Figures 7.3 and 7.4 show two models of four-bar spatial linkages. Figure 7.3 is the well-known Bennett's linkage where the two shafts are at 90° to each other and can both rotate through 360°; it is a double crank and consists of four revolutes. (It was devised in 1903 by Bennett at Cambridge.)

Figure 7.4 is a particular version of the linkage shown diagrammatically in Fig. 7.1 but

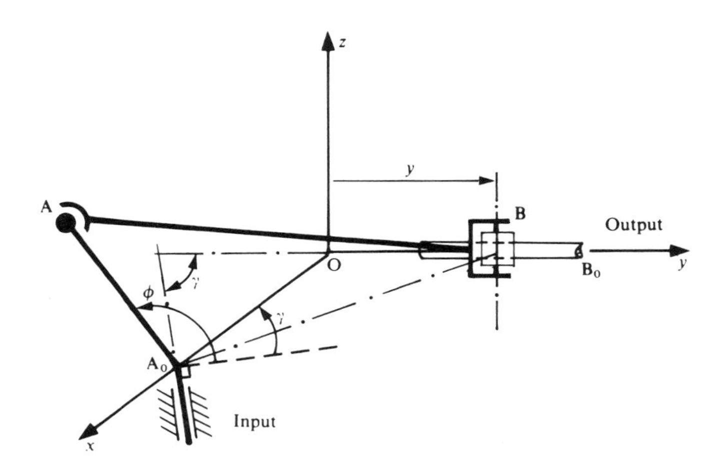

Figure 7.2 Typical spatial linkage consisting of a revolute, a spherical joint, and a slider.

Figure 7.3 Model of the Bennett linkage. (*Courtesy of the University of Bath.*)

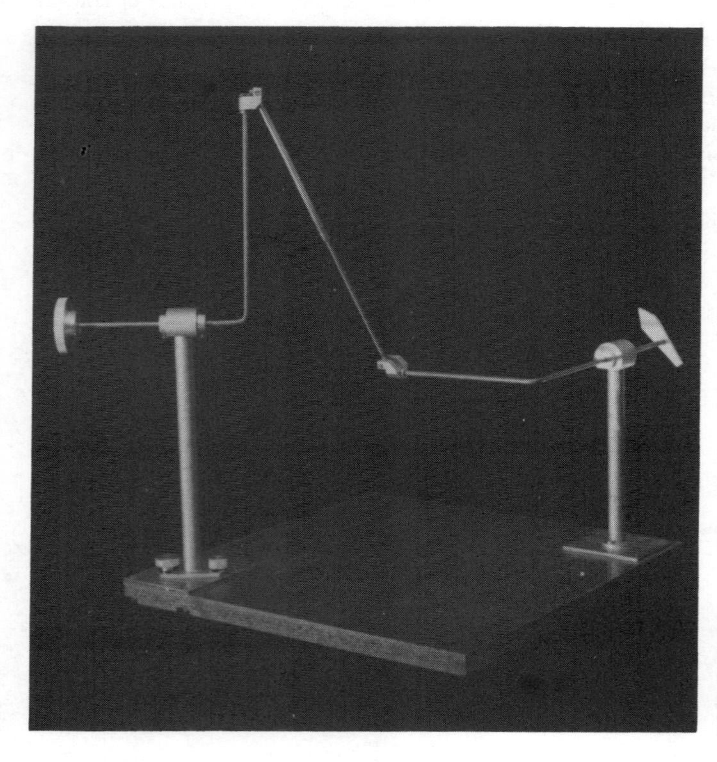

Figure 7.4 Model of a typical four-bar spatial linkage of the R RR RR R type where the spherical joints shown in Fig. 7.1 have been replaced by revolutes. (*Courtesy of the University of Bath.*)

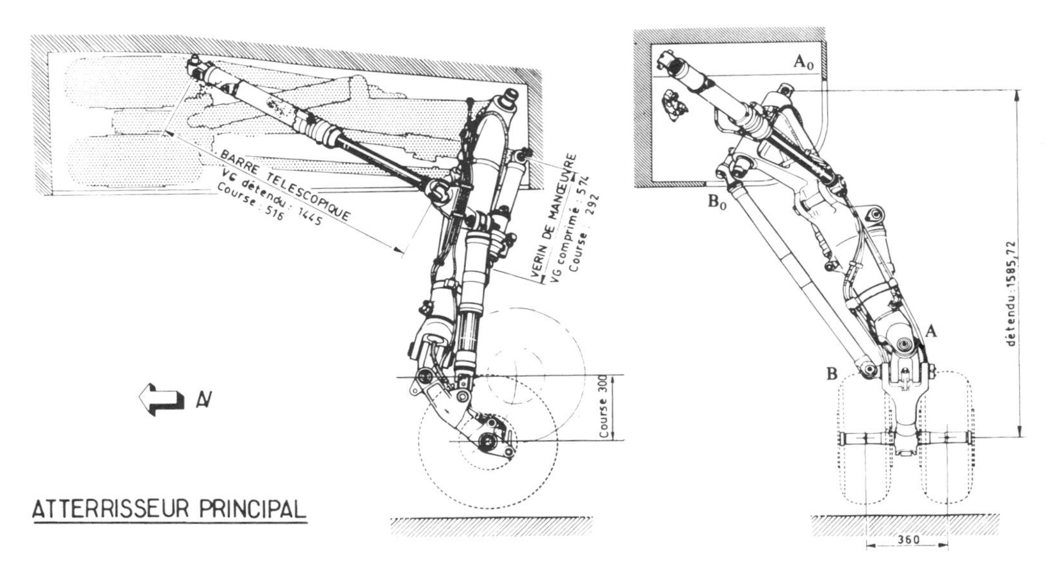

Figure 7.5 Four-bar spatial linkage in practice—the Jaguar aircraft undercarriage. (*Courtesy of Messier-Hispano-Bugatti.*)

Figure 7.6 Main undercarriage of the Jaguar aircraft. (*Courtesy of Messier-Hispano-Bugatti.*)

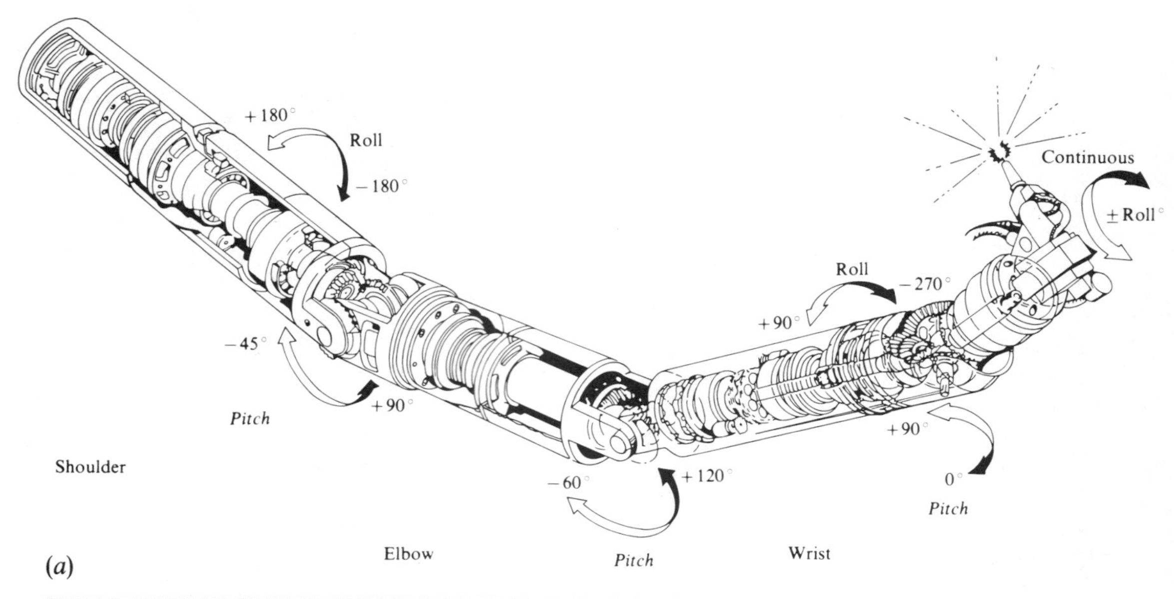

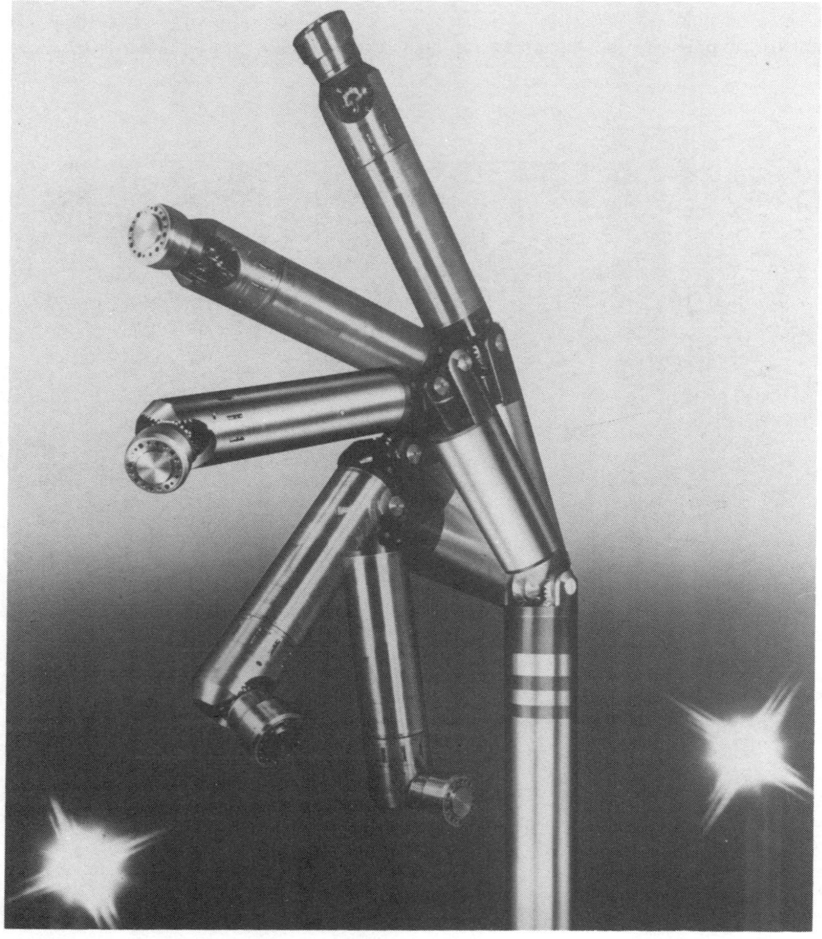

Figure 7.7 Robotic manipulator. (*Courtesy of SAC Hitec Ltd.*)

where the spherical joints have been replaced by revolutes; hence this linkage is an R $\widehat{\mathrm{RR}}$ $\widehat{\mathrm{RR}}$ R type. Furthermore, the input and output shafts are at different levels, and in the model the angle between the shafts and the lengths of the coupler can be altered for demonstration purposes.

Figures 7.5, 7.6, and 7.7 are practical examples of spatial mechanisms. Figure 7.5 is a drawing of the main undercarriage of the Jaguar aircraft. It is a spatial four-bar linkage made up of revolutes at the joints A_0ABB_0, i.e., it is an R R $\widehat{\mathrm{RRR}}$ $\widehat{\mathrm{RR}}$ mechanism. The swinging lever, to which an independent shock absorber is attached, is pivoted on the barrel through a universal joint. The universal joint-swinging lever assembly is guided by a rod connected to the aircraft, thus creating a relative movement between the barrel and the swinging lever on retraction or extension. This allows the wheels to be accommodated inside the fuselage in a horizontal position. Figure 7.6 is a photograph of this undercarriage.

Figure 7.7(*a*) is a cutaway drawing of a manipulator. It is of the 'open-chain' type, consisting of a shoulder, elbow, and wrist, and has six degrees of freedom. All movements are obtained by means of motors driving sets of gears. Its main applications are in nuclear inspections and repair work including arc welding. Figure 7.7(*b*) is a photograph of the manipulator in several positions.

In order to analyse, and synthesize, spatial linkages there are a number of mathematical methods available such as projective geometry, matrix algebra, complex numbers, and vector algebra. In this introduction to the subject we propose to use the vector algebra method because it is straightforward even though the necessary manipulation can be lengthy, a fact due to the nature of spatial motion.

7.2 ELEMENTS OF VECTOR ALGEBRA

7.2.1 Position Vector

Consider any point P in space relative to a fixed point O. To reach P we use a straight line $\mathbf{r}$ directed from O to P. This is the position vector. If we now take Cartesian axes Oxyz we can develop the analysis using vector calculus (Fig. 7.8). A vector may be expressed conveniently in terms of its components and the three unit vectors $\mathbf{i}$, $\mathbf{j}$, and $\mathbf{k}$, whose directions are those of the axes of the Cartesian frame of reference Ox, Oy and Oz. If x, y, and z are the directed lengths of the projections of the vector $\mathbf{r}$ on these axes, we then have the following vector equation:

$$\mathbf{r}=x\mathbf{i}+y\mathbf{j}+z\mathbf{k} \tag{7.1}$$

Furthermore, if the particle is moving along the space curve AB then x, y, and z must be functions of time, i.e.,

$$x=x(t) \qquad y=y(t) \qquad z=z(t)$$

Hence the vector equation for the position of the particle at any instant may be written thus:

$$\mathbf{r}=x(t)\mathbf{i}+y(t)\mathbf{j}+z(t)\mathbf{k} \tag{7.2}$$

The magnitude of the position vector $\mathbf{r}$, denoted by $|\mathbf{r}|$ or simply r, is

$$r=\sqrt{x^2+y^2+z^2}$$

The velocity of the particle is obtained by differentiating Eq. (7.2) with respect to time:

$$\dot{\mathbf{r}}=\dot{x}(t)\mathbf{i}+\dot{y}(t)\mathbf{j}+\dot{z}(t)\mathbf{k}$$

$\dot{\mathbf{i}}=\dot{\mathbf{j}}=\dot{\mathbf{k}}=0$, since the axes are Cartesian axes, fixed in space and of constant magnitude.

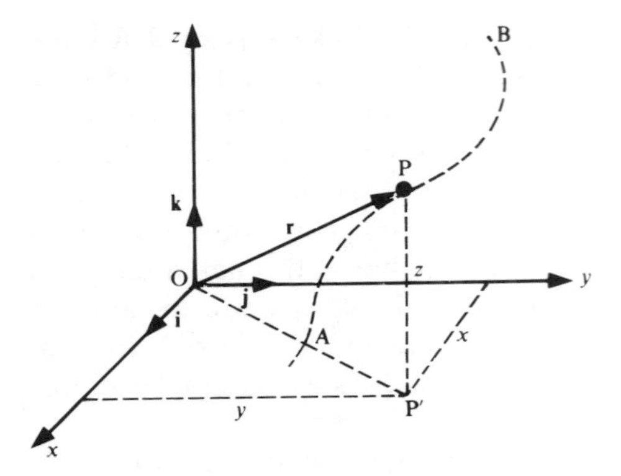

Figure 7.8 Position vector of a point in Cartesian coordinates.

Similarly, the acceleration of the particle is

$$\ddot{\mathbf{r}} = \ddot{x}(t)\mathbf{i} + \ddot{y}(t)\mathbf{j} + \ddot{z}(t)\mathbf{k}$$

It is assumed that the reader is familiar with addition, subtraction, multiplication, and division of vectors by a scalar.

If **A** and **B** are two vectors then the resultant vector **C** obtained by adding **A** and **B** is

$$\mathbf{C} = \mathbf{A} + \mathbf{B}$$

where the positive sign means addition in accordance with the triangle law. Similarly, subtracting **B** from **A** gives a new vector **P**, as follows:

$$\mathbf{P} = \mathbf{A} - \mathbf{B}$$
$$= \mathbf{A} + (-\mathbf{B})$$

i.e., we add to **A** the negative of vector **B**, which is a vector whose magnitude and direction are those of **B** but whose sense is opposite to that of **B**.

7.2.2 Scalar Product of Two Vectors

Let **A** and **B** be two vectors of magnitude A and B originating from point O as shown in Fig. 7.9 and let θ be the angle between their directions, measured positively in the sense of the rotation from **A** to **B**.

The scalar product of the two vectors denoted by **A·B** (pronounced 'A dot B') is defined by the equation

$$\mathbf{A} \cdot \mathbf{B} = AB \cos \theta \tag{7.3}$$

It is a scalar quantity and is measured by the projection of **B** on **A**. From this definition it follows that

$$\mathbf{B} \cdot \mathbf{A} = BA \cos(-\theta) = AB \cos \theta = \mathbf{A} \cdot \mathbf{B}$$

so that the order is immaterial, i.e., the scalar multiplication of vectors is commutative.

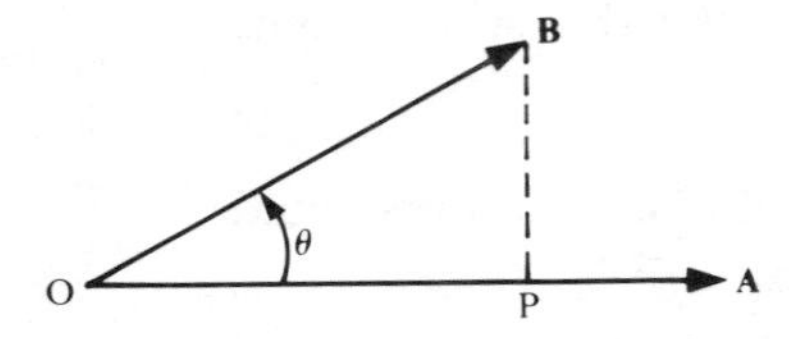

Figure 7.9 Scalar product.

If $90^\circ < \theta < 270^\circ$ then the scalar product is negative and if $\theta = 90^\circ$ then the scalar product is zero.

For the fundamental unit vectors **i**, **j**, and **k** we have

$$\mathbf{i}\cdot\mathbf{i} = \mathbf{j}\cdot\mathbf{j} = \mathbf{k}\cdot\mathbf{k} = 1 \tag{7.4}$$

and

$$\mathbf{i}\cdot\mathbf{j} = \mathbf{j}\cdot\mathbf{k} = \mathbf{k}\cdot\mathbf{i} = 0$$

If

$$\mathbf{A} = a_x\mathbf{i} + a_y\mathbf{j} + a_z\mathbf{k}$$

and

$$\mathbf{B} = b_x\mathbf{i} + b_y\mathbf{j} + b_z\mathbf{k}$$

then using Eq. (7.4) we get

$$\mathbf{A}\cdot\mathbf{B} = a_x b_x + a_y b_y + a_z b_z \tag{7.5}$$

It therefore follows that if $\mathbf{A} = \mathbf{B}$ then

$$\mathbf{A}^2 = a_x^2 + a_y^2 + a_z^2$$

If $\mathbf{A}\cdot\mathbf{B} = 0$, and $\mathbf{A} \neq 0$ and $\mathbf{B} \neq 0$, then the two vectors are at right angles to each other.

7.2.3 Vector Product of Two Vectors

Let **A** and **B** be two vectors of magnitudes A and B respectively and θ be the angle between them measured positively from **A** towards **B** as shown in Fig. 7.10. The vector product $\mathbf{A} \wedge \mathbf{B}$ or $\mathbf{A} \times \mathbf{B}$, pronounced 'A vector B' or 'A cross B', is defined to be a vector whose magnitude is

$$AB \sin\theta \tag{7.6}$$

and whose direction is perpendicular to both vectors and whose sense is in accordance with the right-hand corkscrew rule.

From this definition it follows that the vector multiplication of vectors is not commutative, since

$$BA \sin(-\theta) = -AB \sin\theta$$

Hence

$$\mathbf{B} \wedge \mathbf{A} = -\mathbf{A} \wedge \mathbf{B} \tag{7.7}$$

as illustrated in Fig. 7.10.

When two vectors are parallel their vector product is zero, since $\sin\theta = 0$ or $\sin\pi = 0$. Conversely, if $\mathbf{A} \wedge \mathbf{B} = 0$ then either $\mathbf{A} = 0$ or **A** is parallel to **B**. We should note that two vectors are parallel irrespective of sense if the lines representing them are parallel; thus **P** and $-\mathbf{P}$ are opposite but parallel vectors.

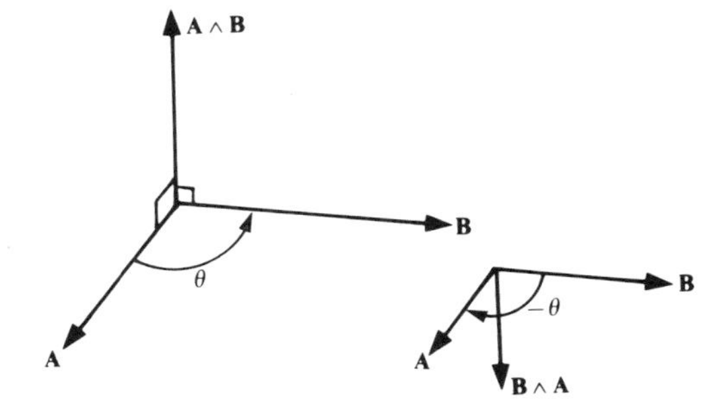

Figure 7.10 Vector product.

Let us now consider the vector products of the unit vectors **i**, **j**, and **k**; from the definition we have

$$\mathbf{i} \wedge \mathbf{i} = \mathbf{j} \wedge \mathbf{j} = \mathbf{k} \wedge \mathbf{k} = 0$$

$$\mathbf{i} \wedge \mathbf{j} = \mathbf{k} \qquad \mathbf{j} \wedge \mathbf{k} = \mathbf{i} \qquad \mathbf{k} \wedge \mathbf{i} = \mathbf{j}$$

Hence for two vectors **A** and **B** we have

$$\begin{aligned} \mathbf{A} \wedge \mathbf{B} &= (a_x\mathbf{i} + a_y\mathbf{j} + a_z\mathbf{k}) \wedge (b_x\mathbf{i} + b_y\mathbf{j} + b_z\mathbf{k}) \\ &= a_x\mathbf{i} \wedge (b_x\mathbf{i} + b_y\mathbf{j} + b_z\mathbf{k}) + a_y\mathbf{j} \wedge (b_x\mathbf{i} + b_y\mathbf{j} + b_z\mathbf{k}) \\ &\quad + a_z\mathbf{k} \wedge (b_x\mathbf{i} + b_y\mathbf{j} + b_z\mathbf{k}) \\ &= (a_yb_z - a_zb_y)\mathbf{i} + (a_zb_x - a_xb_z)\mathbf{j} + (a_xb_y - a_yb_x)\mathbf{k} \end{aligned} \tag{7.8}$$

The above result can be represented more simply by means of a determinant, thus:

$$\mathbf{A} \wedge \mathbf{B} = \begin{vmatrix} \mathbf{i} & \mathbf{j} & \mathbf{k} \\ a_x & a_y & a_z \\ b_x & b_y & b_z \end{vmatrix} \tag{7.8a}$$

The last two rows are in the order of the terms of the product; from the properties of determinants, if these two rows are interchanged, the value of the determinant is unchanged but its sign is changed; thus

$$\mathbf{A} \wedge \mathbf{B} = -\mathbf{B} \wedge \mathbf{A}. \tag{7.9}$$

Vector triple product We shall encounter situations in dynamics where three vectors **A**, **B**, and **C** occur as the product $\mathbf{A} \wedge (\mathbf{B} \wedge \mathbf{C})$. This is referred to as a *vector triple product*, and sometimes we may find it convenient to expand such a product into the following useful expression:

$$\mathbf{A} \wedge (\mathbf{B} \wedge \mathbf{C}) = (\mathbf{A} \cdot \mathbf{C})\mathbf{B} - (\mathbf{A} \cdot \mathbf{B})\mathbf{C} \tag{7.10}$$

7.2.4 Application to Mechanics

(*a*) Work done by a force A force acting on a particle is said to do work when the particle is displaced in a given direction; the work done is a scalar quantity and is proportional to the force and the component of the displacement in the direction of the force.

If **P** and **s** are the force and the displacement respectively, then the work done is

$$Ps \cos\theta = \mathbf{P} \cdot \mathbf{s}$$

where θ is the angle between **P** and **s**. Thus work is the scalar product of the force and the displacement. The work done is zero when **P** and **s** are perpendicular to each other.

If a particle is acted upon by a number of forces $\mathbf{P}_1, \mathbf{P}_2, \mathbf{P}_3, \ldots, \mathbf{P}_n$ then it follows that during a displacement **s** of the particle, each force does work, i.e., $\mathbf{P}_1 \cdot \mathbf{s}$, $\mathbf{P}_2 \cdot \mathbf{s}, \ldots, \mathbf{P}_n \cdot \mathbf{s}$. Hence the total work done is

$$\sum_1^n (\mathbf{P} \cdot \mathbf{s}) = \mathbf{s} \cdot (\Sigma \mathbf{P}) = \mathbf{s} \cdot \mathbf{R}$$

where **R** is the resultant of the forces.

(*b*) Moment of a force Consider a force **P** and AB its line of action as shown in Fig. 7.11. Let

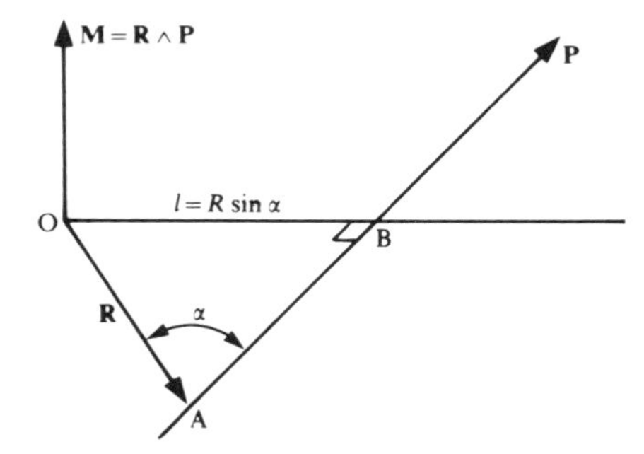

Figure 7.11 Moment of a force in space.

O be any convenient point, and **R** the position vector of any point A on the line of action of the force from O. The moment **M** of the force about the point O is defined by the vector product

$$\mathbf{M} = \mathbf{R} \wedge \mathbf{P}$$

This vector is perpendicular to the plane of **R** and **P**, and therefore to the plane containing the line of action AB of the force and the point O. The magnitude of this moment is lP, where l is the length of the perpendicular OB to the line of action. Since $l = \mathrm{R} \sin \alpha$, the magnitude of the moment is

$$RP \sin \alpha$$

where α is the angle between the direction of the position vector and the line of action of the force.

If a number of forces $\mathbf{P}_1, \mathbf{P}_2, \ldots, \mathbf{P}_n$ act through the same point A, they have a resultant $\mathbf{Q} = \Sigma \mathbf{P}$. The moment of this resultant about O is

$$\begin{aligned} \mathbf{R} \wedge \mathbf{Q} &= \mathbf{R} \wedge (\mathbf{P}_1 + \mathbf{P}_2 + \cdots + \mathbf{P}_n) \\ &= \mathbf{R} \wedge \mathbf{P}_1 + \mathbf{R} \wedge \mathbf{P}_2 + \cdots + \mathbf{R} \wedge \mathbf{P}_n \end{aligned}$$

It is therefore equal to the vector sum of the moments of each force.

7.3 KINEMATICS OF RIGID BODIES

7.3.1 Rotation About a Fixed Point: Velocities

Consider a rigid body rotating about a fixed point A. Let ω be its angular velocity and AX the instantaneous axis of rotation as shown in Fig. 7.12. The angular velocity may be represented by a vector $\boldsymbol{\omega}$ directed along AX in accordance with the right-hand rule, and if P is any point in the body whose position vector from A is **R** then its velocity $\mathbf{v}_{\mathrm{PA}}$ is given by

$$\mathbf{v}_{\mathrm{PA}} = \boldsymbol{\omega} \wedge \mathbf{R} \tag{7.10}$$

If M is the projection of P on AX, the velocity of P is $\omega \mathrm{MP} = \omega R \sin \theta$ at right angles to the plane AXP; hence

$$|V_{\mathrm{PA}}| = \omega R \sin \theta$$

To show that angular velocities obey the rule for the addition of vectors, let the body have a number of angular velocities $\boldsymbol{\omega}_1, \boldsymbol{\omega}_2, \ldots$ about axes through A. Then if $\mathbf{v}_1, \mathbf{v}_2, \ldots$ are their

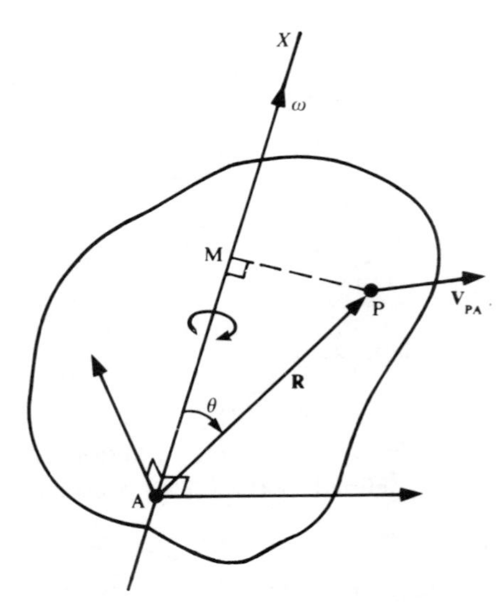

Figure 7.12 Rotation of a rigid body about a fixed point: velocity.

contributions to the velocity $\mathbf{v}$ of the point P then

$$\begin{aligned}\mathbf{v} &= \mathbf{v}_1 + \mathbf{v}_2 + \cdots = \boldsymbol{\omega}_1 \wedge \mathbf{R} + \boldsymbol{\omega}_2 \wedge \mathbf{R} + \cdots \\ &= (\boldsymbol{\omega}_1 + \boldsymbol{\omega}_2 + \cdots) \wedge \mathbf{R} = \boldsymbol{\omega} \wedge \mathbf{R}\end{aligned}$$

where $\boldsymbol{\omega}_1 + \boldsymbol{\omega}_2 + \cdots$ is the resultant of the angular velocities added vectorially.

The point A need not always be a fixed point, but can be a point moving with a velocity $\mathbf{v}_A$ relative to a Newtonian frame, whereupon the absolute velocity $\mathbf{v}$ of the point P is

$$\begin{aligned}\mathbf{v} &= \mathbf{v}_A + \boldsymbol{\omega} \wedge \mathbf{R} \\ &= \mathbf{v}_A + \mathbf{v}_{PA}\end{aligned} \tag{7.11}$$

where $\mathbf{v}_{PA}$ is the velocity of P relative to A and equals $\boldsymbol{\omega} \wedge \mathbf{R}$.

Equation (7.11) is illustrated in Fig. 7.13 where the body might be one of the links in a spatial mechanism, for example, where A would, very likely, be a spherical joint. If Q is any other point of the link, as shown, then its velocity relative to x–y–z frame is

$$\begin{aligned}\mathbf{v}_Q &= \mathbf{v}_A + \mathbf{v}_{QA} \\ &= \mathbf{v}_A + \boldsymbol{\omega} \wedge \mathbf{d}\end{aligned}$$

where $\mathbf{d}$ is the position vector of Q from A.

Example 7.1 If point A of the link in Fig. 7.13 coincides with O, the origin of the x–y–z frame, and the link has an angular velocity $\boldsymbol{\omega} = \mathbf{i} + 1.5\mathbf{j} + 3\mathbf{k}$ rad/s, calculate the velocity components of point P given that $\mathrm{R} = 150\mathbf{i} + 125\mathbf{j} + 275\mathbf{k}$ mm.

SOLUTION (Fig. 7.14) In this case $\mathbf{v}_A = 0$ and $\mathbf{v}_{PA} = \omega \wedge \mathbf{R} = \mathbf{v}$ (say). Let $\boldsymbol{\omega} = \omega_x\mathbf{i} + \omega_y\mathbf{j} + \omega_z\mathbf{k}$ and $\mathbf{R} = x\mathbf{i} + y\mathbf{j} + z\mathbf{k}$ (Fig. 7.14).

Then
$$\boldsymbol{\omega} \wedge \mathbf{R} = \begin{vmatrix} \mathbf{i} & \mathbf{j} & \mathbf{k} \\ \omega_x & \omega_y & \omega_z \\ x & y & z \end{vmatrix}$$

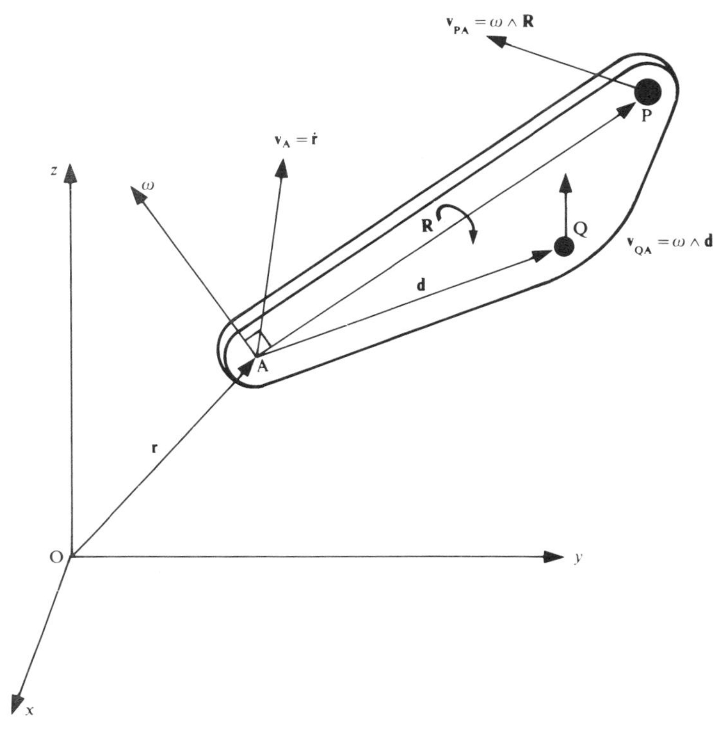

Figure 7.13 Rotation of a link in a spatial mechanism.

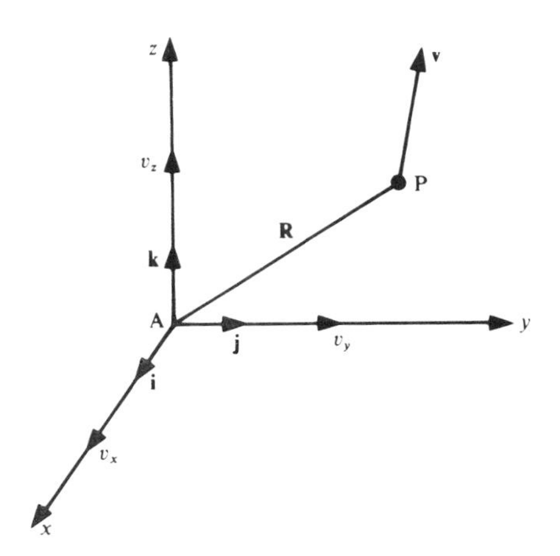

Figure 7.14 Velocity components of the link in Fig. 7.13.

by Eq. (7.9). Hence the components of **v** are

$$
\begin{aligned}
v_x &= z\omega_y - y\omega_z \\
&= 0.275 \times 1.5 - 0.125 \times 3 \\
&= 0.038 \text{ m/s} \\
v_y &= x\omega_z - z\omega_x \\
&= 0.15 \times 3 - 0.275 \times 1 \\
&= 0.175 \text{ m/s}
\end{aligned}
$$

$$v_z = y\omega_x - x\omega_y$$
$$= 0.125 \times 1 - 0.15 \times 1.5$$
$$= -0.100 \text{ m/s}$$

or
$$\mathbf{v}_{PA} = 0.038\mathbf{i} + 0.175\mathbf{j} - 0.100\mathbf{k}$$

7.3.2 Rotation About a Fixed Point: Accelerations

The acceleration of any point of a rigid body is the time derivative of the velocity; thus if we refer to Fig. 7.13 the acceleration **a** of P relative to the fixed frame x–y–z is

$$\mathbf{a} = \frac{d}{dt}(\mathbf{v}) = \frac{d}{dt}(\mathbf{v}_A + \boldsymbol{\omega} \wedge \mathbf{R})$$

Performing the differentiation yields

$$\mathbf{a} = \dot{\mathbf{v}}_A + \dot{\boldsymbol{\omega}} \wedge \mathbf{R} + \boldsymbol{\omega} \wedge \dot{\mathbf{R}}$$

but $\dot{\mathbf{R}}$ is the velocity of **P** relative to **A**. Hence

$$\dot{\mathbf{R}} = \boldsymbol{\omega} \wedge \mathbf{R}$$

by Eq. (7.10). So the acceleration is

$$\mathbf{a} = \mathbf{a}_A + \dot{\boldsymbol{\omega}} \wedge \mathbf{R} + \boldsymbol{\omega} \wedge (\boldsymbol{\omega} \wedge \mathbf{R}) \tag{7.12}$$

The first term is the acceleration of A. The second and third terms represent the acceleration of **P** relative to **A**: $\dot{\boldsymbol{\omega}} \wedge \mathbf{R}$ is the acceleration component perpendicular to the plane of $\dot{\boldsymbol{\omega}}$ and **R**, and $\boldsymbol{\omega} \wedge (\boldsymbol{\omega} \wedge \mathbf{R})$ is the acceleration component perpendicular to $\boldsymbol{\omega}$ and in the plane of $\boldsymbol{\omega}$ and **R** (Fig. 7.15). Also $\dot{\boldsymbol{\omega}}$ is the angular acceleration of the body; if the body rotates at constant angular velocity, $\dot{\boldsymbol{\omega}} = 0$ and $\dot{\boldsymbol{\omega}} \wedge \mathbf{R} = 0$.

As a simple illustration of the application of Eqs (7.11) and (7.12) consider the following example.

Example 7.2 Two sliders A and B are constrained to move in slots at right angles to each other, as shown in Fig. 7.16, and are connected by the rigid link AB of length 450 mm. At

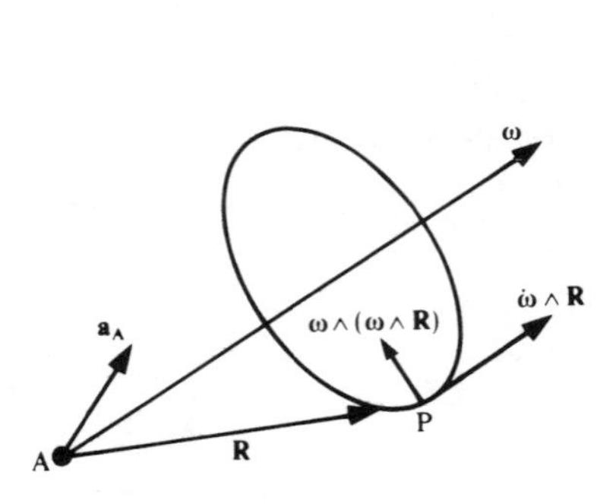

Figure 7.15 Rotation of a rigid body about a fixed point: acceleration.

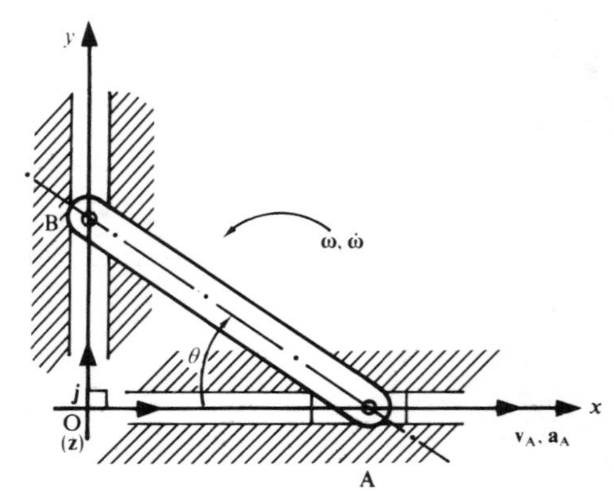

Figure 7.16 Sliders.

the instant when $\theta = 30°$ slider A is moving with a velocity of 0.6 m/s and an acceleration of 1.2 m/s^2 in the direction shown. Calculate the velocity and acceleration of the slider B at that instant and the angular velocity and acceleration of the link AB.

Solution We take the xOy frame of reference to coincide with the slots. The Oz axis is towards the reader. Let $\boldsymbol{\omega}$ and $\dot{\boldsymbol{\omega}} = \boldsymbol{\alpha}$ be respectively the angular velocity and angular acceleration of the link AB, and $\mathbf{v}_B$ and $\mathbf{a}_B$ be the velocity and acceleration of slider B.

We are given $\mathbf{v}_A = 0.6\mathbf{i}$ $\quad \mathbf{a}_A = 1.2\mathbf{i}$

Also let $\mathbf{v}_B = v\mathbf{j}$ $\quad \mathbf{a}_B = a\mathbf{j}$

and $\boldsymbol{\omega} = \omega\mathbf{k}$ $\quad \boldsymbol{\alpha} = \alpha\mathbf{k}$

since the link is moving in the x–y plane.

From Eq. (7.11) we have

$$\mathbf{v}_B = \mathbf{v}_A + \boldsymbol{\omega} \wedge \mathbf{R}$$

i.e.,

$$v\mathbf{j} = 0.6\mathbf{i} + \omega\mathbf{k} \wedge (-0.45 \cos 30°\ \mathbf{i} + 0.45 \sin 30°\ \mathbf{j})$$
$$= 0.6\mathbf{i} - 0.39\omega\mathbf{j} - 0.225\omega\mathbf{i}$$

Hence, comparing the $\mathbf{i}$ and $\mathbf{j}$ components,

$$0 = 0.6 - 0.225\omega \qquad \omega = 2.67 \text{ rad/s}$$

and

$$v = -0.39 \times 2.67 \qquad v = 1.04 \text{ m/s}$$

From Eq. (7.12) we have

$$\mathbf{a}_B = \mathbf{a}_A + \boldsymbol{\alpha} \wedge \mathbf{R} + \boldsymbol{\omega} \wedge (\boldsymbol{\omega} \wedge \mathbf{R})$$

where $\boldsymbol{\omega}$ is the angular velocity and $\boldsymbol{\alpha}$ is the angular acceleration of the link.

Substituting numerical values yields

$$a\mathbf{j} = 1.2\mathbf{i} + \alpha\mathbf{k} \wedge (-0.39\mathbf{i} + 0.225\mathbf{j}) + 2.67\mathbf{k} \wedge [2.67\mathbf{k} \wedge (-0.39\mathbf{i} + 0.225\mathbf{j})]$$
$$= 1.2\mathbf{j} - 0.39\alpha\mathbf{j} - 0.225\alpha\mathbf{i} + 2.78\mathbf{i} - 1.604\mathbf{j}$$
$$= (3.98 - 0.225\alpha)\mathbf{i} - (0.39\alpha + 1.604)\mathbf{j}$$

Hence, comparing the $\mathbf{i}$ and $\mathbf{j}$ components,

$$\alpha = 3.98/0.225 = 17.69 \text{ rad/s}^2$$

and

$$a = -0.39 \times 17.69 - 1.604 = -8.5 \text{ m/s}^2$$

Note: The reader can verify these answers graphically using the instant-centre method and an acceleration diagram.

7.4 KINEMATICS OF A TYPICAL FOUR-BAR SPATIAL LINKAGE

Figure 7.17 shows such a four-bar RSSR spatial linkage in which the input crank A_0A rotates about the Y-axis so that A moves in a circular path in the X–Z plane. The output link B_0B rotates about an axis parallel to the X-axis in the X–Y plane and oscillates through an angle

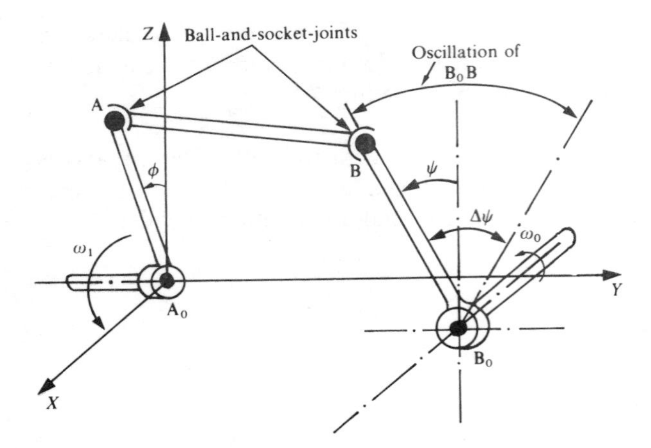

Figure 7.17 Typical four-bar spatial linkage.

$\Delta\psi$ in a plane parallel to the Y–Z plane giving the coupler AB motion in three dimensions.

To analyse such a linkage we replace the links by the vectors **a**, **b**, **c**, and **d** as shown in Fig. 7.18.

Writing the vector or loop equation for the linkage we have

$$\mathbf{a}+\mathbf{b}=\mathbf{c}+\mathbf{d} \tag{7.13}$$

which upon differentiating becomes

$$\dot{\mathbf{a}}+\dot{\mathbf{b}}=\dot{\mathbf{c}}$$

since **d** is a vector of constant magnitude and direction.

If ω_1 is the input angular velocity of the link A_0A, then in vector form we have

$$\boldsymbol{\omega}_1=\omega_1\mathbf{j}$$

Similarly, the output angular velocity of the link B_0B is

$$\boldsymbol{\omega}_0=\omega_0\mathbf{i}$$

Let V_A be the velocity of the point A as a point on A_0A, then

$$\mathbf{V}_A=\boldsymbol{\omega}_1\wedge\mathbf{a}=\omega_1\mathbf{j}\wedge\mathbf{a}$$

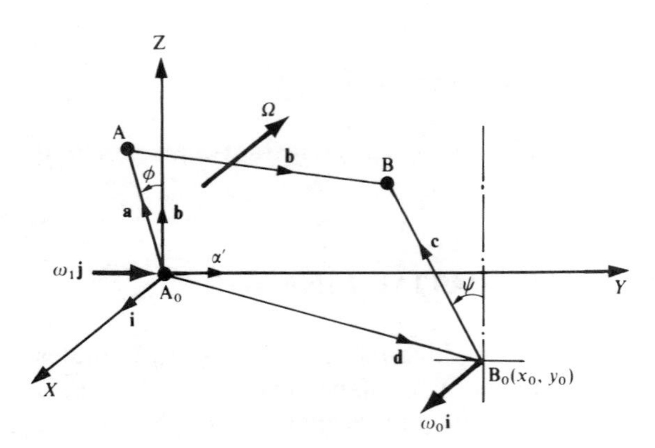

Figure 7.18 Vector representation of the linkage in Fig. 7.17.

Also, if $\mathbf{V}_B$ is the velocity of a point B as a point on B_0B, then

$$\mathbf{V}_B = \boldsymbol{\omega}_0 \wedge \mathbf{c} = \omega_0 \mathbf{i} \wedge \mathbf{c}$$

But the velocity $\mathbf{V}_B$ of a point B on the coupler AB is given by

$$\mathbf{V}_B = \mathbf{V}_A + \mathbf{V}_{BA}$$

or

$$\mathbf{V}_B = \mathbf{V}_A + \boldsymbol{\Omega} \wedge \mathbf{b} \qquad \text{by Eq. (7.11)}$$

where $\boldsymbol{\Omega}$ is the angular velocity of the coupler AB which can be expressed in terms of its components ω_x, ω_y, and ω_z, i.e.,

$$\boldsymbol{\Omega} = \omega_x \mathbf{i} + \omega_y \mathbf{j} + \omega_z \mathbf{k}$$

Hence

$$\boldsymbol{\Omega} \wedge \mathbf{b} = \mathbf{V}_B - \mathbf{V}_A = \omega_0 \mathbf{i} \wedge \mathbf{c} - \omega_1 \mathbf{j} \wedge \mathbf{a} \tag{7.14}$$

Since $\mathbf{V}_{BA}$, i.e., the velocity of B on AB as seen by an observer at A, is perpendicular to the coupler AB then by taking the 'dot' (scalar) product of $\mathbf{V}_{BA}$ with $\mathbf{b}$ we must have

$$\mathbf{V}_{BA} \cdot \mathbf{b} = 0$$

Substituting for $\mathbf{V}_{BA}$ yields

$$(\omega_0 \mathbf{i} \wedge \mathbf{c} - \omega_1 \wedge \mathbf{a}) \cdot \mathbf{b} = 0 \tag{7.15}$$

To solve for ω_0 we need to express **a**, **b**, and **c** in terms of the unit vectors **i**, **j**, and **k**, the input angle ϕ and the output angle ψ.

Referring to Fig. 7.18 we have

$$\left.\begin{aligned} \mathbf{a} &= a \sin \phi \mathbf{i} + \mathbf{a} \cos \phi \mathbf{k} \\ \mathbf{c} &= -c \sin \psi \mathbf{j} + c \cos \psi \mathbf{k} \\ \mathbf{d} &= x_0 \mathbf{i} + y_0 \mathbf{j} \end{aligned}\right\} \tag{7.16}$$

x_0 and y_0 are the coordinates of B_0.

Solving for **b** from Eq. (7.13) we get

$$\mathbf{b} = \mathbf{c} + \mathbf{d} - \mathbf{a}$$

and substituting for **a**, **c**, and **d** from (7.16) yields

$$\mathbf{b} = (x_0 - a \sin \phi)\mathbf{i} + (y_0 - c \sin \psi)\mathbf{j} + (c \cos \psi - a \cos \phi)\mathbf{k} \tag{7.17}$$

Substituting Eq. (7.17) in Eq. (7.15) and performing the dot product will yield an expression for the output angular velocity ω_0. The problem is thus reduced to one of algebraic manipulation of the vector equations.

An alternative approach is to consider the length $b = AB$ of the coupler and make use of the fact that since the link is rigid $\dot{b} = 0$, as follows.

From Eq. (7.17),

$$b^2 = (x_0 - a \sin \phi)^2 + (y_0 - c \sin \psi)^2 + (c \cos \psi - a \cos \phi)^2$$

Differentiating yields

$$\begin{aligned} O = (x_0 - a \sin \phi)(-a \cos \phi \dot{\phi}) + (y_0 - c \sin \psi)(-c \cos \psi)\dot{\psi} \\ + (c \cos \psi - a \cos \phi)(-c \sin \psi \dot{\psi} + a \sin \phi \dot{\phi}) \end{aligned}$$

Expanding and collecting terms the angular velocity ω_0 of the output is given by

$$\omega_0 = \frac{a(x_0 \cos\phi - c\sin\phi\cos\psi)}{c(a\cos\phi\sin\psi - y_0\cos\psi)} \tag{7.18}$$

Example 7.3 Figure 7.19 shows a crank A_0A rotating in the x–y plane at a constant angular velocity $\omega_1 = 10$ rad/s and driving the slider B on a rod PQ by means of the link AB. The rod is in the y–z plane and parallel to A_0y. Calculate the velocity of the slider and the angular velocity of the link AB when $\phi = 90°$.

SOLUTION When $\phi = 90°$ the mechanism is in the position shown in Fig. 7.20:

$$\mathbf{V}_B = V_{B\mathbf{j}}$$

$$\mathbf{V}_A = \omega_1 \mathbf{j} \wedge \mathbf{a}$$

$$= 10\mathbf{j} \wedge 50\mathbf{i} = -500\mathbf{k}$$

1. *Velocity of the slider* Let Ω be the angular velocity of the link AB, then by Eq. (7.11) we have

$$\mathbf{V}_B = \mathbf{V}_A + \mathbf{V}_{BA} = \mathbf{V}_A + \Omega \wedge \mathbf{b}$$

where

$$\mathbf{b} = \mathbf{R}_B - \mathbf{a} = 150\mathbf{j} + 75\mathbf{k} - 50\mathbf{i}.$$

Hence

$$V_B\mathbf{j} = -500\mathbf{k} + \begin{vmatrix} \mathbf{i} & \mathbf{j} & \mathbf{k} \\ \omega_x & \omega_y & \omega_z \\ -50 & 150 & 75 \end{vmatrix} \quad \text{by Eq. (7.8}a\text{)}$$

$$V_B\mathbf{j} = (75\omega_y - 150\omega_z)\mathbf{i} + (-75\omega_x - 50\omega_z)\mathbf{j} + (150\omega_x + 50\omega_y)\mathbf{k} - 500\mathbf{k}$$

Equating the coefficients of the unit vectors, we get

$$75\omega_y - 150\omega_z = 0 \tag{7.19}$$

$$-75\omega_x - 50\omega_z = V_B \tag{7.20}$$

$$150\omega_x + 50\omega_y = 500 \tag{7.21}$$

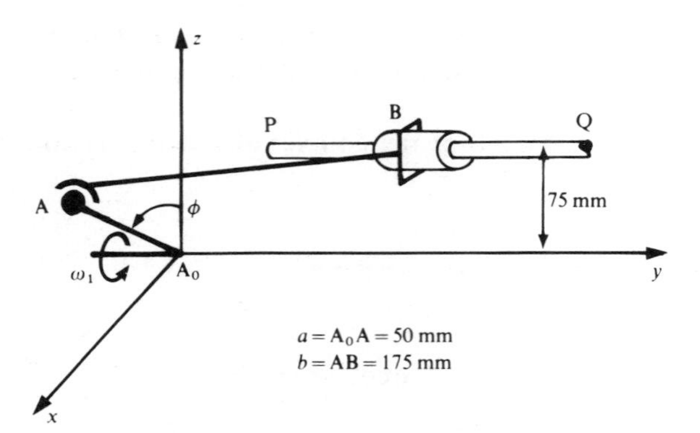

Figure 7.19

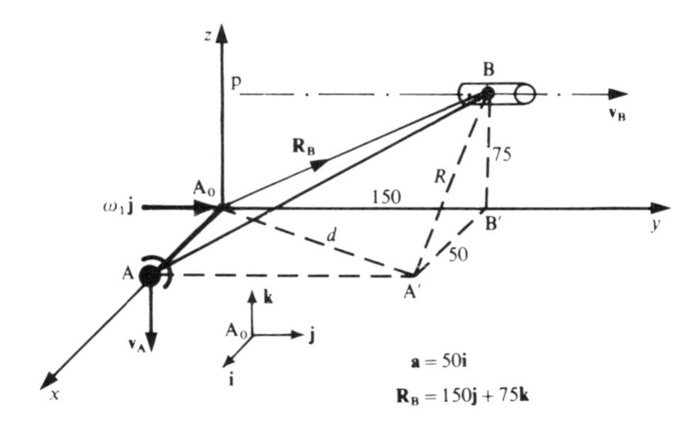

Figure 7.20

Multiplying Eq. (7.19) by -2, Eq. (7.20) by 6, Eq. (7.21) by 3, and adding yields

$$\underline{V_B = -250 \text{ mm/s}}$$

i.e., the collar is moving towards P at that particular instant.

2. *Angular velocity of the link AB* An examination of Eqs (7.19) to (7.21) reveals that the determinant of the coefficients of the components of the angular velocity is zero; we cannot, therefore, calculate their values. We therefore require another equation to take into account the constraint at B. The connection at B is such that the slider cannot rotate about $A'B = \mathbf{R}$, say, so that the projection of Ω on $\mathbf{R}$ must be zero, i.e., $\Omega \cdot \mathbf{R} = 0$.

From Fig. 7.20 we see that

$$\mathbf{R} + \mathbf{R}_B = \mathbf{d}$$

Hence

$$\mathbf{R} = 50\mathbf{i} + 150\mathbf{j} - 150\mathbf{j} - 75\mathbf{k}$$
$$= 50\mathbf{i} - 75\mathbf{k}$$

and

$$(\omega_x\mathbf{i} + \omega_y\mathbf{j} + \omega_z\mathbf{k}) \cdot (50\mathbf{i} - 75\mathbf{k}) = 0$$

i.e.,

$$50\omega_x - 75\omega_z = 0 \tag{7.22}$$

Solving Eqs (7.19) to (7.22) yields

$$\underline{V_B = -250 \text{ mm/s}} \qquad \underline{\omega_x = 57.7 \text{ rad/s}} \qquad \underline{\omega_y = 76.9 \text{ rad/s}} \qquad \underline{\omega_z = 38.5 \text{ rad/s}}$$

Note: Since the determinant of Eqs (7.19) to (7.21) is zero, we conclude that there are an infinite number of solutions for ω_x, ω_y, and ω_z. We could therefore assign any value to any one of the components of the angular velocity of the link. For example, by letting $\omega_y = 0$ we find $V_B = -250$ mm/s as before. The reader can verify that by letting $\omega_x = 0$, say, the answer will be the same. The reason for this is due to the fact that whatever the motion of the link AB about its own axis, if any, it cannot influence the output velocity or acceleration of the slider; this was already mentioned in the introduction to this chapter.

Example 7.4 Figure 7.21 shows a four-bar spatial linkage where the input and output shafts are at 90° to each other. Double revolutes are used at A and B as shown by the photograph

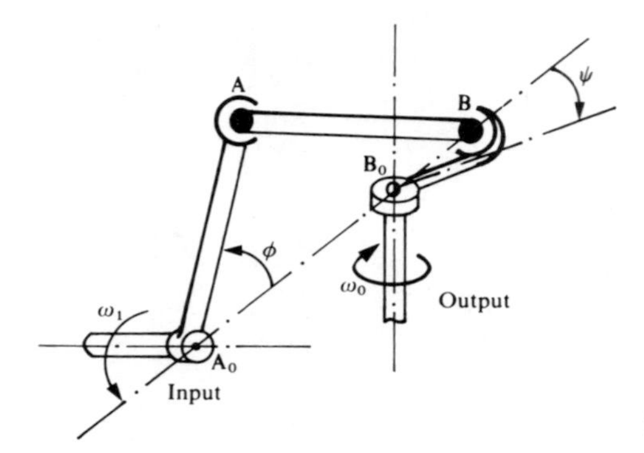

Figure 7.21

(Fig. 7.22) of a demonstration model of this linkage. The input shaft at A_0 has a constant angular velocity ω_1 in a clockwise direction on looking along the shaft towards A_0. Obtain expressions for the angular velocity ratio and for the acceleration. Using the data given below, tabulate the output angle, velocity ratio and acceleration for one complete revolution of A_0A. Data: $A_0A = a = 102$ mm; $AB = b = 381$ mm; $B_0B = c = 254$ mm; $A_0B_0 = d = 314$ mm.

When the input speed is 240 rev/min calculate (*a*) the maximum values of the output velocity and acceleration, (*b*) the angular velocity of the coupler AB for $\phi = 60°$.

SOLUTION 1. *Output velocity and acceleration* Figure 7.23 shows the linkage with the links expressed as vectors given by

$$\mathbf{a} = -a\cos\phi\mathbf{i} + \sin\phi\mathbf{k}$$
$$\mathbf{c} = -c\cos\psi\mathbf{i} + c\sin\psi\mathbf{j}$$
$$\mathbf{d} = -d\mathbf{i}$$

Since $\mathbf{a}+\mathbf{b}=\mathbf{c}+\mathbf{d}$, then

$$\mathbf{b} = \mathbf{c} + \mathbf{d} - \mathbf{a}$$

Substituting for **c**, **d**, and **a** we get

$$\mathbf{b} = (-c\cos\psi - d + a\cos\phi)\mathbf{i} + (c\sin\psi)\mathbf{j} - (a\sin\phi)\mathbf{k} \tag{7.23}$$

For the velocity of B we have $\mathbf{V}_B = \mathbf{V}_A + \mathbf{V}_{BA}$ by Eq. (7.11).

Therefore
$$\mathbf{V}_{BA} = \mathbf{V}_B - \mathbf{V}_A = -\boldsymbol{\omega}_0 \wedge \mathbf{c} - \boldsymbol{\omega}_1 \wedge \mathbf{a}$$
$$= -(\omega_0\mathbf{k} \wedge \mathbf{c} + \omega_1\mathbf{j} \wedge \mathbf{a})$$

where ω_0 is the angular velocity of the output link BB_0.

Since $\mathbf{V}_{BA}$, i.e., the velocity of B relative to A, is perpendicular to AB then

$$\mathbf{V}_{BA} \cdot \mathbf{b} = 0$$

Hence

$$(\omega_0\mathbf{k} \wedge \mathbf{c} + \omega_1\mathbf{j} \wedge \mathbf{a}) \cdot \mathbf{b} = 0$$

but
$$\mathbf{V}_B = -\omega_0\mathbf{k} \wedge \mathbf{c} = -\omega_0\mathbf{k} \wedge (c\cos\psi\mathbf{i} + c\sin\psi\mathbf{j}) = \omega_0 c\cos\psi\mathbf{j} + \omega_0 c\sin\psi\mathbf{i}$$

and
$$\mathbf{V}_A = \omega_1\mathbf{j} \wedge \mathbf{a} = \omega_1\mathbf{j} \wedge (-a\cos\phi\mathbf{i} + a\sin\phi\mathbf{k}) = \omega a\cos\phi\mathbf{k} + \omega_1 a\sin\phi\mathbf{i}$$

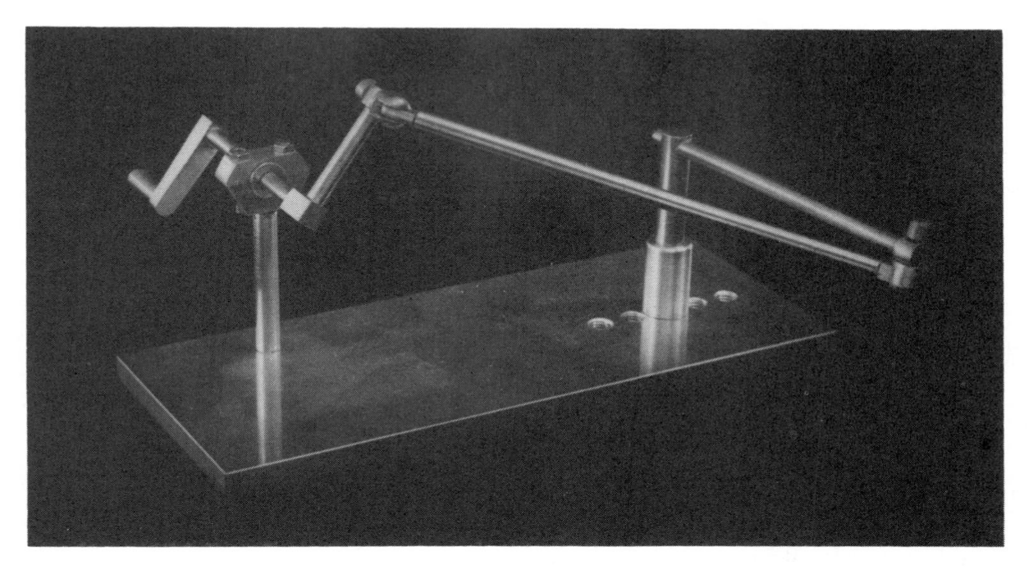

Figure 7.22 Model of the four-bar spatial linkage showing the double revolutes that replace the two spherical joints. (*Courtesy of the University of Bath.*)

It follows that

$$\mathbf{V}_{BA} = (\omega_0 c \sin\psi - \omega_1 a \sin\phi)\mathbf{i} + \omega_0 c \cos\psi\mathbf{j} - \omega_1 a \cos\phi\mathbf{k}$$

Upon performing the dot product with **b** from (7.23) we get

$$-\omega_0 c^2 \sin\psi\cos\psi + (\omega_1 a \sin\phi - \omega_0 c \sin\psi)(a\cos\phi - c\cos\psi - d) - \omega_1 a^2 \sin\phi\cos\phi = 0$$

which upon expanding and solving for the velocity ratio yields

$$\frac{\omega_0}{\omega_1} = \frac{a\sin\phi(d + c\cos\psi)}{c\sin\psi(d - a\cos\phi)} \tag{7.24}$$

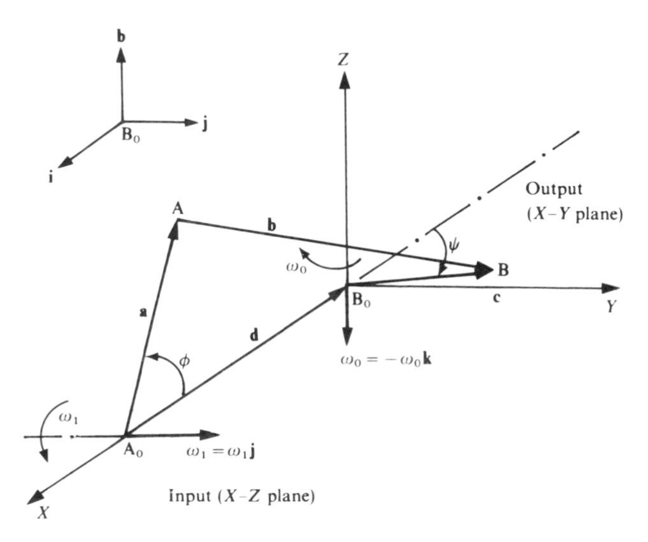

Figure 7.23

In order to solve for this ratio we need the value of the output angle ψ (note the similarity with the planar four-bar linkage).

From Eq. (7.23) we have

$$b^2=(-c\cos\psi+a\cos\phi-d)^2+c^2\sin^2\psi+a^2\sin^2\phi \tag{7.25}$$

Expanding and solving for $\cos\psi$ yields

$$\cos\psi=\frac{b^2-a^2-c^2-d^2+2ad\cos\phi}{2c(d-a\cos\phi)} \tag{7.26}$$

To obtain an expression for the acceleration α_0 of the output BB_0 it is best to differentiate Eq. (7.24).

Let

$$u=a\sin\phi(d+c\cos\psi)$$
$$v=c\sin\psi(d-a\cos\phi)$$

then

$$\omega_0=\frac{u}{v}\omega_1 \tag{7.27}$$

Differentiating (7.27) with respect to time, we get

$$\alpha_0=\frac{vu'-uv'}{v^2}\omega_1^2+\frac{u}{v}\alpha_1 \tag{7.28}$$

We also note that

$$\omega_0=\frac{d\psi}{dt}=\frac{d\psi}{d\phi}\frac{d\phi}{dt}=\frac{u}{v}\omega_1$$

Hence

$$u'=a\cos\phi(d+c\cos\psi)-ac\sin\phi\sin\psi\left(\frac{u}{v}\right)$$

$$v'=c\cos\psi(d-a\cos\phi)\left(\frac{u}{v}\right)+ac\sin\phi\sin\psi$$

In this case $\alpha_1=0$, i.e., the input is not accelerating. Table 7.1 shows the result of the computation for all values of ϕ from 0 to 360° in steps of 15°.

Figure 7.24 is a plot of the velocity ratio and of output acceleration. We notice that the mechanism with the pillar in the particular position shown has a constant deceleration for a 150° movement of the input.

The graph also shows the cyclic nature of the velocity and acceleration just like the planar mechanism.

From Table 7.1 we find that the total angular movement of the output is

$$\Delta\psi=115.93-70.64=45.29^\circ \quad \text{for} \quad \Delta\phi \text{ of } 180^\circ$$
$$\omega_1=2\pi\times 240/60=25.1 \text{ rad/s}$$

Maximum velocity, $\omega_{0_{max}}=0.427\times 25.1=\underline{10.7 \text{ rad/s}}$

Maximum acceleration, $\alpha_{0_{max}}=0.8\times 25.1^2=\underline{504 \text{ rad/s}^2}$

2. *Angular velocity of the coupler* There are two situations to consider:

(*a*) the magnitude only of the angular velocity of the coupler;
(*b*) the components of the angular velocity of the coupler parallel to the axes.

Table 7.1

ϕ,°	ψ,°	ω_0/ω_1	α_0/ω_1^2
0	70.64	0	0.8
15	72.2	0.199	0.68
30	76.32	0.343	0.41
45	82.08	0.414	0.143
60	88.42	0.427	−0.045
75	94.62	0.397	−0.153
90	100.24	0.349	−0.205
105	105.06	0.293	−0.225
120	109.0	0.233	−0.229
135	112.05	0.173	−0.226
150	114.21	0.115	−0.222
165	115.50	0.057	−0.219
180	115.93	0	−0.218
195	115.50	−0.057	−0.219
210	114.21	−0.115	−0.222
225	112.05	−0.173	−0.226
240	109.0	−0.233	−0.229
255	105.06	−0.293	−0.225
270	100.24	−0.349	−0.205
285	94.62	−0.397	−0.153
300	88.42	−0.427	−0.045
315	82.08	−0.414	0.143
330	76.32	−0.343	0.41
345	72.20	−0.199	0.68
360	70.64	0	0.8

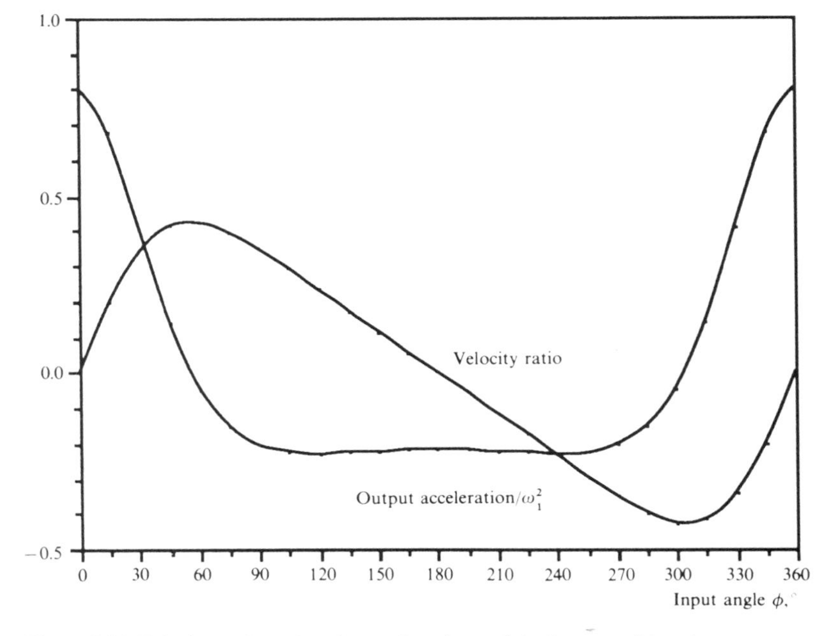

Figure 7.24 Velocity and acceleration as functions of the input position ϕ.

Case (a). From Eq. (7.11) we have

$$\begin{aligned}\mathbf{V}_{BA} &= \mathbf{V}_B - \mathbf{V}_A \\ &= -\boldsymbol{\omega}_0 \wedge \mathbf{c} - \boldsymbol{\omega} \wedge a, \text{ from solution, part 1 above} \\ &= (\omega_0 c \sin\psi - \omega_1 a \sin\phi)\mathbf{i} + \omega_0 c \cos\psi \mathbf{j} - \omega_1 a \cos\phi \mathbf{k}\end{aligned}$$

The magnitude V_{BA} of $\mathbf{V}_{BA}$ is given by

$$V_{BA} = \sqrt{(\omega_0 c \sin\psi - \omega_1 a \sin\phi)^2 + (\omega_0 c \cos\psi)^2 + (\omega_1 a \cos\phi)^2}$$

If Ω is the magnitude of the angular velocity of the coupler then

$$\Omega = V_{BA}/b$$

Since we already have expressions for ψ and ω_0 we can readily calculate Ω.

When $\phi = 60°$, $\psi = 88.42°$, $\omega_0 = 10.7$ rad/s, $b = 0.381$ m.

Substitution in the above equation yields

$$\Omega = \frac{1}{0.381}\sqrt{\begin{matrix}(10.7 \times 0.254 \sin 88.42 - 25.1 \times 0.102 \sin 60)^2 + \\ (10.7 \times 0.254 \cos 88.42)^2 + (25.1 \times 0.102 \cos 60)^2)\end{matrix}}$$

Hence $\underline{\Omega = 3.61 \text{ rad/s.}}$

Case (b). From Eq. (7.11) we have

$$\boldsymbol{\Omega} \wedge \mathbf{b} = \mathbf{V}_B - \mathbf{V}_A \tag{7.29}$$

Substituting for **b**, $\mathbf{V}_B$, and $\mathbf{V}_A$ we get

$$\begin{vmatrix}\mathbf{i} & \mathbf{j} & \mathbf{k} \\ \omega_x & \omega_y & \omega_z \\ (a\cos\phi - c\cos\psi - d) & (c\sin\psi) & (-a\sin\phi)\end{vmatrix} \begin{matrix} = (\omega_0 c \sin\psi - \omega_1 a \sin\phi)\mathbf{i} \\ + \omega_0 c \cos\psi \mathbf{j} \\ + -\omega_1 a \cos\phi \mathbf{k}\end{matrix}$$

Expanding the determinant and equating the coefficients of **i**, **j**, and **k** yields the following equations:

$$\begin{bmatrix} 0 & -a\sin\phi & -c\sin\psi \\ a\sin\phi & 0 & (a\cos\phi - c\cos\psi - d) \\ c\sin\psi & -(a\cos\phi - c\cos\psi - d) & 0 \end{bmatrix} \begin{Bmatrix}\omega_x \\ \omega_y \\ \omega_z\end{Bmatrix} = \begin{Bmatrix}\omega_0 c \sin\psi - \omega_1 a \sin\phi) \\ \omega_0 c \cos\psi \\ -\omega_1 a \cos\phi\end{Bmatrix} \tag{7.30}$$

i.e.

$$\mathbf{A\Omega} = \mathbf{B}$$

An examination of these equations reveals that the matrix **A** is singular, i.e., the determinant of the matrix is zero, hence this set of equations has either an infinite number of solutions or none at all.

Since this is a real problem the trivial case of $\Omega = 0$ cannot be considered and the fact that Ω can have an infinite number of solutions is due to the extra degree of freedom of the coupler which is its ability to rotate about its own axis, theoretically.

To obtain the angular velocity $\boldsymbol{\Omega}$ we can proceed as follows. Let us consider Eq. (7.29) above, i.e.,

$$\boldsymbol{\Omega} \wedge \mathbf{b} = \mathbf{B}_{BA}$$

If we pre-multiply by **b** we get

$$\mathbf{b} \wedge (\boldsymbol{\Omega} \wedge \mathbf{b}) = \mathbf{b} \wedge \mathbf{V}_{\mathrm{BA}}$$

and expanding the left-hand side yields Eq. (7.10) (see 7.2.3)

$$\mathbf{b} \wedge (\boldsymbol{\Omega} \wedge \mathbf{b}) = (\mathbf{b} \cdot \mathbf{b})\boldsymbol{\Omega} - (\mathbf{b} \cdot \boldsymbol{\Omega})\mathbf{b} \tag{7.31}$$

The first term of Eq. (7.31) equals $b^2\boldsymbol{\Omega}$ and the second term is zero because the components of $\boldsymbol{\Omega}$ taken along body axes, i.e., axes fixed to the coupler AB at A, as shown in Fig. 7.25, are perpendicular to **b**, except for ω'_y which is along AB and may be equated to zero since any spin of AB about its own axis does not affect the output motion of the mechanism.

Thus we have

$$b^2\boldsymbol{\Omega} = \mathbf{b} \wedge \mathbf{V}_{\mathrm{BA}} = \mathbf{b} \wedge (\mathbf{V}_{\mathrm{B}} - \mathbf{V}_{\mathrm{A}})$$

Substituting for $\boldsymbol{\Omega}$, **b**, $\mathbf{V}_{\mathrm{B}}$ and $\mathbf{V}_{\mathrm{A}}$ and solving for $\boldsymbol{\Omega}$ yields

$$(\omega_x\mathbf{i} + \omega_y\mathbf{j} + \omega_z\mathbf{k}) = \frac{1}{b^2}\begin{vmatrix} \mathbf{i} & \mathbf{j} & \mathbf{k} \\ (a\cos\phi - c\cos\psi - d) & c\sin\psi & -a\sin\phi \\ (\omega_0 c\sin\psi - \omega_1 a\sin\phi) & \omega_0 c\cos\psi & -\omega_1 a\cos\phi \end{vmatrix}$$

Equating the coefficients of **i**, **j**, and **k** and using the numerical values of case (*a*) we find $\omega_x = -2.239$ rad/s, $\omega_y = -2.685$ rad/s, and $\omega_z = -1.013$ rad/s.

Hence $$\Omega = \sqrt{\omega_x^2 + \omega_y^2 + \omega_z^2} = \sqrt{2.239^2 + 2.685^2 + 1.013^2} = \underline{3.64 \text{ rad/s}}$$

which is close to that obtained for case (*a*).

Note: To show once again that giving any arbitrary value to one of the components of $\boldsymbol{\Omega}$ in Eq. (7.30) does not affect the output, let $\omega_x = 0$; then solving Eq. (7.30) for ω_y and ω_z and hence the velocity ratio yields

$$\frac{\omega_0}{\omega_1} = \frac{a\sin\phi(c\cos\psi + d)}{c\sin\psi(d - a\cos\phi)} \qquad \text{as Eq. (7.20)}$$

From a practical point of view such as the design of the links of a spatial mechanism we shall require the components of velocity and acceleration of the coupler, particularly in the case of a high-speed mechanism, where the inertia forces and torques may be high and will add to the forces and torques being transmitted. In such situations we will have to adopt

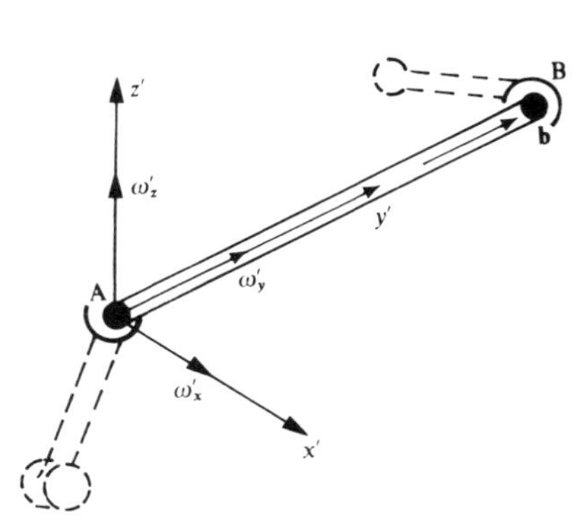

Figure 7.25

the procedure discussed in case (*b*) for the angular velocity of the coupler and a similar procedure for its angular acceleration, as illustrated in Example 7.5.

Example 7.5 Calculate the angular acceleration of the coupler AB of Example 7.4 when the input is rotating at 1200 rev/min and accelerating at 3000 rev/min/s in the sense of ω_1 at the instant when $\phi = 0$

SOLUTION Recalling Eq. (7.12) for the acceleration of a point P on a link AP, we have

$$\mathbf{a}_p = \mathbf{a}_A + \boldsymbol{\alpha} \wedge \mathbf{R} + \boldsymbol{\omega} \wedge (\boldsymbol{\omega} \wedge \mathbf{R})$$

where $\mathbf{a}_A$ = acceleration of the point A
$\boldsymbol{\alpha}$ = angular acceleration of the link
$\boldsymbol{\omega}$ = angular velocity of the link
$\mathbf{R}$ = length of the link

The diagram of the mechanism is reproduced in Fig. 7.26 for convenience. Let us consider each link, as follows:

1. *Point A on link A_0A*

$$\mathbf{a} = -a \cos \phi \mathbf{i} + a \sin \phi \mathbf{k}$$
$$= -0.102\mathbf{i} \qquad \text{since } \phi = 0$$

$$\mathbf{V}_A = \omega_1 \mathbf{j} \wedge \mathbf{a}$$
$$= 125.7\mathbf{j} \wedge (-0.102\mathbf{i})$$
$$= 12.82\mathbf{k}$$

$$\mathbf{a}_A = \alpha_1 \mathbf{j} \wedge \mathbf{a} + \omega_1 \mathbf{j} \wedge (\omega_1 \mathbf{j} \wedge \mathbf{a})$$

Tangential component   Centripetal component

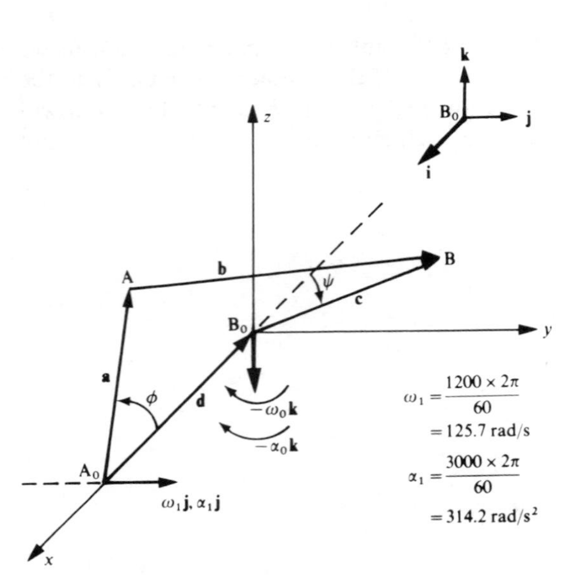

Figure 7.26

But $\omega_1 \mathbf{j} \wedge \mathbf{a} = \mathbf{V}_A$

We have

$$\alpha_1 \mathbf{j} \wedge \mathbf{a} = 314.2\mathbf{j} \wedge (-0.102\mathbf{i}) = 32.05\mathbf{k}$$

$$\omega_1 \mathbf{j} \wedge \mathbf{V}_A = 125.7\mathbf{j} \wedge 12.82\mathbf{k} = 1611\mathbf{i}$$

Hence

$$\underline{\mathbf{a}_A = 1611\mathbf{i} + 32.05\mathbf{k}} \tag{7.32}$$

2. *Point B on link B_0B*

$$\begin{aligned}\mathbf{c} &= -c\cos\psi\mathbf{i} + c\sin\psi\mathbf{j}\\ &= -0.254\cos 70.64\mathbf{i} + 0.254\sin 70.64\mathbf{j}\\ &= -0.0842\mathbf{i} + 0.240\mathbf{j}\end{aligned}$$

Since

$$\psi = 70.64^\circ \text{ when } \phi = 0 \text{ (see Table 7.1)}$$

$$\begin{aligned}\mathbf{V}_B &= -\omega_0 \mathbf{k} \wedge \mathbf{c}\\ &= 0 \text{ since } \omega_0/\omega_1 = 0 \text{ when } \phi = 0 \text{ (see Table 7.1)}\end{aligned}$$

$$\mathbf{a}_B = -\alpha_0 \mathbf{k} \wedge \mathbf{c} - \omega_0 \mathbf{k} \wedge \mathbf{V}_B = -\alpha_0 \mathbf{k} \wedge \mathbf{c}$$

$$= \begin{vmatrix} \mathbf{i} & \mathbf{j} & \mathbf{k} \\ 0 & 0 & -\alpha_0 \\ -0.0842 & 0.240 & 0 \end{vmatrix} = \underline{0.240\alpha_0\mathbf{i} + 0.0842\alpha_0\mathbf{j}} \tag{7.33}$$

3. *Point B on link AB*

$$\mathbf{b} = \mathbf{c} + \mathbf{d} - \mathbf{a} = (-c\cos\psi - d + a\cos\phi)\mathbf{i} + c\sin\psi\mathbf{j} - a\sin\phi\mathbf{k}$$

Substituting values we get

$$\begin{aligned}\mathbf{b} &= (-0.254\cos 70.64 - 0.314 + 0.102)\mathbf{i} + 0.254\sin 70.64\mathbf{j}\\ &= -0.296\mathbf{i} + 0.240\mathbf{j}\end{aligned}$$

$$\mathbf{a}_{BA} = \alpha_c \wedge \mathbf{b} + \omega_c \wedge (\omega_c \wedge \mathbf{b}) = \alpha_c \wedge \mathbf{b} + \omega_c \wedge \mathbf{V}_{BA}$$

where $\boldsymbol{\omega}_c$ = angular velocity of the coupler $= \omega_x\mathbf{i} + \omega_y\mathbf{j} + \omega_z\mathbf{k}$

$\boldsymbol{\alpha}_c$ = angular acceleration of the coupler

$= \boldsymbol{\alpha}_x\mathbf{i} + \boldsymbol{\alpha}_y\mathbf{j} + \boldsymbol{\alpha}_z\mathbf{k}$

From Example 7.4, case (*b*), we have

$$\omega_x\mathbf{i} + \omega_y\mathbf{j} + \omega_z\mathbf{k} = \begin{vmatrix} \mathbf{i} & \mathbf{j} & \mathbf{k} \\ -0.296 & 0.24 & 0 \\ 0 & 0 & -12.82 \end{vmatrix}$$

hence $\omega_x = -21.2$, $\omega_y = -26.13$, and $\omega_z = 0$.

From Example 7.4, case (*a*),

$$\mathbf{V}_{BA} = -\omega_1 a\cos\phi\mathbf{k} = -125.7 \times 0.102\mathbf{k} = -12.82\mathbf{k}$$

$$\boldsymbol{\omega}_c \wedge \mathbf{V}_{BA} = \begin{vmatrix} \mathbf{i} & \mathbf{j} & \mathbf{k} \\ -21.2 & -26.13 & 0 \\ 0 & 0 & -12.82 \end{vmatrix} = 335\mathbf{i} - 272\mathbf{j} \tag{7.34}$$

$$\alpha_c \wedge \mathbf{b} = \begin{vmatrix} \mathbf{i} & \mathbf{j} & \mathbf{k} \\ \alpha_x & \alpha_y & \alpha_z \\ -0.296 & 0.240 & 0 \end{vmatrix} = -0.240\alpha_z\mathbf{i} - 0.296\alpha_z\mathbf{j} + (0.240\alpha_x + 0.296\alpha_y)\mathbf{k} \quad (7.35)$$

Writing the vector equation for the acceleration of the point B on AB we have

$$\mathbf{a}_B = \mathbf{a}_A + \mathbf{a}_{BA},$$

i.e.,

$$\mathbf{a}_{Bt} + \mathbf{a}_{Bn} = \mathbf{a}_{At} + \mathbf{a}_{An} + \mathbf{a}_{BAt} + \mathbf{a}_{BAn} \quad (7.36)$$

where the suffices 't' and 'n' stand for the tangential and normal components of acceleration respectively.

Before we can solve for α_c we need to introduce the condition that $\boldsymbol{\alpha}_c \cdot \mathbf{b} = 0$ to eliminate the possible spin of the coupler about its axis. This leads to the following equation:

$$(\alpha_x\mathbf{i} + \alpha_y\mathbf{j} + \alpha_z\mathbf{k}) \cdot (-0.296\mathbf{i} + 0.240\mathbf{j}) = 0$$

i.e.,

$$-0.296\alpha_x + 0.240\alpha_y = 0 \quad (7.37)$$

Substituting Eqs (7.32) to (7.35) in Eq. (7.36), we get

$$0.240\alpha_0\mathbf{i} + 0.0842\alpha_0\mathbf{j} = 1611\mathbf{i} + 32.05\mathbf{k} - 0.240\alpha_z\mathbf{i} - 0.296\alpha_z\mathbf{j} + 0.2400\alpha_x\mathbf{k} + 0.296\alpha_y\mathbf{k} + 335\mathbf{i} - 272\mathbf{j}$$

Equating the coefficients of **i**, **j**, and **k** yields

$$0.240\alpha_0 + 0.240\alpha_z = 1611 + 335 = 1946 \quad (7.38)$$

$$0.0842\alpha_0 + 0.296\alpha_z = -292 \quad (7.39)$$

$$0.240\alpha_x + 0.296\alpha_y = -32.05 \quad (7.40)$$

and

$$-0.296\alpha_x + 0.240\alpha_y = 0 \quad (7.37)^*$$

From Eqs (7.38) and (7.39),

$$\alpha_0 = 12\,607 \text{ rad/s}^2 \quad \text{in the assumed direction}$$

$$\alpha_z = -4497 \text{ rad/s}^2$$

From Eq. (7.40), $\alpha_x = 0.811\alpha_y$

Substituting in Eq. (7.37) yields

$$\alpha_y = -32.05/(0.24 \times 0.811 + 0.296) = -65.3 \text{ rad/s}^2$$

and

$$\alpha_x = -53 \text{ rad/s}^2$$

Hence at the instant when $\phi = 0°$, the angular acceleration of the output $\boldsymbol{\alpha}_0 = -12\,607\mathbf{k}$ rad/s^2, and that of the coupler $\boldsymbol{\alpha}_c = -53\mathbf{i} - 65.3\mathbf{j} - 4497\mathbf{k}$ rad/s^2.

The angular acceleration α_0 of the output is also given by Eq. (7.28). When $\phi = 0$, $\omega_0 = 0$, and, hence, $u/v = 0$. Equation (7.28) becomes $\alpha_0 = (u'/v)\omega_1^2$; from Table 7.1 we see that $u'/v = 0.8$, and the acceleration of the output is:

$$\alpha_0 = 0.8 \times 125.7^2 = \underline{12\,640 \text{ rad/s}^2}$$

which is, near enough, the value obtained by the above method.

If the acceleration of the coupler is not required, then the method described in the solution of Example 7.4 is to be preferred as it is very much shorter.

EXERCISES

7.1 Figure 7.27 shows two rods AA_1 and BB_1 constrained to move in guides G_1 and G_2 respectively. They are connected by means of ball-and-socket joints at A and B.

Taking a set of coordinates as shown and with A moving at a constant velocity of 1.25 m/s, calculate at the instant when the coordinates of A and B are as indicated: (*a*) the velocity and acceleration of B; (*b*) the angular velocity and angular acceleration of AB.

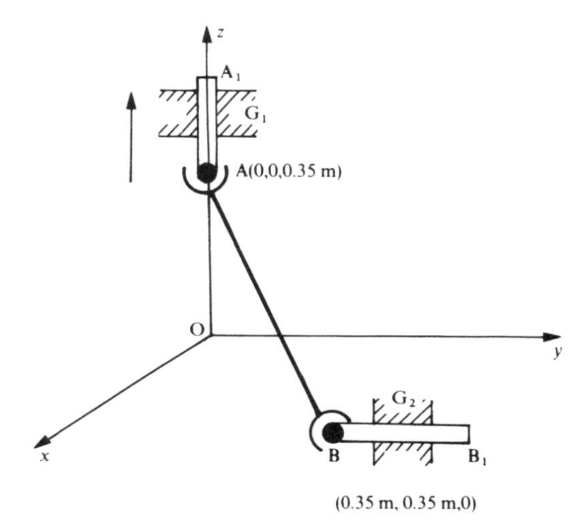

Figure 7.27

7.2 The four-bar spatial linkage shown in Fig. 7.28 has two shafts at 90° to each other. The input shaft A_0x which rotates about the axis A_0x and the output shaft By which rotates about an axis parallel to A_0y. Data: $A_0A = 96$ mm; $AB = 360$ mm; $BB_0 = 240$ mm; $x = 72$ mm; $y = 288$ mm; $\phi = 45°$; $\omega_1 = 60$ rad/s.

Calculate: (*a*) the angular velocity ω_0 and angular acceleration α_0 of the output shaft; (*b*) the angular velocity and angular acceleration of the coupler AB.

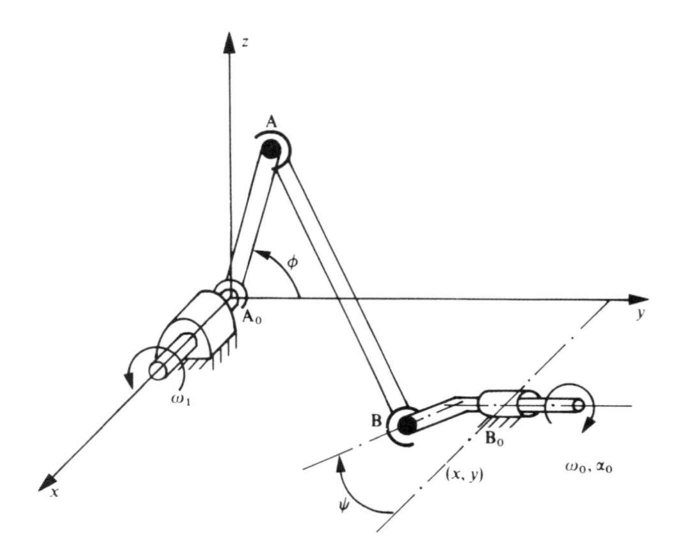

Figure 7.28

7.3 Referring to Example 7.4 on page 199, plot the output angular displacement when $b = d$, $a = c = b/\sqrt{2}$ and with $b = 300$ mm for one complete revolution of the input crank.

What is the total angular displacement of the output link B_0B?

Plot also the angular velocity ratio (ω_0/ω_1) and the angular acceleration ratio, α_0/ω_1^2, for one complete rotation of the input.

Note: With the above link ratios the linkage is that due to Bennett (see Fig. 7.3).

7.4 For the typical four-bar RSSR spatial linkage shown in Fig. 7.29, where $A_0A = a$, the input link, $AB = b$, the coupler, and $B_0B = c$, the output link whose axis is in the x–y plane and at an angle θ with A_0y. B_0 has coordinates x, y and the input and output angles are ϕ and ψ respectively in the direction shown, **e** and **t** are unit vectors along and perpendicular to the output shaft.

Show that the velocity ratio is given by:

$$\frac{\omega_0}{\omega_1} = \frac{a[\sin\phi(x - c\cos\psi\sin\theta) + c\cos\phi\sin\psi]}{c[\sin\psi(x\sin\theta + a\sin\theta\cos\phi - y\cos\theta) - a\sin\phi\cos\psi]}$$

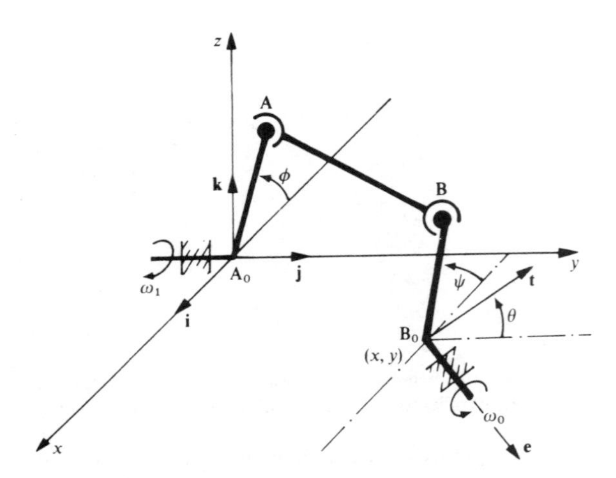

Figure 7.29

7.5 Figure 7.30 shows the general arrangement of the spatial four-bar RSSR linkage. The input and output shafts are inclined at an angle α to each other. For convenience a reference frame $Oxyz$ has been chosen such that the input shaft

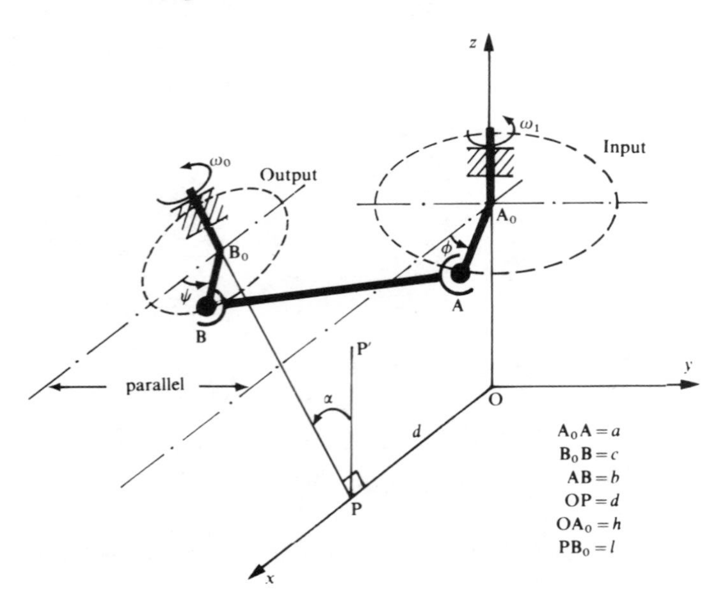

Figure 7.30

lies along the Oz-axis, PB_0 is perpendicular to Ox and PP' is parallel to Oz. $A_0A = a$; $B_0B = c$; $AB = b$; $OP = d$; $OA_0 = h$; $PB_0 = l$.

Derive an expression for the output position ψ as a function of the input angle ϕ and show that the velocity ratio ω_0/ω_1 is given by

$$\frac{\omega_0}{\omega_1} = \frac{a\{c(\cos\phi\sin\psi\cos\alpha - \sin\phi\cos\psi) - d\sin\phi - l\cos\phi\sin\alpha\}}{c\{a(\cos\phi\sin\psi - \sin\phi\cos\psi\cos\alpha) - d\sin\psi - h\sin\alpha\cos\psi\}}$$

7.6 The axes $Oxyz$ shown in Fig. 7.31 are rectangular. Arm OA rotates about the y-axis with an angular velocity of 10 rad/s in the direction shown. Arm PB is constrained to rotate about an axis parallel to Ox. Pivot P lies on the y-axis 0.5 m from O. Arm length OA is 0.2 m, and PB is 0.3 m. Calculate the angular velocity of PB at the instant shown, when OA is 25° before the direction Ox, and PB makes an angle of 75° to Oy. What is the component of the linear acceleration of B in the direction BP?

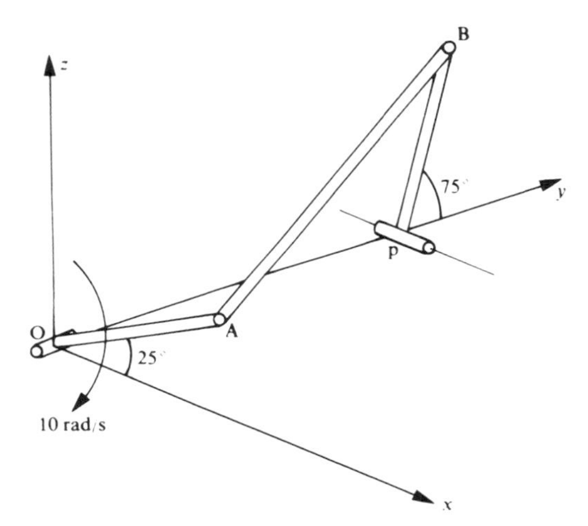

Figure 7.31

7.7 Figure 7.32 shows a universal joint known as a Hooke's joint. It consists of a driving yoke A which is fixed to the input shaft rotating at ω_1. The axis of the driven yoke B makes an angle α with the input shaft and rotates at ω_2. Show that the velocity ratio is given by:

$$\frac{\omega_2}{\omega_1} = \frac{\cos\alpha}{1 - \sin^2\phi\sin^2\alpha}$$

where ϕ is the displacement of the input and ψ the corresponding output displacement, and $\omega_1 = \dot{\phi}$, $\omega_2 = \dot{\psi}$. The joints at c, c′, d, and d′ are revolutes.

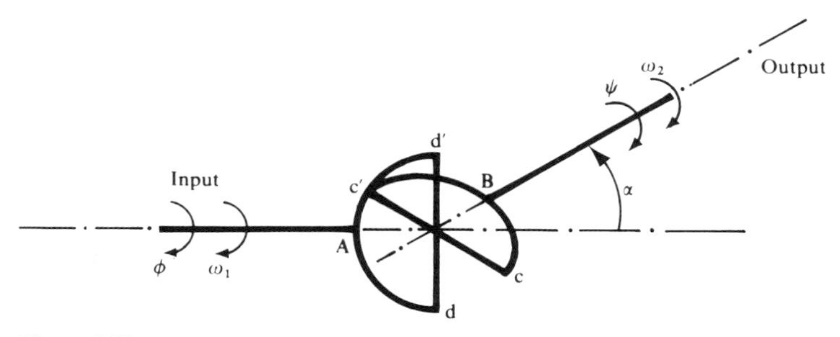

Figure 7.32

7.8 The cylinder shown in Fig. 7.33 rotates at a constant angular velocity Ω about a vertical axis OZ. At the same time it is being elevated at a constant angular velocity ω in a vertical plane while the piston rod is moving outwards with a constant velocity u.

Using the data given, calculate (*a*) the velocity of B, and (*b*) the acceleration of B. (Data: OB = 1.75 m, $u = 0.5$ m/s, $\Omega = 0.75$ rad/s, $\omega = 0.4$ rad/s and $\theta = 60°$.)

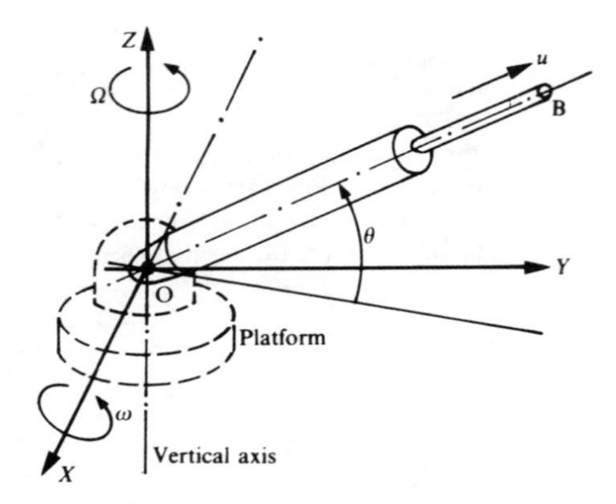

Figure 7.33

CHAPTER

EIGHT

FORCES IN MECHANISMS

8.1 INTRODUCTION

A mechanism is a device that transforms motion and in so doing transmits power by means of forces and torques in its elements. Consider, for example, the transmission system shown in Fig. 8.1, where a motor drives a load by means of a reduction gearbox, a four-bar linkage, and geared segments. The linkage is used in this case to provide an oscillatory output which is then amplified by the geared segments. The torque provided by the motor must be sufficient to overcome the load torque demand at B_0' throughout the cycle of operation as well as the inertia forces, inertia torques and friction.

The input torque is transformed into a force at the point of contact between the teeth of the meshing gears, resulting in a torque being applied to the link A_0A that is integral with the gear. This new torque provides a force at the joint A to the coupler AB, which in turn applies a force at B to the link B_0B. The force at B now generates a torque on the geared segment, which is then transmitted to the load by another force of contact between the gears.

It follows then that to design the elements of such a system or any other system we must first evaluate the torques and forces involved. In the case of the four-bar linkage these will be the forces in the links, as well as forces on the pins and bearings at A_0, A, B, and B_0. In some designs the pins and bearings may be the critical elements because they are also subjected to wear caused by rubbing due to the relative velocity between the links, clearances, and jerks.

In some mechanisms the inertia forces and torques due to the accelerations can be of the same order of magnitude as the static forces and torques. In certain cases such as high-speed machinery, inertia forces and torques can be very much greater than the static ones.

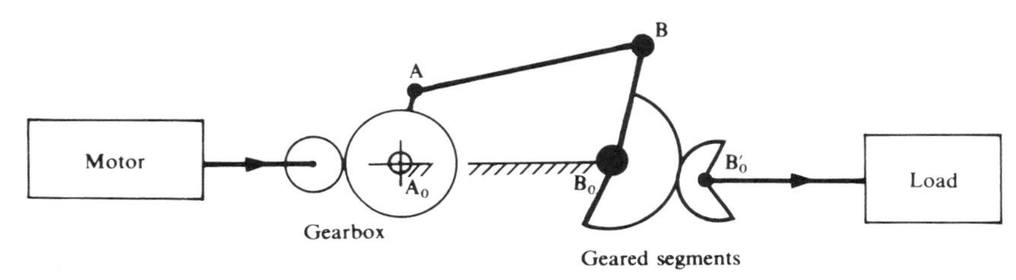

Figure 8.1 Typical transmission system: use of a linkage between motor and load.

The task of the designer is to ensure that all the elements of the system have been properly designed so that they will not fail in service.

The above remarks apply whether the mechanism is a precision one such as the one discussed in Chapter 6, the stitching machine on page 4, the aircraft undercarriage mentioned in Sec. 7.1, or the crane in Fig. 1.7.

In the following sections we investigate the static and dynamic forces and torques.

8.2 TRANSMISSION OF FORCES, STATIC EQUILIBRIUM

Let us consider the transmission of forces in mechanisms neglecting inertia forces and friction for the time being, in which case the forces and torques will be in static equilibrium. We know from our study of 'statics' that for a body to be in equilibrium we must have

$$\Sigma \text{ forces} = 0 \tag{8.1}$$

and

$$\Sigma \text{ torques} = 0 \tag{8.2}$$

For example, Fig. 8.2(*a*) shows two meshing gears. A is the driver rotating at ω_1, and B the driven gear rotating at ω_2; T_1 is the applied torque and T_2 the resisting torque. If the system is not accelerating it must be in equilibrium.

The free-body diagrams of Fig. 8.2(*b*) show the forces acting at the point of contact P between the gears and at each bearing; the force F is inclined at an angle ϕ, the pressure angle of the gears, friction being neglected.

Considering gear A, for equilibrium we must have

$$\left.\begin{aligned} F\cos\phi - Y_1 &= 0 \\ F\sin\phi - X_1 &= 0 \end{aligned}\right\} \quad \text{by (8.1)}$$

and

$$T_1 - F\cos\phi R_1 = 0 \quad \text{by (8.2)}$$

X_1 and Y_1 are the reactions at the bearings O_1.

Similarly for gear B:

$$\left.\begin{aligned} F\cos\phi - Y_2 &= 0 \\ F\sin\phi - X_2 &= 0 \end{aligned}\right\} \quad \text{by (8.1)}$$

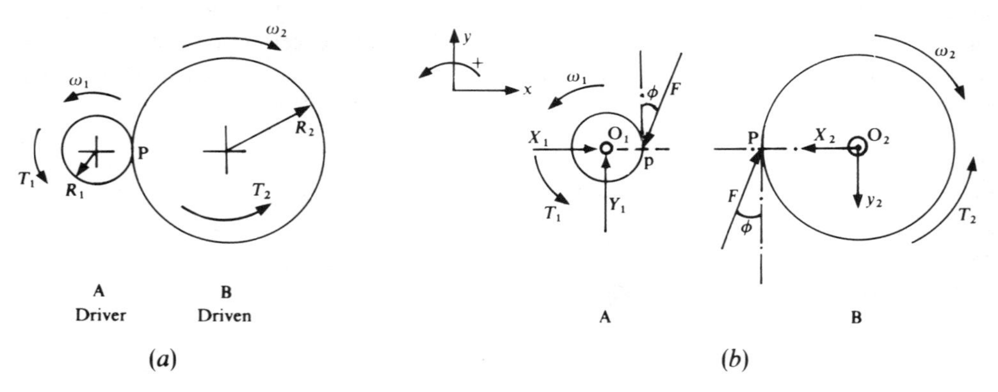

Figure 8.2 Forces in a simple gear train.

and $$T_2 - F \cos \phi R_2 = 0 \quad \text{by (8.2)}$$

X_2 and Y_2 are the reactions at the bearing O_2. Also for the system as a whole it follows that $X_1 - X_2 = 0$ and $Y_1 - Y_2 = 0$. Hence the contact force F is given by

$$F = T_1 / \cos \phi R_1$$

and $X_1 = X_2 = T_1 \tan \phi / R_1$, $Y_1 = Y_2 = T_1 / R_1$ by substitution for F in the above equations.

Furthermore, since

$$T_1 = F \cos \phi R_1 \quad \text{and} \quad T_2 = F \cos \phi R_2$$

then $$T_2 / T_1 = R_2 / R_1$$

From the kinematics of the system, if V is the velocity at the point of contact, then

$$V = \omega_1 R_1 = \omega_2 R_2$$

therefore $$R_2 / R_1 = \omega_1 / \omega_2$$

Hence $$T_2 / T_1 = \omega_1 / \omega_2 \quad \text{or} \quad T_1 \omega_1 = T_2 \omega_2$$

This last relationship tells us that in the absence of friction the power delivered by the motor is equal to the power absorbed by the load.

Before proceeding further let us recall the fact that for a body to be in equilibrium under the action of two forces F_1 and F_2 applied at A and B respectively, as shown in Fig. 8.3, these two forces must be equal in magnitude, opposite in sense and directed along the line joining A to B; i.e., the forces must be collinear. In the case of a body subjected to three forces F_1, F_2, and F_3 applied at A, B, and C respectively as shown in Fig. 8.4, then for static equilibrium the line of action of the forces must be concurrent and meet at some point P.

If a couple is applied to a body then this body will be in static equilibrium if it is acted upon by an equal and opposite couple as shown in Fig. 8.5. In the figure couple Fd formed by the two equal, opposite, and parallel forces F at A and B is balanced by the couple $F_1 d_1$ formed by two equal, opposite, and parallel forces applied at C and D.

Hence $Fd - F_1 d_1 = 0$ for equilibrium. If on the other hand a pure couple C is applied to a body, then that body will be in static equilibrium if another couple C is applied in the opposite direction or by the application of two equal, opposite, and parallel forces applied at two points in the body as in Fig. 8.5.

8.3 STATIC FORCES IN MECHANISMS

In order to calculate the forces and torques in any mechanism it is necessary to isolate each element of the mechanism as a free body and then apply the equilibrium conditions, Eqs (8.1) and (8.2).

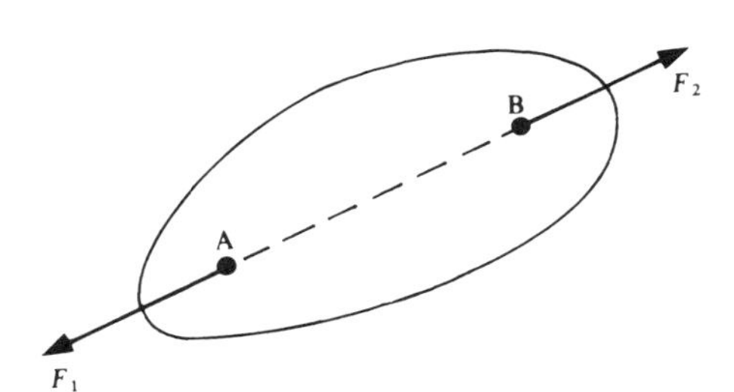

Figure 8.3 Equilibrium of a body under the action of two forces.

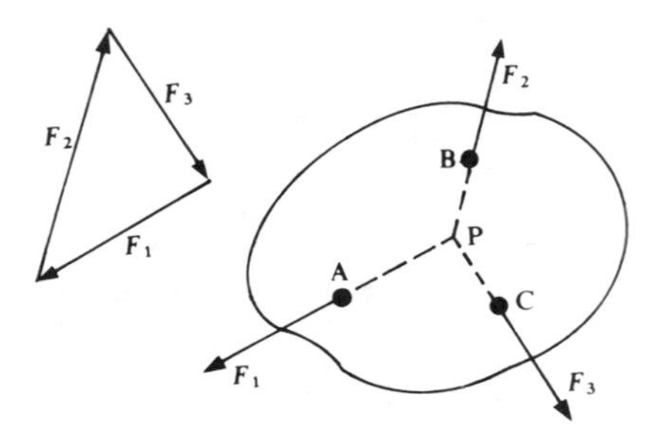

Figure 8.4 Body in equilibrium under the action of three forces: their lines of action must be concurrent.

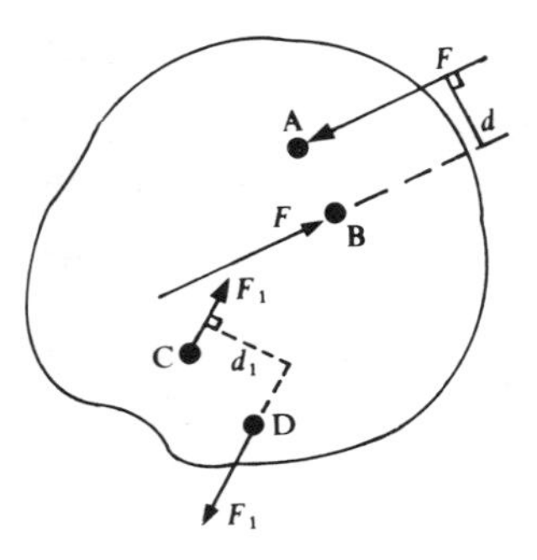

Figure 8.5 Equilibrium of a body under the action of two couples.

Let us consider the four-bar linkage shown in Fig. 8.6. An input torque T is applied to the shaft at A_0 to overcome a force P at C on the coupler ABC. We require the value of this torque when the linkage is in the position defined by $\phi = \phi_0$, and the load P is at an angle β to the line CC_1 parallel to A_0B_0.

Figure 8.7 shows the free-body diagrams and the forces acting. We should note that the forces at B and B_0 must be collinear, as discussed in Sec. 8.2 (Fig. 8.3). Taking the Cartesian frame of reference shown with origin at A_0, we see that the reactions at each joint are as indicated neglecting the weights of the links.

The equilibrium conditions lead to three equations for each element of the linkage as follows:

(*a*) Link A_0A

$$X_0 + X_A = 0$$

$$Y_0 + Y_A = 0$$

$$T - X_A a \sin \phi_0 + Y_A a \cos \phi_0 = 0 \qquad \text{by taking moments about } A_0$$

(*b*) Link ABC

$$P \cos \beta + R_0 \cos \psi_1 - X_A = 0$$

$$R_0 \sin \psi_1 - P \sin \beta - Y_A = 0$$

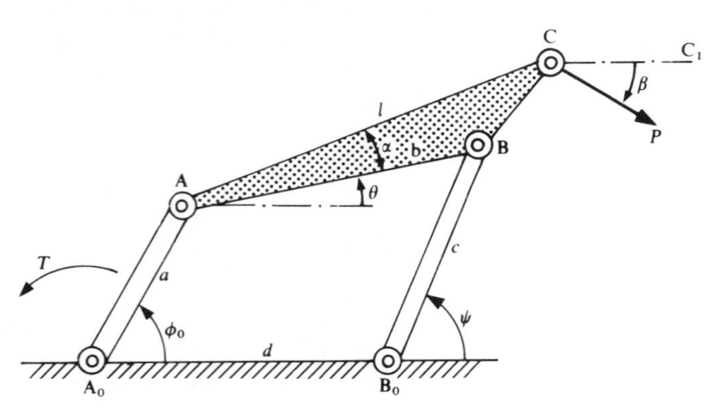

Figure 8.6 Four-bar linkage under the action of a force and a couple.

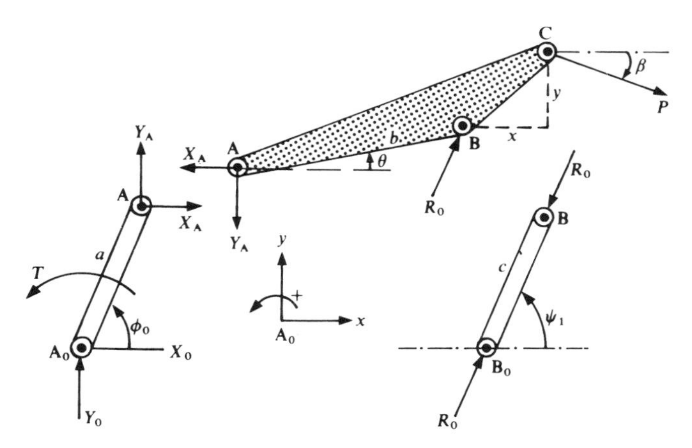

Figure 8.7 Free-body diagram of each member of the four-bar linkage shown in Fig. 8.6.

Taking moments about B, we have

$$Y_A b \cos\theta - X_A b \sin\theta - P\cos\beta y - P\sin\beta x = 0$$

It can be seen that there are eight unknowns (X_0, Y_0, T, X_A, Y_A, R_0, ψ_1, and θ); hence the six equilibrium equations together with the kinematic relationships between ϕ_0, θ, and ψ_1 will enable us to calculate all of them.

The values could also be obtained graphically by first drawing the linkage in the required position; θ and ψ_1 are therefore automatically defined. Then, considering the link ABC, R_0, X_A, and Y_A are obtained from a single diagram since the link is subjected to three forces, namely P, R_0, and R_A (whose components are X_A and Y_A), which must be concurrent as shown in Fig. 8.8, hence R_0 and R_A are found. From the equations in (*a*) above we have $X_0 = -X_A$ and $Y_0 = -Y_A$, and T is obtained from the third equation. In general, however, we shall require the values of the forces and the torques for all values of ϕ within the range of operation of the linkage. This would require many diagrams, and it would therefore be more efficient to derive analytical expressions for all the unknowns as functions of the input ϕ and use a pocket or desk computer to obtain the desired numerical values.

Example 8.1 Figure 8.9 shows a gripper G consisting of a four-bar linkage whose proportions are such that the point C on the coupler ABC follows a straight path C_1C_2, i.e., parallel to A_0B_0 for $90° \leqslant \phi \leqslant 270°$.

Using the data given below calculate the value of the force P applied in the direction of motion of C to overcome the torque T applied at A_0 by the gripper when $\phi = 90°$.

Calculate also the forces in all the links neglecting friction and inertia forces.

Data: $a = 120$ mm; $b = 300$ mm; $c = 300$ mm; BC $= 300$ mm; $d = 240$ mm; $T = 25$ N m.

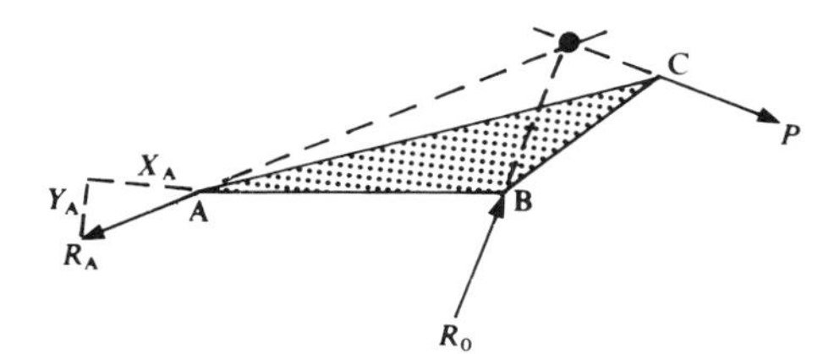

Figure 8.8 Equilibrium of the coupler.

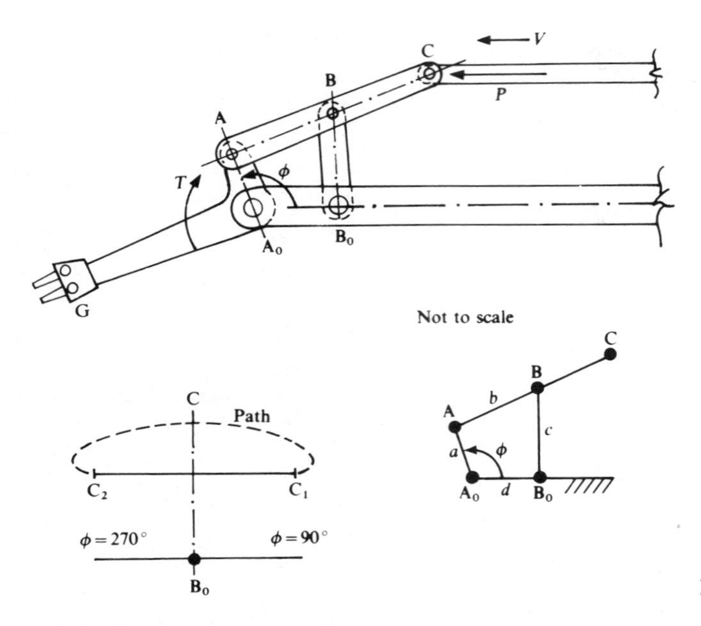

Figure 8.9

SOLUTION 1. *Value of P*
We can obtain an expression for P by equating the power in to the power out since friction is neglected. This gives

$$T\omega = PV$$

where ω is the constant angular velocity of the output link A_0A; hence

$$P = T\omega/V$$

We see, therefore, that what is needed is the kinematic relationship between ω and V. In the required position, i.e., $\phi = 90°$, we can obtain the desired relationships quite simply as shown in Fig. 8.10. In that position both A_0A and BB_0 are perpendicular to A_0B_0; hence the instantaneous centre for AB is at infinity, so that

$$a\omega = V$$

therefore $\omega/V = 1/a$ and $P = T/a = 25/0.12 = \underline{208.3\ \text{N}}.$

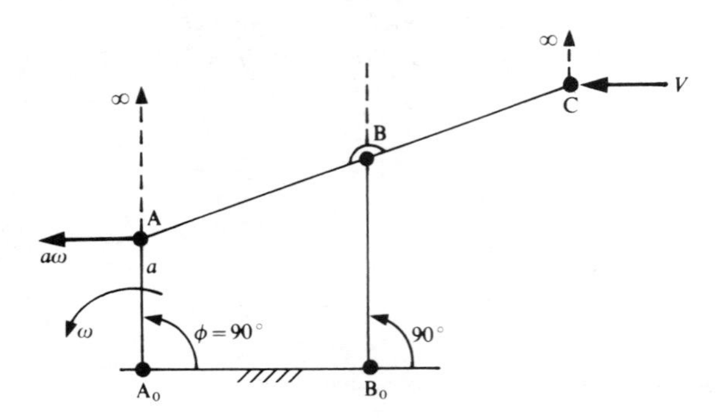

Figure 8.10

2. *Forces in the links*

The free-body diagram for each link is shown in Fig. 8.11. Since link ABC, the coupler, is subjected to three forces, then for equilibrium their lines of action must meet at a point, Q in this case, as was explained in Sec. 8.2. The line of action of F_B must be along BB_0, hence the point Q is located and by joining AQ the line of action of F_A is defined as shown in Fig. 8.11. If we know F_A then the forces in the link AA_0 follow.

To calculate the values of these forces we could draw a force diagram for link ABC since P is known from solution, part 1. Alternatively we can proceed as follows.

Let θ be the angle the coupler makes with A_0B_0 and let X_A and Y_A be the components of F_A acting on link A_0A (Fig. 8.12). Then taking moments about A_0 we have

$$X_A a - T = 0$$

hence $X_A = T/a = 25/0.120 = \underline{208.3\text{ N.}}$

Considering link ABC we have for equilibrium

$$X_A - P = 0 \qquad Y_A - F_B = 0$$

Taking moments about A yields

$$P2b \sin\theta - F_B d = 0$$

Hence

$$F_B = 2b \sin P/d.$$

It follows that

$$P = X_A = \underline{208.3\text{ N}}$$
$$F_B = 2 \times 300 \times 0.6 \times 208.3/240 = \underline{312.4\text{ N}}$$

and

$$Y_A = F_B = \underline{312.4\text{ N}}$$

In this case we could have obtained the values of the forces by considering the equilibrium of the linkage as a whole as shown in Fig. 8.13. Let X_A, Y_A, and F_B be the reactions at A_0 and B_0; then for equilibrium we have

$$X_A - P = 0 \tag{8.3}$$

$$Y_A - F_R = 0 \tag{8.4}$$

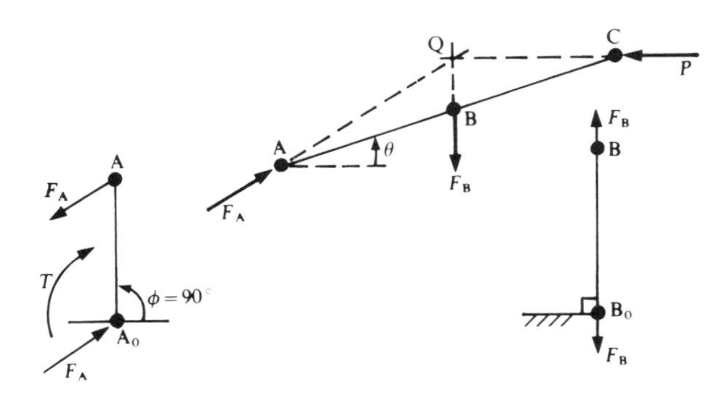

Figure 8.11

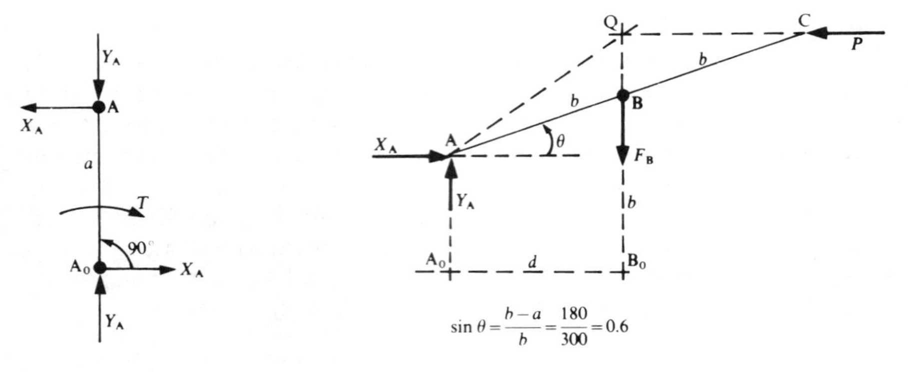

Figure 8.12

Moments about A_0:

$$P(2b\sin\theta + a) - F_B d - T = 0 \tag{8.5}$$

Hence from Eq. (8.5),

$$F_B = \frac{P(2b\sin\theta + a) - T}{d}$$

But $\sin\theta = 0.6$. Substituting numerical values yields

$$F_B = \frac{208.3(2\times 0.3\times 0.6 + 0.120) - 25}{0.24} = \underline{312.4\text{ N}}$$

From Eq. (8.3),

$$X_A = P = \underline{208.3\text{ N}}$$

and from (8.4)

$$Y_A = F_B = \underline{312.4\text{ N}}$$

Example 8.2 Figure 8.14 shows a slider crank mechanism. At the instant when $\phi = 45°$ the force on the slider is 500 N. Calculate the torque T which must be exerted on the crank A_0A in order to maintain equilibrium and the corresponding forces in all the members if $a = 65$ mm and $b = 200$ mm. Neglect friction and inertia forces.

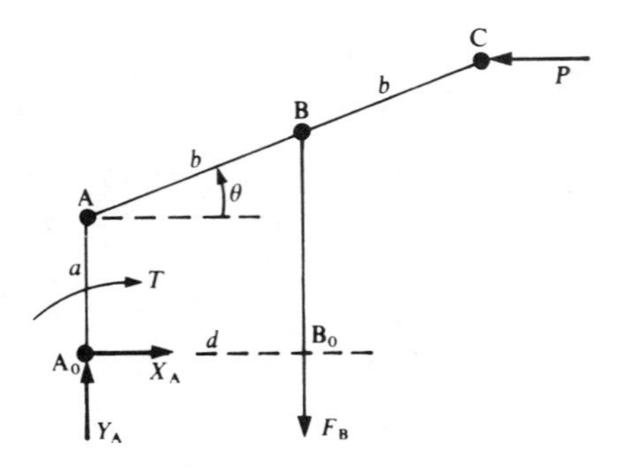

Figure 8.13

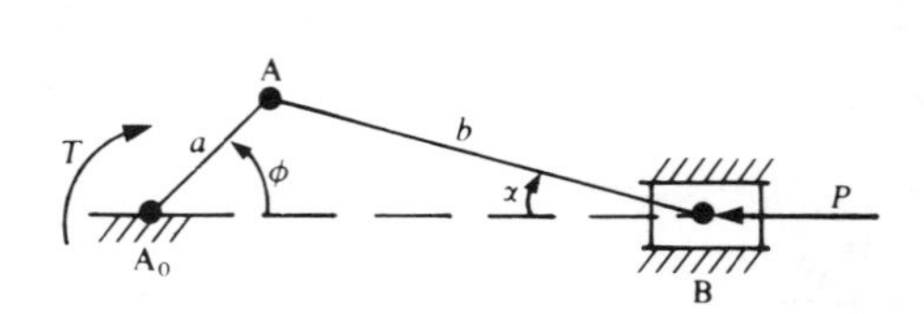

Figure 8.14

SOLUTION We first draw free-body diagrams for all the members of the mechanisms as shown in Fig. 8.15. The forces F_A in the connecting rod must be equal, opposite in sense, and collinear as mentioned before. The slider is subjected to three forces: P which is known in magnitude and direction and F_A and F_0 which are known in direction only.

The angle α made by the connecting rod can be calculated from the geometry of the mechanism, thus:

$$a \sin \phi = b \sin \alpha$$

since $\phi = 45°$ it follows that

$$\alpha = \arcsin\left(\frac{a}{b}\sin\phi\right) = \arcsin\left(\frac{65}{200}\sin 45\right) = 13.29°$$

Applying the equilibrium equations to each member in turn for the three unknowns, F_A, F_0, and T, we have

$$F_A \cos\alpha - P = 0 \tag{8.6}$$

$$F_0 - F_A \sin\alpha = 0 \tag{8.7}$$

$$F_A a \sin(\phi + \alpha) - T = 0 \tag{8.8}$$

From Eq. (8.6) $F_A = P/\cos\alpha = 500/\cos 13.29 = \underline{513.75\ \text{N}}$
From Eq. (8.7) $F_0 = F_A \sin\alpha = 513.75 \sin 13.29 = \underline{118\ \text{N}}$
From Eq. (8.8) $T = F_A a \sin(\phi + \alpha) = 513.75 \times 0.065 \times \sin 58.29 = \underline{28.4\ \text{N m}}$

8.4 INERTIA FORCES AND TORQUES

When a mechanism is accelerating the resulting forces and torques are referred to as inertia forces and inertia torques. They are also known as dynamic forces and torques. These can be very high, particularly in high-speed mechanisms, e.g., the slider crank mechanism used in internal-combustion engines, packaging machines, stitching machines, etc. Such forces and torques will add, vectorially, to the static forces and torques discussed in the previous sections, and in some cases these inertia effects can be many times the static ones. It is therefore important to be able to calculate them in order to take them into account in the detail designs of all the members of the mechanisms, i.e., links, slides, and bearings.

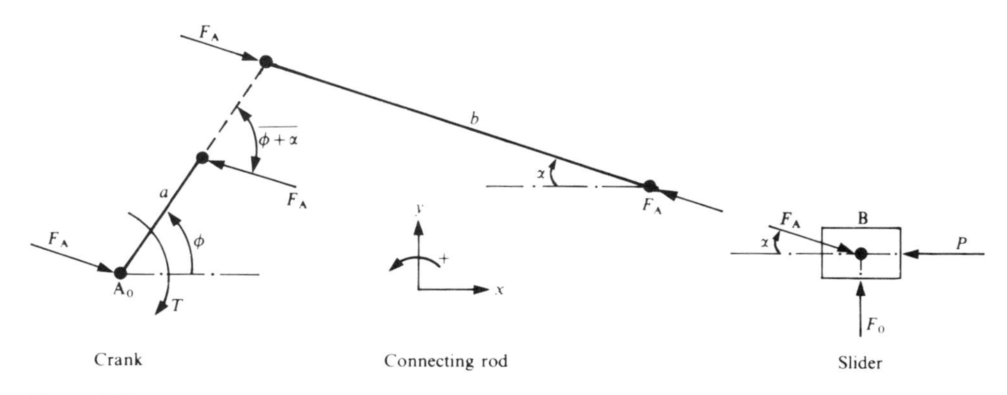

Figure 8.15

We usually know from kinematics the acceleration **a** of the centre of mass and the angular acceleration $\boldsymbol{\alpha}$ of a rigid body; we therefore know from dynamics that there must be a force **F** and a torque **T** acting on the body such that

$$\mathbf{F}=M\mathbf{a} \tag{8.9}$$

and

$$\mathbf{T}=I\boldsymbol{\alpha} \tag{8.10}$$

where M is the mass of the body and I is its moment of inertia about an axis through the centre of mass.

When calculating forces and torques in mechanisms graphically it may be useful to replace a force and a torque by a single force which is a well-known principle in mechanics and is illustrated in Fig. 8.16. In Fig. 8.16(*a*) the body is shown subjected to a force $\mathbf{F}=m\mathbf{a}$ and a torque $\mathbf{T}=I\boldsymbol{\alpha}$, whereas in Fig. 8.16(*b*) two extra forces have been added, one at G opposite to the applied force and another at some point 0 at a distance d. For the two systems to be dynamically equivalent we must have $Fd=T=I\alpha$. The two forces at G cancel and hence the single force at O replaces the original force and couple. The force at O must therefore be applied at a distance d from the centre of mass G such that

$$d=\frac{I\alpha}{F}=\frac{I\alpha}{ma} \tag{8.11}$$

In a given mechanism if we know the (mass) × (acceleration) and (moment of inertia) × (angular acceleration) of a particular link as inertia force and inertia torque respectively, Eqs (8.3) and (8.4) can be expressed as follows:

$$\Sigma(\text{forces})=0 \tag{8.12}$$

and

$$\Sigma(\text{torques})=0 \tag{8.13}$$

These equations may be considered as equations of static equilibrium in which Σ(forces) includes the inertia forces $-\Sigma$ma and Σ(torques) includes the inertia torques $-\Sigma I\alpha$. Thus the dynamic problem is reduced to one of statics, known as D'Alembert's principle, and is particularly useful in the case of a graphical calculation of forces and torques.

As an example of the method let us consider the four-bar linkage shown in Fig. 8.17 in which the input crank A_0A rotates at a constant angular velocity ω, the linkage being of the crank–

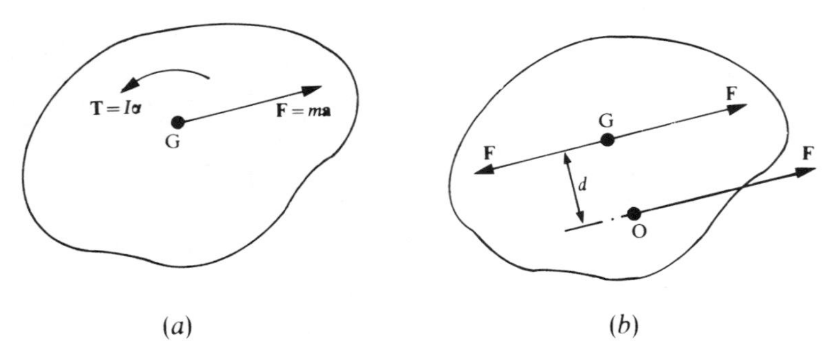

Figure 8.16 Inertia force and inertia torque.

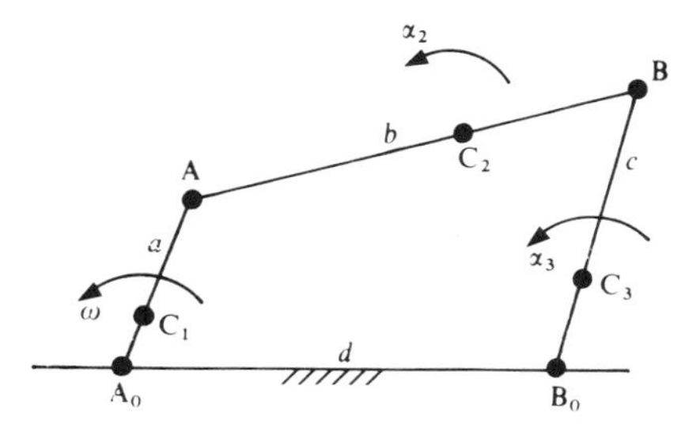

Figure 8.17 Four-bar linkage whose input crank rotates at a constant angular velocity.

rocker type. Let C_1, C_2, and C_3 be the centres of mass of links A_0A, AB, and B_0B. We require to calculate the torque T which must be applied at the shaft A_0 to produce the motion.

If the problem is to be analysed graphically we must first draw an acceleration diagram to a suitable scale S; this is shown in Fig. 8.18 and is discussed in Sec. 3.3.1. From this diagram the actual accelerations of the centres of mass are:

$$a_1 = (a_0c_1)S \text{ for } C_1$$
$$a_2 = (a_0c_2)S \text{ for } C_2$$
$$a_3 = (b_0c_3)S \text{ for } C_3$$

and the angular accelerations are:

$$\alpha_2 = (bb''S)/b \text{ for AB}$$
$$\alpha_3 = (b_0b'S)/c \text{ for } B_0B$$

The senses are determined from the diagram and are shown in Fig. 8.17. α_1 for link A_0A is zero since ω is constant.

Figure 8.19 shows the three links together with the inertia forces and torques shown dotted opposite to the directions of the accelerations. In the case of the coupler, and the follower, we can replace the inertia force and torque by a single force at a distance d from the centre of mass

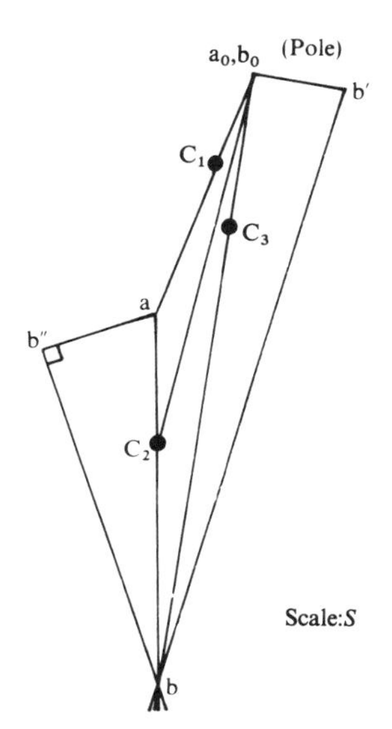

Figure 8.18 Acceleration diagram.

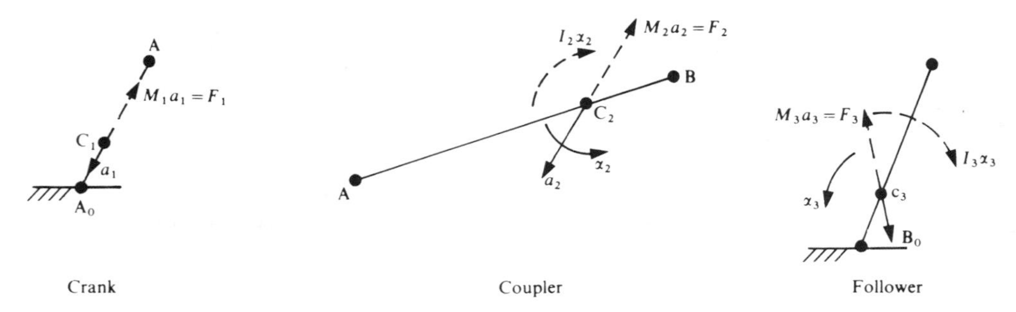

Figure 8.19 Inertia forces and inertia torques on each element of the linkage of Fig. 8.18.

given by Eq. (8.11). The result is shown in Fig. 8.20, where

$$d_2 = I_2\alpha_2/M_2a_2 \qquad \text{and} \qquad d_3 = I_3\alpha_3/M_3a_3$$

We are now in a position to draw a free-body diagram for each link as shown in Fig. 8.21. The inertia forces F_1, F_2, and F_3 are considered to be external forces which must be in equilibrium with the reactions at each joint. The directions of these reactions are at this stage unknown. Starting with joint B on link B_0B, the follower, let R_n and R_t be the normal and transverse reactions. By taking moments about B_0, R_t is obtained in magnitude and sense.

Next we consider link AB, the coupler, and take moments about A to determine R_n in magnitude and sense.

At this stage since we now know R_n and R_t we can draw a polygon of forces for the coupler to determine R_2 as shown in Fig. 8.22(*a*); in the diagram O is the pole, the reference point.

At A on link A_0A, the crank, the force is the vector sum of F_1 and R_2, i.e., $\mathbf{R}'_2 = \mathbf{F}_1 + \mathbf{R}_2$, and it follows that for equilibrium of forces $R_1 = R'_2$, i.e., R_1 must be equal in magnitude and direction but of opposite sense to R'_2 in Fig. 8.22(*b*).

Taking moments about A_0 we obtain the magnitude of the required torque T, i.e.,

$$T = R'_2 h$$

It only remains to calculate the reaction R_3 at B_0. This is obtained quite simply by drawing a polygon of forces for the follower since R_n and R_t are known or by means of a polygon of forces for the linkage as a whole as shown in Fig. 8.22(*c*).

Note: The diagrams of forces in Fig. 8.22 have not been drawn to any scale.

It should be pointed out that the vector sum of the inertia forces, i.e., $\mathbf{S} = \mathbf{F}_1 + \mathbf{F}_2 + \mathbf{F}_3$, commonly referred to as the shaking force, will result in vibrations of the linkage. Such vibrations may, in some instances, be unacceptable and attempts to balance the linkage will be

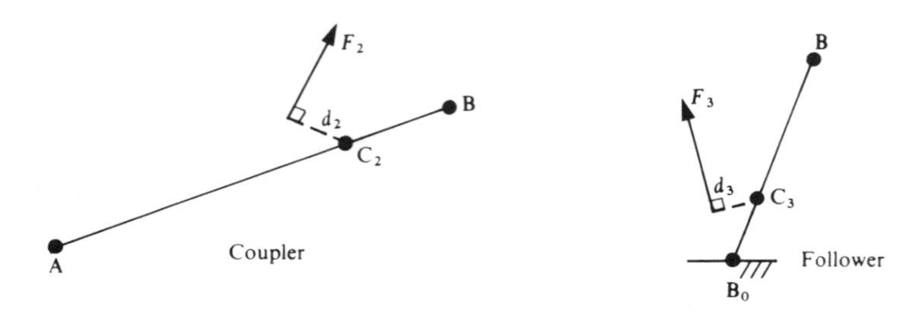

Figure 8.20 Replacing inertia force and torque by a single inertia force.

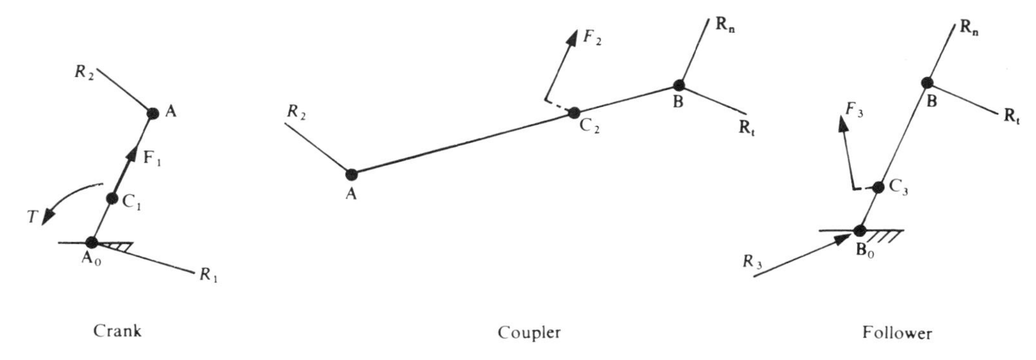

Figure 8.21 Free-body diagrams of the members of the four-bar linkage shown in Fig. 8.19 under the action of their respective inertia force.

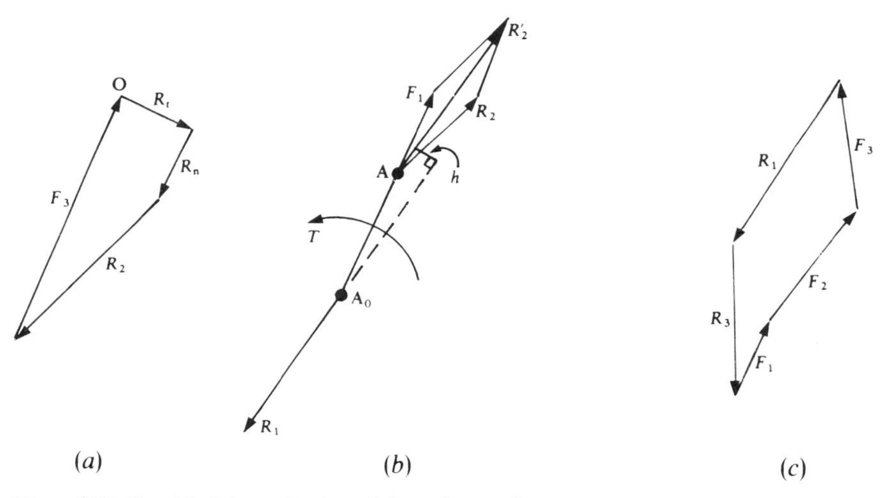

Figure 8.22 Graphical determination of the unknown forces.

necessary to avoid damage to the mechanism and to reduce the transmission of these vibrations to the surroundings.

Example 8.3 Referring to Example 8.2, calculate the torque T required to overcome inertia forces when the crank is rotating at 3000 rev/min and $\phi = 30°$, given the following:

Mass of the crank, $m_1 = 1.75$ kg.
Mass of the connecting rod, $m_2 = 3.25$ kg.
Mass of the slider or piston, $m_3 = 1.25$ kg.
Moment of inertia of the crank, $I_1 = 2.5 \times 10^{-3}$ kg m^2.
Moment of inertia of the connecting rod, $I_2 = 4.3 \times 10^{-2}$ kg m^2.

The location of the centres of mass are shown in Fig. 8.23.

SOLUTION The first step is to calculate the linear and angular accelerations of each member of the mechanism. For this we require the angular velocities of the crank and the connecting

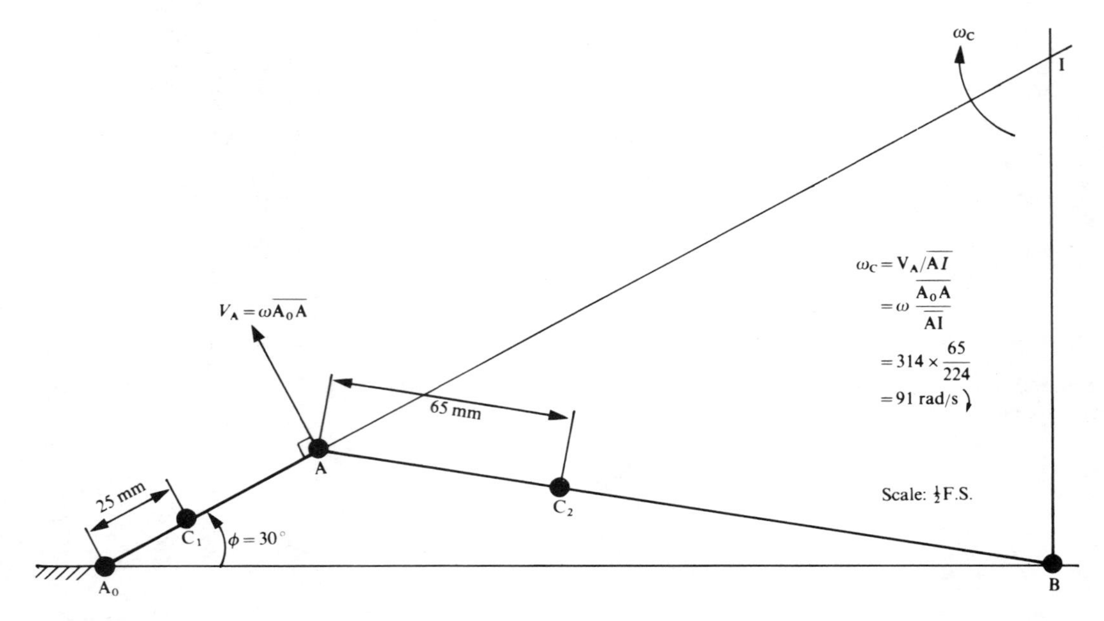

Figure 8.23

rod. The angular velocitity ω of the crank is given by

$$\omega = 2\pi \times 3000/60 = 314 \text{ rad/s anticlockwise}$$

The angular velocity ω_c of the connecting rod is obtained from the instant-centre diagram shown as Fig. 8.23 (see Sec. 2.2).

The next step is the calculation of the accelerations.

1. *Crank A_0A*:
   Centripetal acceleration, $a_A = \omega^2\overline{A_0A} = 314^2 \times 0.065 = 6408 \text{ m/s}^2 \downarrow$.
2. *Connecting rod AB*:
   Centripetal acceleration, $a_n = \omega_0^2\overline{AB_0} = 91^2 \times 0.2^2 = 1656 \text{ m/s}^2 \nwarrow$.

The acceleration diagram can now be drawn (see Sec. 8.3) as shown in Fig. 8.24.

The inertia forces are:

$$F_1 = m_1 a_1 = 1.75 \times 2400 = 4200 \text{ N}$$
$$F_2 = m_2 a_2 = 3.15 \times 6300 = 20\,475 \text{ N}$$
$$F_3 = m_3 a_3 = 1.25 \times 6690 = 8363 \text{ N}$$

The inertia torque on the connecting rod is given by:

$$T_2 = I_2 \alpha_2 = 4.3 \times 10^{-2} \times 14\,700 = 632 \text{ N m}$$

Hence for the connecting rod the force F_2 and the torque T_2 are equivalent to the single force F_2 at a distance d_2 given by

$$d_2 = I_2\alpha_2 / M_2\alpha_2 = 632/20\,475 = 0.031 \text{ m}$$

Next we draw the free body diagrams as shown in Fig. 8.25; there are two unknowns for the connecting rod, R_2 and R_3.

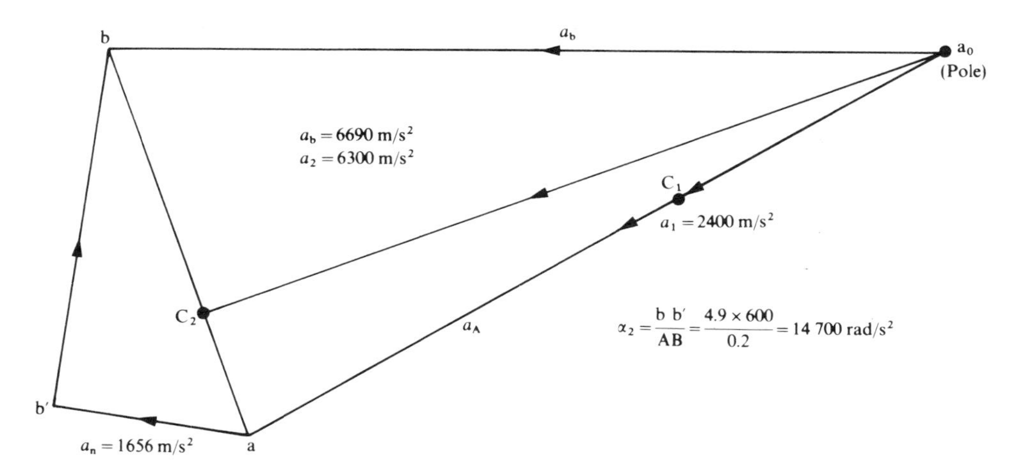

Figure 8.24

Taking moments about A we have

$$-F_2h + F_3e + R_3l = 0$$

Solving for R_3 yields, $R_3 = (20\,475 \times 0.008 - 8363 \times 0.036)/0.194 = -708$ N, i.e., $R_3 = 708$ N $\downarrow$
R_2 is obtained from the polygon of forces shown in Fig. 8.26. By scaling we find $R_2 = 29\,000$ N.

The free-body diagram for the crank is shown in Fig. 8.27. For equilibrium we have

$$T - R_1d_1 = 0$$

Hence $T = R_1d_1 = 32\,000 \times 0.018 = \underline{576 \text{ N m}}$

Although in the preceeding pages we have analysed the static and dynamic effects separately to illustrate the method, in a particular case we can combine the two effects in a single analysis, which will require one set of diagrams instead of two. A possible difficulty may arise if the inertia forces are very much greater than the static forces, in which case accuracy may be lost in drawing the force polygons.

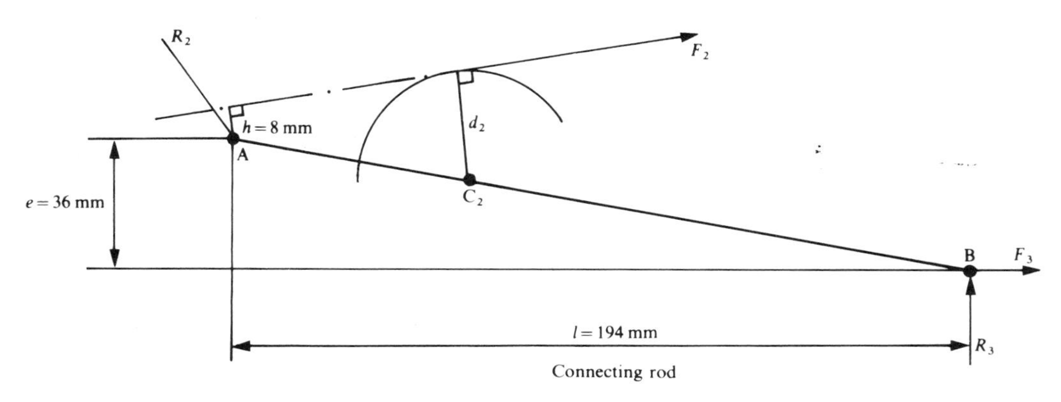

Figure 8.25

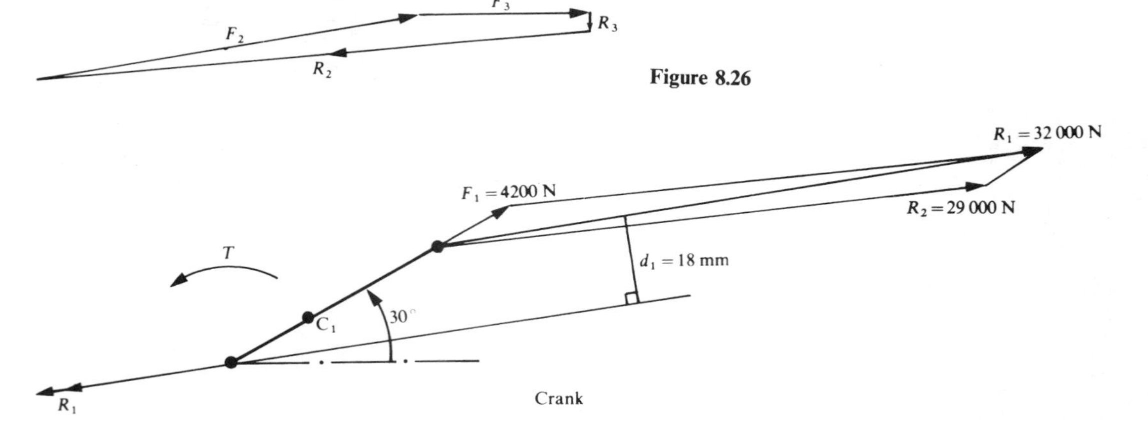

Figure 8.26

Figure 8.27

8.5 FORCES IN MECHANISMS USING THE METHOD OF VIRTUAL WORK

The last two sections were devoted to the use of the principle of equilibrium of forces using analytical or graphical methods to calculate forces and torques. We now consider a second method that has certain advantages over the above because it simplifies the analysis; this method is that of virtual work.

The principle of virtual work states that if a system is in equilibrium under the action of forces and torques, the total work done by these forces and torques is equal to zero for small displacements compatible with the constraints.

As a reminder, the work done δU by a force F during a displacement δr is defined to be (Fig. 8.28)

$$\delta U = F \cos \theta \delta r$$

Thus the work is the product of the component of the force along the displacement. It is in fact the scalar product of the force $\mathbf{F}$ and the displacement $\delta\mathbf{r}$, i.e.,

$$\delta U = \mathbf{F} \cdot \delta\mathbf{r}$$

From this definition it can be seen that work may be positive, negative or zero, depending upon the value of the angle θ.

Similarly, if a torque $\mathbf{T}$ is applied to a body, then for a small rotation $\delta\boldsymbol{\theta}$ of the body it can be shown that the work done $\delta U = \mathbf{T} \cdot \delta\boldsymbol{\theta}$. If a system such as the one shown in Fig. 8.29 is in equilibrium under the action of externally applied forces and torques, then for small displacements the total virtual work must be zero, i.e.,

$$\Sigma\mathbf{F} \cdot \delta\mathbf{r} + \Sigma\mathbf{T} \cdot \delta\boldsymbol{\theta} = 0 \qquad (8.14)$$

The small displacements referred to are also called *virtual* displacements since they do not actually occur in the real system; we simply imagine the system to be slightly displaced from its position of equilibrium. It must be remembered that since reactions at joints occur in pairs and are equal and opposite then when the system is taken as a whole these reactions will not appear in the equations.

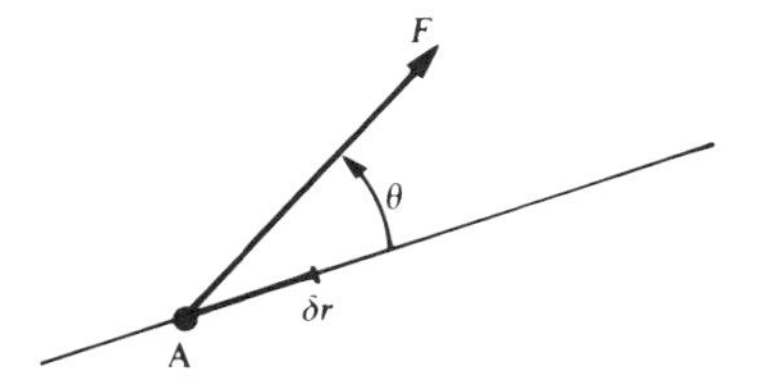

Figure 8.28 Action of a force along a given direction.

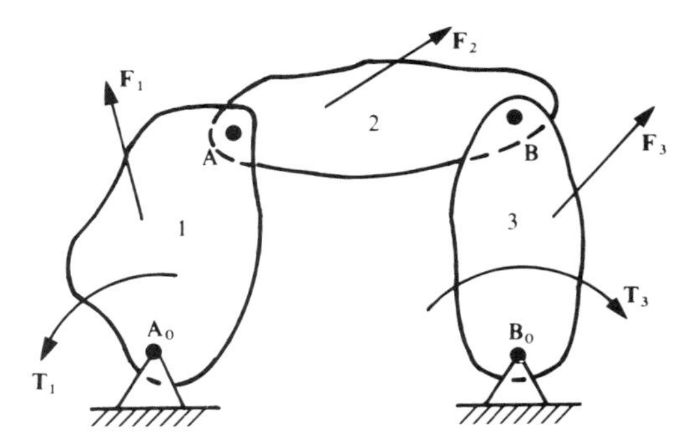

Figure 8.29 Typical linkage under the action of a number of forces and torques.

Furthermore, since such small displacements take place during the same interval of time we can divide Eq. (8.14) by the time δt, and as $\delta t \to 0$ we get

$$\Sigma \mathbf{F} \cdot \mathbf{v} + \Sigma \mathbf{T} \cdot \omega = 0 \tag{8.15}$$

Hence the virtual work is proportional to the velocities of the points of application of the forces.

Equation (8.15) can be modified to include all inertia forces and torques thus:

$$\Sigma \mathbf{F} \cdot \mathbf{v} + \Sigma \mathbf{F}_i \cdot \mathbf{v} + \Sigma \mathbf{T} \cdot \omega + \Sigma \mathbf{T}_i \cdot \omega = 0 \tag{8.16}$$

where $\mathbf{F}_i = -m\mathbf{a}$ (inertia force) and $\mathbf{T}_i = -I\boldsymbol{\alpha}$ (inertia torque). Note the sign!

Equation (8.16) may be solved graphically or analytically; the latter is on the whole preferable. We now propose to solve Example 8.3 by the method of virtual work.

Example 8.4 Referring to Example 8.3, calculate the torque T required to overcome inertia forces when the crank is rotating at 3000 rev/min and $\phi = 30°$, given the following:

Mass of the crank, $m_1 = 1.75$ kg.
Mass of the connecting rod, $m_2 = 3.25$ kg.
Mass of the slider or piston, $m_3 = 1.25$ kg.
Moment of inertia of the crank, $I_1 = 2.5 \times 10^{-3}$ kg m^2.
Moment of inertia of the connecting rod, $I_2 = 4.3 \times 10^{-2}$ kg m^2.

The locations of the centres of mass are shown in Fig. 8.30.

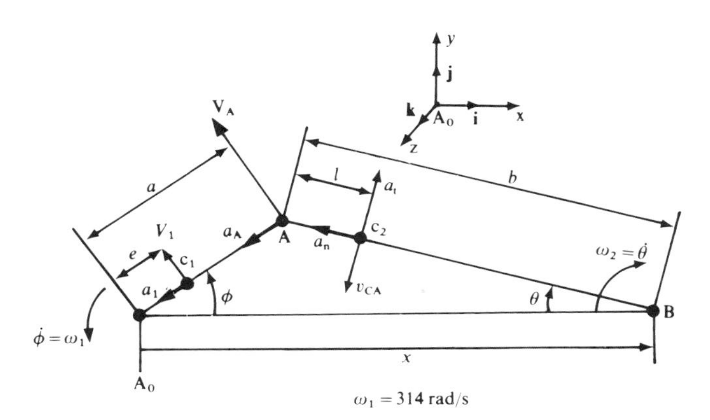

Figure 8.30

SOLUTION 1. Taking the frame of reference shown in Fig. 8.30 with the mechanism in the x–y plane, we have:

$$x = a\cos\phi + b\cos\theta \tag{8.17}$$

$$a\sin\phi = b\sin\theta \tag{8.18}$$

These are the only two equations required to solve the problem by the method of virtual work.

From Eq. (8.18) we get

$$\theta = \arcsin(a\sin\phi/b) = \arcsin(0.065\sin 30/0.2) = 9.35^\circ$$

Differentiating Eq. (8.18) yields

$$\dot\theta = \omega\, a\cos\phi/b\cos\theta = 314\times 0.065\cos 30/0.2\cos 9.35 = \underline{89.6\ \text{rad/s}}\ \circlearrowright$$

Differentiating once more, remembering that in this case $\ddot\phi = 0$, we get

$$\begin{aligned}\ddot\theta &= (b\sin\theta\dot\theta^2 - a\sin\phi\dot\phi^2)/b\cos\theta\\ &= (0.2\sin 9.35\times 89.6^2 - 0.065\sin 30\times 314^2)/0.2\cos 9.35\\ &= -14\,932\ \text{rad/s}^2 = \underline{14\,932\ \text{rad/s}^2}\ \circlearrowleft\end{aligned}$$

Differentiating Eq. (8.17) twice, we find

$$\dot x = -a\sin\phi\dot\phi - b\sin\theta\dot\theta = -0.065\sin 30\times 314 - 0.2\sin 9.35\times 89.6 = \underline{-13.12\ \text{m/s}}$$

and

$$\ddot x = -a\cos\phi\dot\phi^2 - b(\cos\theta\dot\theta^2 + \sin\theta\ddot\theta)\qquad \text{since } \ddot\phi = 0$$

Hence

$$\ddot x = -0.065\cos 30\times 314^2 - 0.2(\cos 9.35\times 89.6^2 + \sin 9.35\,(-14\,932)) = \underline{-6655\ \text{m/s}^2}$$

2. *Velocities*

Let $\mathbf{V}_1$ = velocity of C_1 on A_0A

$\mathbf{V}_A$ = velocity of A on A_0A

$\mathbf{V}_2$ = velocity of C_2 on AB

$\mathbf{V}_{2A}$ = velocity of C_2 relative to A

then $\mathbf{V}_2 = \mathbf{V}_A + \mathbf{V}_{2A}$

also $\boldsymbol{\omega}_1 = 314\mathbf{k}$, $\boldsymbol{\omega}_2 = -89.6\mathbf{k}$

$\mathbf{V}_1 = \omega e = 314\times 0.025 = 7.85$ m/s

Hence $\mathbf{V}_1 = -V_1\sin\phi\mathbf{i} + V_1\cos\phi\mathbf{j} = -7.85\sin 30\mathbf{i} + 7.85\cos 30\mathbf{j}$

$= \underline{-3.927\mathbf{i} + 6.8\mathbf{j}}$

$\mathbf{V}_A = -\omega a\sin\phi\mathbf{i} + \omega a\cos\phi\mathbf{j}$

$= \underline{-10.21\mathbf{i} + 17.68\mathbf{j}}$

$V_{CA} = \dot\theta l = 89.6\times 0.065 = 5.824\swarrow$

$\mathbf{V}_{2A} = -5.824\sin 9.35\mathbf{i} - 5.824\cos 9.3\mathbf{j} = \underline{-0.946\mathbf{i} - 5.747\mathbf{j}}$

Hence $\underline{\mathbf{V}_2 = -11.156\mathbf{i} + 11.933\mathbf{j}}$

Also $\underline{\mathbf{V}_3 = \dot x\mathbf{i} = -13.12\mathbf{i}}$

3. *Accelerations*

Let $\mathbf{a}_1$ = acceleration of C_1 on A_0A

$\mathbf{a}_A$ = acceleration of A

$\mathbf{a}_n$ = centripetal acceleration of C_2 on AB relative to A

$\mathbf{a}_t$ = tangential acceleration of C_2 on AB relative to A

$\mathbf{a}_2$ = acceleration C_2 on AB

then

$$\mathbf{a}_2 = \mathbf{a}_A + \mathbf{a}_n + \mathbf{a}_t$$

$$\mathbf{a}_1 = -e\omega^2 \cos \phi\mathbf{i} - e\omega^2 \sin \phi\mathbf{j}$$

$$= -0.025 \times 314^2 \cos 30\mathbf{i} - 0.025 \times 314^2 \sin 30\mathbf{j}$$

$$= \underline{-2137\mathbf{i} - 1233\mathbf{j}}$$

$$\mathbf{a}_A = -a\omega^2 \cos \phi\mathbf{i} - a\omega^2 \sin \phi\mathbf{j}$$

$$= \underline{-5556\mathbf{i} - 3208\mathbf{j}}$$

$$\mathbf{a}_n = -l\dot{\theta}^2 \cos \theta\mathbf{i} + l\dot{\theta}^2 \sin \theta\mathbf{j}$$

$$= -0.065 \times 89.6^2 \cos 9.35\mathbf{i} + 0.065 \times 89.6^2 \sin 9.35\mathbf{j}$$

$$= \underline{-515\mathbf{i} + 84.8\mathbf{j}}$$

$$\mathbf{a}_t = l\ddot{\theta} \cos \theta\mathbf{i} + l\ddot{\theta} \sin \theta\mathbf{j}$$

$$= 0.065 \times 14\,932 \sin 9.35\mathbf{i} + 0.065 \times 14\,932 \cos 9.35\mathbf{j}$$

$$= \underline{158\mathbf{i} + 958\mathbf{j}}$$

Hence

$$\underline{\mathbf{a}_2 = -5913\mathbf{i} - 2165\mathbf{j}}$$

Also

$$\underline{\mathbf{a}_3 = -6655\mathbf{i}}$$

and

$$\underline{\boldsymbol{\alpha}_2 = 14\,932\mathbf{k}}$$

Applying Eq. (8.16), we get

$$\mathbf{T}_1 \cdot \omega_1 + \mathbf{F}_{i1} \cdot \mathbf{V}_1 + \mathbf{F}_{i2} \cdot \mathbf{V}_2 + \mathbf{T}_{i2} \cdot \omega_2 + \mathbf{F}_{i3} \cdot \mathbf{V}_3 = 0 \tag{8.19}$$

Substituting numerical values yields the followng results:

(a) *For the crank* (Fig. 8.31)

$$\mathbf{T}_1 \cdot \omega_1 = T_1\mathbf{k} \cdot 314\mathbf{k} = 314T_1$$

$$\mathbf{F}_{i1} \cdot \mathbf{V}_1 = -m_1\mathbf{a}_1 \cdot \mathbf{V}_1$$

$$= 0 \text{ since } \mathbf{F}_{i1} \text{ and } \mathbf{V}_1 \text{ are } 90^\circ \text{ to each other}$$

Note: Substituting numerical values does not give zero exactly; this is due to the rounding-off of the intermediate values.

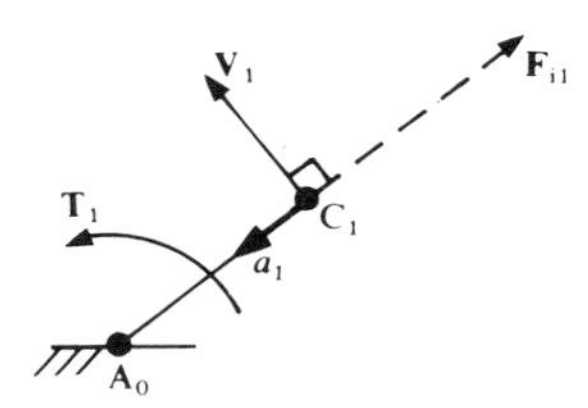

Figure 8.31

(b) *For the connection rod* (Fig. 8.32)

$$\begin{aligned}\mathbf{F}_{i2}\cdot\mathbf{V}_2 &= -m_2\mathbf{a}_2\cdot\mathbf{V}_2\\ &= -3.25(-5913\mathbf{i}-2165\mathbf{j})\cdot(-11.156\mathbf{i}+11.933\mathbf{j})\\ &= -130\,424\end{aligned}$$

$$\begin{aligned}\mathbf{T}_{i2}\cdot\omega_2 &= -I_2\alpha_2\cdot\omega_2\\ &= -4.3\times10^{-3}\times14\,932\mathbf{k}\cdot(-89.6\mathbf{k})\\ &= 57\,530\end{aligned}$$

(c) *For the slider* (Fig. 8.33)

$$\begin{aligned}\mathbf{F}_{i3}\cdot\mathbf{V}_3 &= -m_3\mathbf{a}\cdot\mathbf{V}_3\\ &= -1.25(-6655\mathbf{i})\cdot(-13.12\mathbf{i})\\ &= -10\,9142\end{aligned}$$

Substituting in Eq. (8.19) and solving for T_1 yields

$$T_1=(130\,424+109\,142-57\,530)/314=\underline{580\text{ N m}}$$

This shows that the value of 576 N m obtained by the graphical method is quite acceptable.

In a practical situation we would usually require the variation in the value of the torque needed to overcome the inertia loading during one complete cycle of operation of the mechanism, 360° in the above example. To obtain such a variation it would be advisable to program a microcomputer or pocket computer, which would be more efficient and accurate than the graphical method; furthermore, the latter would need many diagrams, e.g., $12\times5=60$ diagrams, taking 30° increments in the input, five constructions being needed for each position.

The principle of superposition can also be used to calculate the effect of static and dynamic loading, i.e., in the above example the total torque required on the input will be the algebraic sum of the static and the dynamic torque as illustrated in Example 8.5.

Example 8.5 Calculate the total torque required to overcome a static load of 2500 N on the slider and the inertia loading of Example 8.4 when $\phi=30°$.

SOLUTION 1. We know from Example 8.4 that the torque T_1 to overcome the inertia is equal to 580 N m anticlockwise.

2. To calculate the static torque T_s required to overcome the static load $P=2500$ N consider the mechanism in the position shown in Fig. 8.34. Let x be the position of the slider B when A_0A makes an angle ϕ with A_0B. For the mechanism to be in static

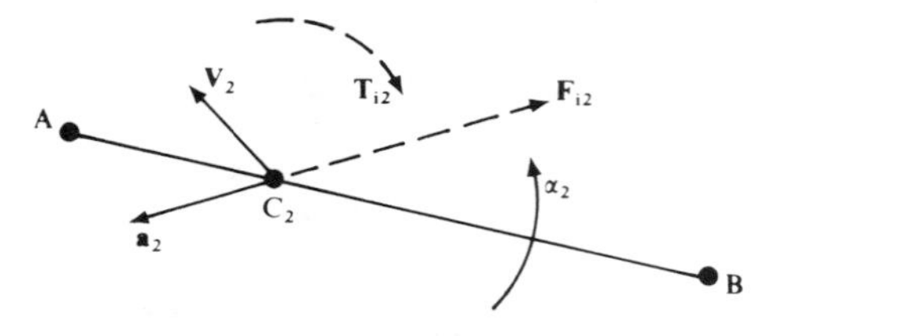

Figure 8.32

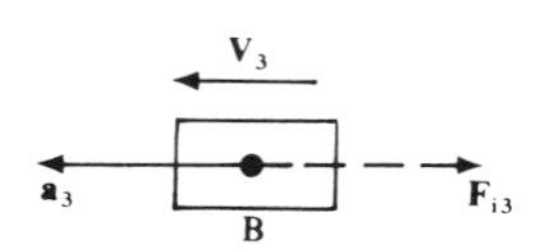

Figure 8.33

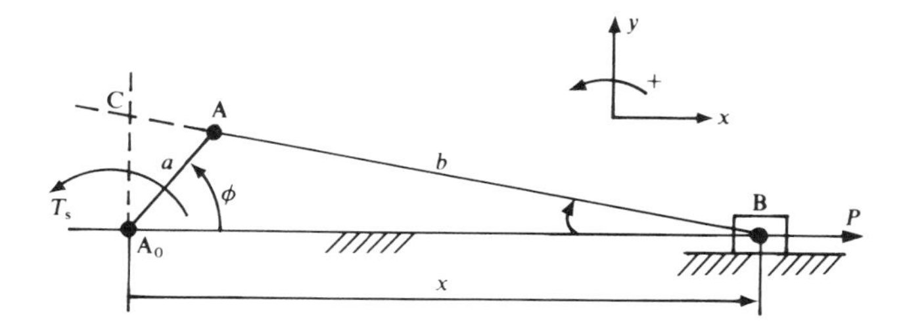

Figure 8.34

equilibrium the total virtual work done must be equal to zero, hence

$$T_s\,\delta\phi + P\,\delta x = 0$$

since

$$x = a\cos\phi + b\cos\theta \tag{8.17)*}$$

$$\delta x = -a\sin\phi\,\delta\phi - b\sin\theta\,\delta\theta \tag{8.20}$$

From Eq. (8.18) we have

$$\sin\theta = \frac{a}{b}\sin\phi$$

Hence

$$\cos\theta\,\delta\theta = \frac{a}{b}\cos\phi\,\delta\phi \tag{8.21}$$

Substituting Eq. (8.21) in (8.20) gives

$$\delta x = -a(\sin\phi + \tan\theta\cos\phi)\,\delta\phi = -a\,\delta\phi\,\sin(\phi+\theta)/\cos\theta$$

It follows that the required torque is given by

$$T_s = Pa\sin(\phi+\theta)/\cos\theta$$

Substituting numerical values yields

$$T_s = 2500 \times 0.065\sin(30+9.35)/\cos 9.35 = 104\ \text{N m}$$

The total torque T required at the input when $\phi = 30°$ is

$$T = 580 + 104 = \underline{684\ \text{N m}}$$

The reader should notice that in that position the ratio of the dynamic torque to the static torque is 5.58 to 1, a point made at the beginning of this section.

Note: From a graphical point of view the static torque can be calculated quite easily from a scale diagram; it is given by

$$T_s = P \times \overline{A_0C}$$

From the geometry it can be shown that

$$\overline{A_0C} = a\sin(\phi+\theta)/\cos\theta \qquad \text{since} \quad \angle A_0AC = (\phi+\theta) \text{ and } \angle A_0CA = (90-\theta)$$

8.6 KINETIC ENERGY AND EQUIVALENT INERTIA

If a link in a mechanism is in motion then we know from the dynamics of rigid bodies that its kinetic energy T_i is given by

$$T_i = \tfrac{1}{2}m_i v_i^2 + \tfrac{1}{2}I_i\omega_i^2$$

where m_i is its mass, I_i its moment of inertia, v_i the linear velocity of its centre of mass, and ω_i its angular velocity.

Hence for a mechanism consisting of N elements (links, slides) its total kinetic energy will be

$$T = \sum_i^N T_i = \sum \tfrac{1}{2}m_i v_i^2 + \sum \tfrac{1}{2}I_i\omega_i^2$$

If ω_1 is the angular velocity of the input to the mechanism then we can write

$$v_i = k_i\omega_1 \qquad \text{and} \qquad \omega_i = K_i\omega_i$$

Hence the kinetic energy can be expressed as follows:

$$T = \tfrac{1}{2}I_e\omega_1^2$$

where

$$I_e = \sum (m_i k_i^2 + I_i K_i^2) \tag{8.22}$$

I_e is referred to as the equivalent inertia, reflected inertia, or generalized inertia. It represents the inertia of the mechanism as seen by the input.

k_i and K_i depend on the configuration of the mechanism at a particular instant. It follows, therefore, that I_e will vary during the cycle of operation of the mechanism. For example, consider the four-bar linkage shown in Fig. 8.35; we saw in Sec. 5.4 that

$$\frac{\omega_3}{\omega_1} = \frac{\sin(\phi - \psi) - d \sin \phi / c}{\sin(\phi - \psi) + d \sin \psi / a}$$

Hence

$$K_3 = \frac{\sin(\phi - \psi) - d \sin \phi / c}{\sin(\phi - \psi) + d \sin \psi / a}$$

Alternatively,

$$K_3 = \frac{a \sin(\theta - \phi)}{c \sin(\psi - \theta)}$$

from Sec. 5.6.

The latter form may be preferable in view of the fact that we shall in most cases require the value of the angle θ when we consider the effecive inertia of the coupler AB.

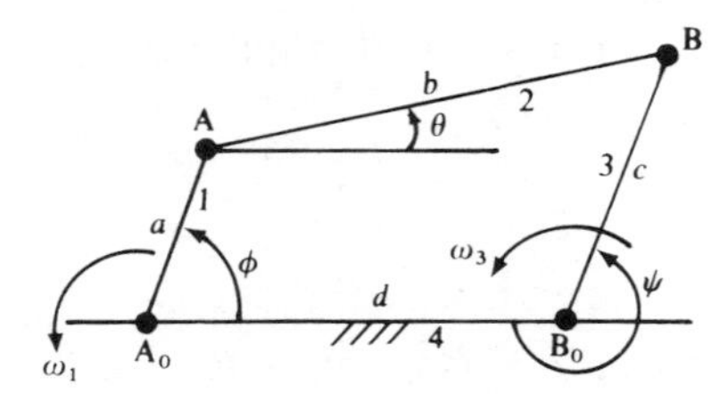

Figure 8.35

Thus for the coupler

$$K_2 = \frac{a \sin(\phi - \psi)}{b \sin(\psi - \theta)}$$

If C_1 is the torque required to overcome the inertia effects in a mechanism, then from dynamics we know that

$$\text{Power} = \text{rate of change of kinetic energy} \tag{8.23}$$

which can be expressed as follows for an assemblage of links:

$$C_1\omega_1 = \frac{d}{dt}\Sigma\tfrac{1}{2}mv^2 + \frac{d}{dt}\Sigma\tfrac{1}{2}I\omega^2 \tag{8.24}$$

It is independent of the accelerations.

Expressing the right-hand side of this equation in terms of the equivalent inertia I_e, we have

$$C_1\omega_1 = \frac{d}{dt}(\tfrac{1}{2}I_e\omega_1^2) \tag{8.25}$$

Integrating yields

$$\int_{\phi_1}^{\phi_2} C_1 \, d\phi = \Delta(\tfrac{1}{2}I_e\omega_1^2) \tag{8.26}$$

i.e.,

$$\text{Work done by the couple} = \text{change in kinetic energy}$$

To use Eq. (8.26), C_1 must be constant or a known function of the input angle ϕ. If C_1, the input couple, is constant (as in many practical cases) then the integral becomes:

$$C_1\,\Delta\phi = \Delta(\tfrac{1}{2}I_e\omega_1^2) = \tfrac{1}{2}(I_{en+1}\dot{\phi}_{n+1}^2 - I_{en}\dot{\phi}_n^2) \tag{8.27}$$

where $\Delta\phi = \phi_{n+1} - \phi_n$.

Equations (8.26) or (8.27) enable us to investigate the dynamic response of a mechanism, which we propose to illustrate by means of the following example.

Example 8.6 A constant torque of 3 N m is applied to the input shaft A_0 of the four-bar linkage shown in Fig. 8.36. Calculate the input speed variation over one cycle, assuming it started from rest at $\phi = 0$ and with zero load torque. Data: $A_0A = 50$ mm; $AB = 200$ mm; $BB_0 = 150$ mm; $A_0B_0 = 205$ mm; $m_1 = 0.125$ kg; $I_{AO} = 1 \times 10^{-4}$ kg m^2; $m_2 = 0.5$ kg; $I_2 = 1.7 \times 10^{-3}$ kg m^2; $m_3 = 0.375$ kg; $I_{BO} = 2.8 \times 10^{-3}$ kg m^2. The centres of mass of the links are half-way along the lengths.

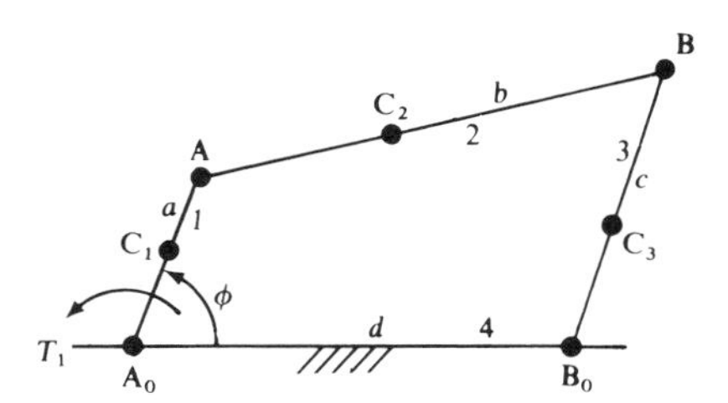

Figure 8.36

SOLUTION Figure 8.37 shows the mechanism in any position defined by the input angle ϕ. $\mathbf{V}_1$, $\mathbf{V}_2$, and $\mathbf{V}_3$ are the linear velocities of the centres of mass C_1, C_2, and C_3 respectively. ω_1, ω_2, and ω_3 are the angular velocities, and θ and ψ the coupler and follower angular positions respectively.

The kinetic energies of the links are as follows:

1. *For* A_0A, *link* 1:

$$T_1 = \tfrac{1}{2}m_1 v_1^2 + \tfrac{1}{2}I_1\omega_1^2 = \tfrac{1}{2}I_{A_0}\omega_1^2$$

where $\omega_1 = \dot{\phi}$ and I_{A_0} = moment of inertia about an axis through A_0

2. *For* AB, *link* 2:

$$T_2 = \tfrac{1}{2}m_2 v_2^2 + \tfrac{1}{2}I_2\omega_2^2$$

where I_2 = moment of inertia of AB about the centre of mass C_2.

3. *For* BB_0, *link* 3:

$$T_3 = \tfrac{1}{2}m_3 v_3^2 + \tfrac{1}{2}I_3\omega_3^2 = \tfrac{1}{2}I_{B_0}\omega_3^2 = \tfrac{1}{2}I_{B_0}K_3^2\omega_1^2$$

Let us consider link 2, the coupler. Taking an x–y frame of reference, the coordinates of the centre of mass are $x_2 = a\cos\phi + l\cos\theta$, where l is the distance of C from A

$$y_2 = a\sin\phi + l\sin\theta$$

Differentiating yields

$$\begin{aligned}\dot{x}_2 &= -a\sin\phi\dot{\phi} - l\sin\theta\dot{\theta}\\ &= -(a\sin\phi + lK_2\sin\theta)\omega_1\\ \dot{y}_2 &= a\cos\phi\dot{\phi} + l\cos\theta\dot{\theta}\\ &= (a\cos\phi + lK_2\cos\theta)\omega_1\end{aligned}$$

where $K_2 = a\sin(\phi-\psi)/b\sin(\psi-\theta)$;
also

$$V_2^2 = \dot{x}_2^2 + \dot{y}_2^2$$

Therefore the kinetic energy of link 2 becomes

$$T_2 = \tfrac{1}{2}\omega_1^2[m_2(a^2 + l^2K_2^2 + 2alK_2\cos(\phi-\theta) + I_2K_2^2]$$

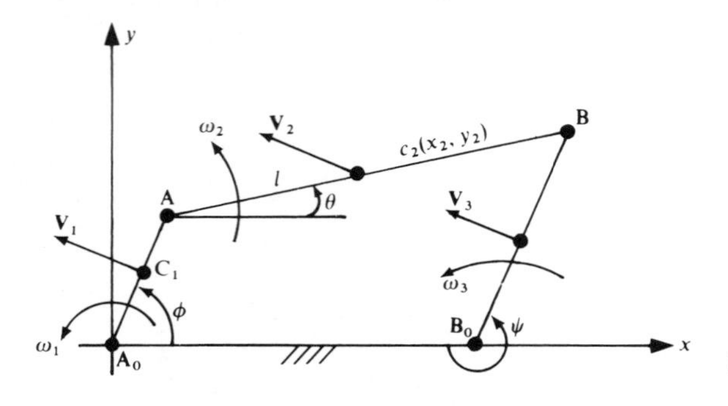

Figure 8.37

The total kinetic energy of the linkage is

$$T=T_1+T_2+T_3=\tfrac{1}{2}(I_{A_0}+m_2(\dot{x}_2^2+\dot{y}_2^2)+I_2K_2^2+I_{b_0}K_3^2)\omega_1^2=\tfrac{1}{2}I_e\omega_1^2$$

where I_e, the equivalent inertia as seen by the input, is given by

$$I_e=I_{A_0}+I_{B_0}K_3^2+I_2K_2^2+m_2(a^2+l^2K_2^2+2alK_2\cos(\phi-\theta))$$

By giving ϕ values from 0 to 360° and substituting the data in the above equation, the variation in I_e is obtained as shown in Fig. 8.38.

The variation in the input speed in the absence of a resisting torque on the output side is obtained by solving Eq. (8.27) for $\dot{\phi}_{n+1}$, giving

$$\dot{\phi}_{n+1}=\sqrt{\left(\frac{2C_1(\phi_{n+1}-\phi_n)+I_{en}\dot{\phi}_n^2}{I_{en+1}}\right)} \qquad n=1 \text{ to } N \text{ steps} \qquad (8.27)$$

where in this case $C_1=3$N m.

On starting with $\phi_1=0$ and using 15° steps for the input position, the fluctuation in the input speed $\dot{\phi}$ is shown in Fig. 8.39, illustrating clearly the effect of the variation of the inertia I_e as seen by the input prime mover. It therefore follows that the system is constantly accelerating and decelerating as the speed increases.

For comparison, the increase in the input speed shown by the straight line is obtained by assuming the inertia of the system is constant and equal to the mean value of I_e, i.e.,

$$I_{\text{average}}=1.165\times10^{-3}\text{ kg m}^2$$

Using the torque–angular-acceleration equation for a rigid body rotating about a fixed axis, we have:

$$C_1=I_{av}\alpha_1$$

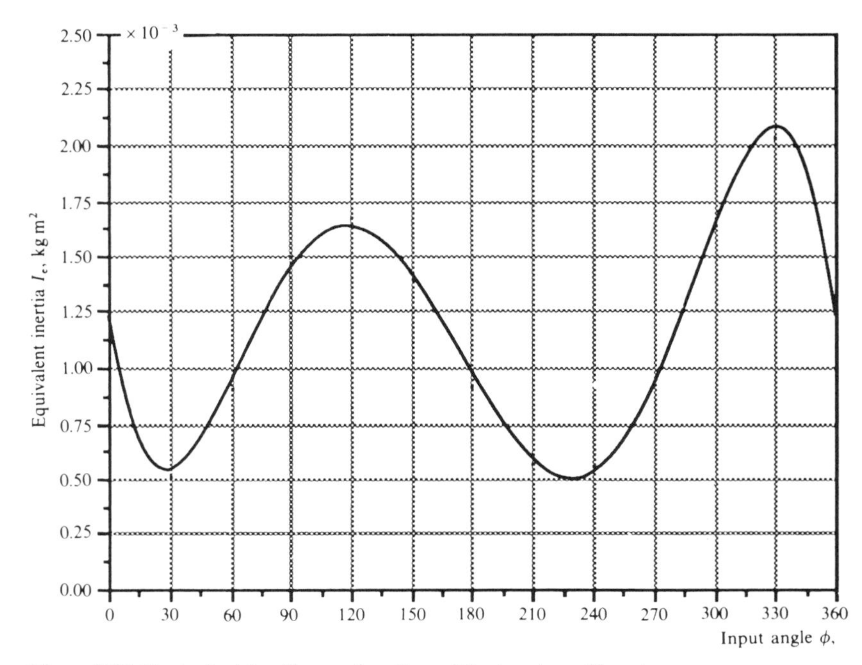

Figure 8.38 Equivalent inertia as a function of the input position ϕ.

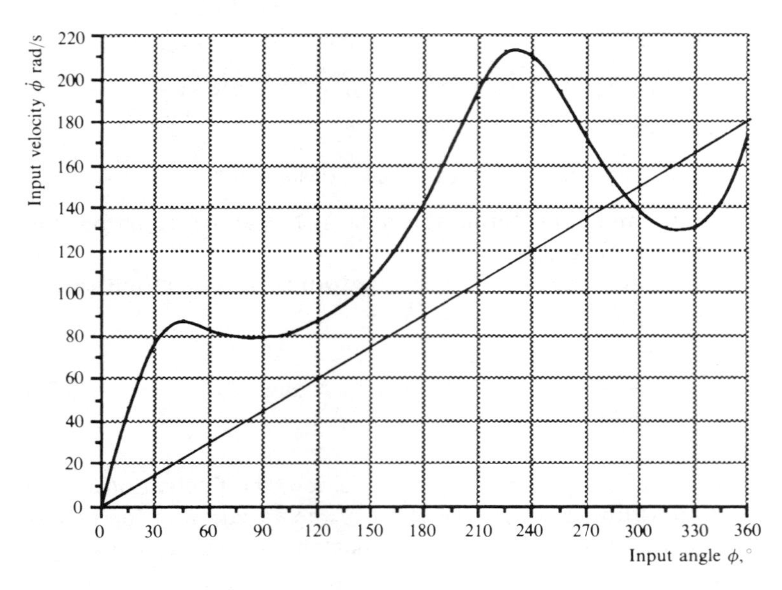

Figure 8.39 Variation in the input velocity as a function of the input position ϕ.

where α_1 is the angular acceleration of the input.

Hence
$$\alpha_1 = \frac{C_1}{I_{av}} = \frac{3}{1.165 \times 10^{-3}} = 2575 \text{ rad/s}^2$$

The speed of the input when $\phi = 360°$ is therefore given by:

$$\omega_1 = \sqrt{2\alpha_1 \phi} = \sqrt{2 \times 2575 \times 2\pi} = 180 \text{ rad/s}$$

Alternatively, from Eq. (8.27),

$$\omega_1 = \sqrt{2C_1 \Delta\phi / \mathrm{I}_{av}} = \sqrt{2 \times 3 \times 2\pi / 1165 \times 10^{-3}} = 180 \text{ rad/s}$$

In practical situations, in order to minimize undesirable speed fluctuations, a large flywheel could be installed on the output shaft of the prime mover.

Thus if I is added inertia the total inertia as seen by the prime mover will be equal to $I + I_e$. For example, with $I = 5 \times 10^{-3}$ kg m^2 and solving Eq. (8.27) we obtain the new input speed as shown in Fig. 8.40, indicating quite clearly that the fluctuations without added inertia have virtually disappeared, and that the acceleration of the system between 35 and 240° is the same as that with a total inertia $I + I_{eav}$.

Example 8.7 If in Example 8.6 the input angular velocity is kept constant at 425 rad/s, calculate the total torque required to overcome the inertia effects, the weights of the links, and a load torque of 10 N m, neglecting friction.

Solution 1. From Eq. (8.14) we have

$$C_1 \omega_1 = \frac{\mathrm{d}}{\mathrm{d}t}\left(\tfrac{1}{2} I_e \omega_1^2\right) = \omega_1 \dot{\omega}_1 I_e + \tfrac{1}{2}\omega_1^2 \frac{\mathrm{d}I_e}{\mathrm{d}t}$$

$$= \omega_1 \dot{\omega}_1 I_e + \tfrac{1}{2}\omega_1^2 \frac{\mathrm{d}I_e}{\mathrm{d}\phi}\frac{\mathrm{d}\phi}{\mathrm{d}t}$$

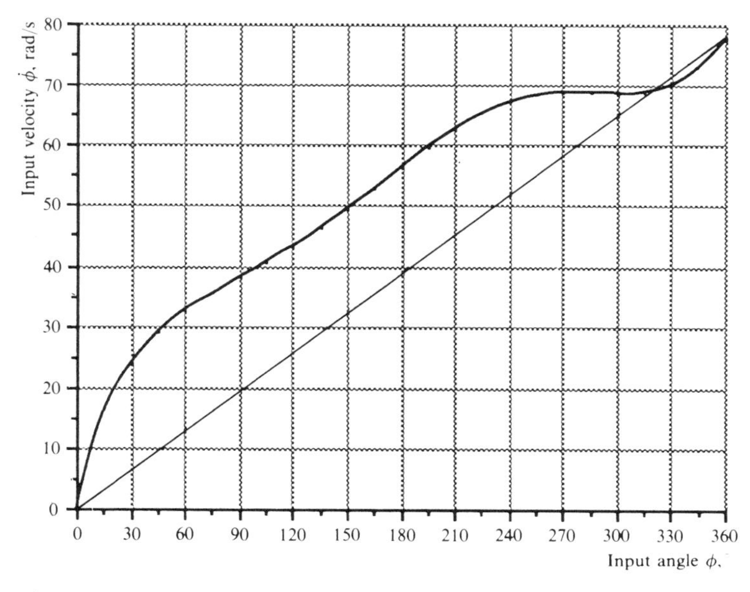

Figure 8.40 Effect of adding inertia on the input velocity.

therefore

$$C_1 = \dot{\omega}_1 I_e + \tfrac{1}{2}\omega_1^2 \frac{dI_e}{d\phi} \tag{8.28}$$

Since in this case ω_1 is constant, $\dot{\omega}_1 = 0$, hence the dynamic torque required is given by:

$$C_1 = \tfrac{1}{2}\omega_1^2 \frac{dI_e}{d\phi} \tag{8.29}$$

Upon performing the differentiation (which we leave as an exercise for the reader) and for ϕ from 0 to 360° and with $\omega_1 = (2\pi \times 425)/60 = 44.5$ rad/s, we obtain the variation in the value of the input torque (dynamic) required to overcome the inertia effects as shown in Fig. 8.41.

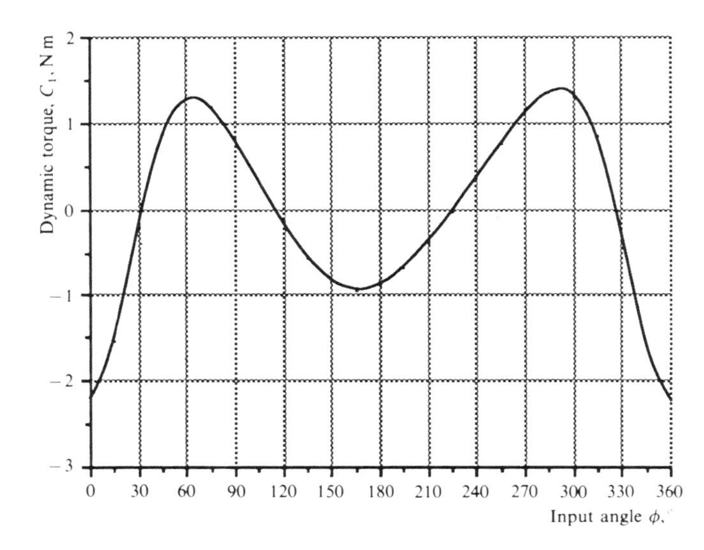

Figure 8.41 Dynamic torque.

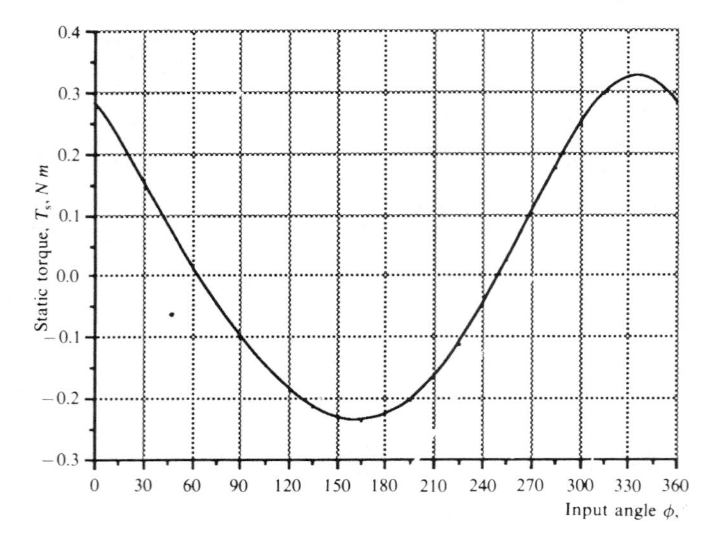

Figure 8.42 Static torque.

2. The torque (static) necesary to overcome the weights of the links is shown in Fig. 8.42. This is obtained by considering the equilibrium of the linkage as a function of the weights and the input crank angle position as explained in Sec. 8.3.

3. If T_1 is the torque required at the input to overcome the load torque T_o at the output, then neglecting friction and equating powers we have

$$T_1\omega_1 = T_o\omega_o$$

i.e.,

$$T_1 = \frac{\omega_o}{\omega_1} T_o = K_3 T_o$$

where

$$K_3 = \frac{a \sin\{\theta - \phi\}}{c \sin\{\psi - \theta\}} \quad \text{(see Sect. 8.6)}$$

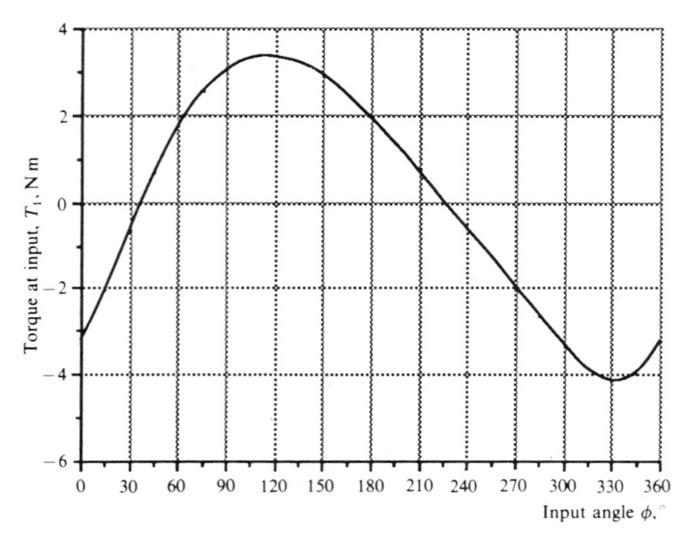

Figure 8.43 Torque to overcome load torque.

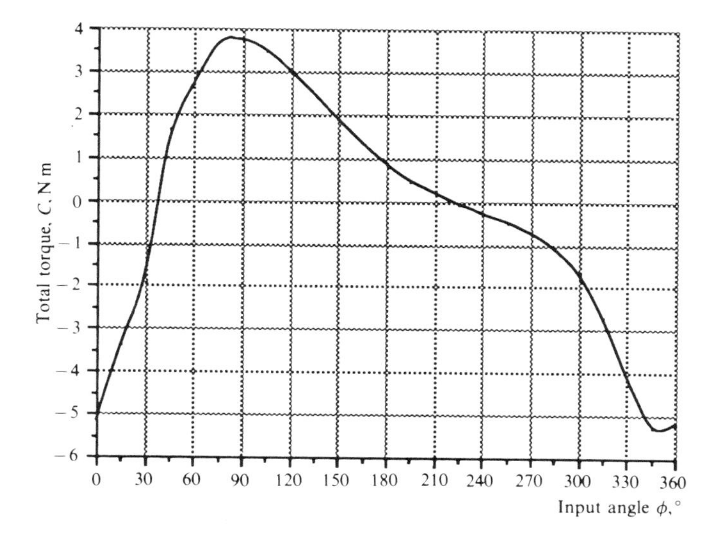

Figure 8.44 Input torque.

Figure 8.43 shows the result of the computation. It therefore follows that the total torque C required at the input is the algebraic sum of the three torques, namely

$$C = C_1 + T_s + T_1$$

The variation in C is shown in Fig. 8.44 for one complete revolution of the input crank.

8.7 FRICTION IN MECHANISMS

In mechanisms we are particularly concerned with the effect of friction at joints. In most cases the friction force is of the dry type, even if the surfaces are slightly lubricated; we can therefore write that the friction force F is given by

$$F = \mu R$$

where μ is the coefficient of kinetic friction and R is the reaction at a joint.

Associated with the coefficient of friction is the friction angle ϕ, defined by $\mu = \tan \phi$.

Consider a link L as shown in Fig. 8.45 connected to another link to which a pin is attached.

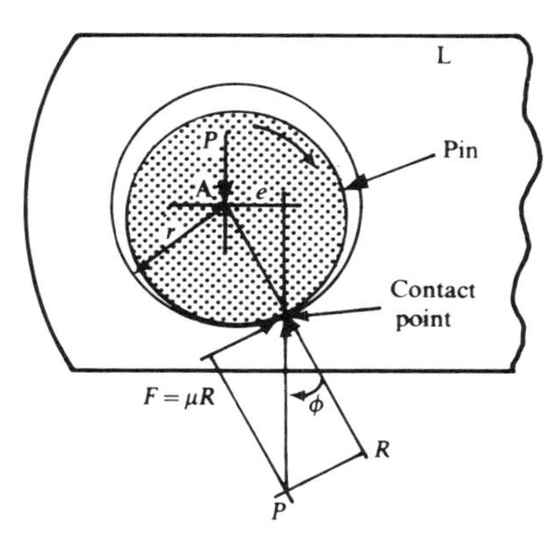

Figure 8.45 Forces on a pin rotating in the hole of a link due to friction.

If there is relative rotation between the links as indicated in the diagram, then the pin will ride up the bearing and assume the position shown. The clearance that must exist between the bearing and the pin is shown greatly exaggerated for clarity.

At the contact point there are two forces: a reaction R and a friction force $F=\mu R$ due to the load P, hence there is a friction torque as follows:

$$T_f = Fr = Pe$$

where $e = r \sin \phi$ is referred to as the *friction radius* and the circle drawn with that radius as the *friction circle.*

In many cases in practice both μ and r are small, therefore e will also be small; consequently the direction of the forces in a link will not change significantly as a result of friction at a joint. The direction of the forces in a link such as the coupler of a four-bar linkage will be tangential to the friction circles as shown in Fig. 8.46, i.e., they will be inclined to the centreline AB (and not through the pin centres) by an angle β given by

$$\beta = \arcsin\left(\frac{e_1 + e_2}{b}\right)$$

where $b = \overline{AB}$,

$$\text{or by } \beta = \arcsin\left(\frac{r_1 \sin \phi_1 + r_2 \sin \phi_2}{b}\right)$$

To appreciate the effect of friction consider the case of a typical link of length $b = 150$ mm, $r_1 = r_2 = 7.5$ mm, and $\mu = 0.2$. Substituting in the expression for β yields

$$\beta = \arcsin\left(\frac{2 \times 7.5 \sin(\arctan 0.2)}{150}\right) = 1.12°$$

As a second example consider the coupler of a four-bar linkage used in a mechanical digger where $b = 350$ mm, $r_1 = r_2 = 30$ mm, and $\mu = 0.5$, typically. The value of β is given by

$$\beta = \arcsin[2 \times 30 \sin(\arctan 0.5)/350] = 4.40°$$

Hence the inclination of the line of action of the forces to the centreline AB will not affect the force analysis by any significant amount. We therefore conclude that in most cases we can neglect friction effects except in special mechanisms such as those that rely on a toggle principle for their operation. Friction forces will also have to be included in mechanisms with slides if lubrication is sparse.

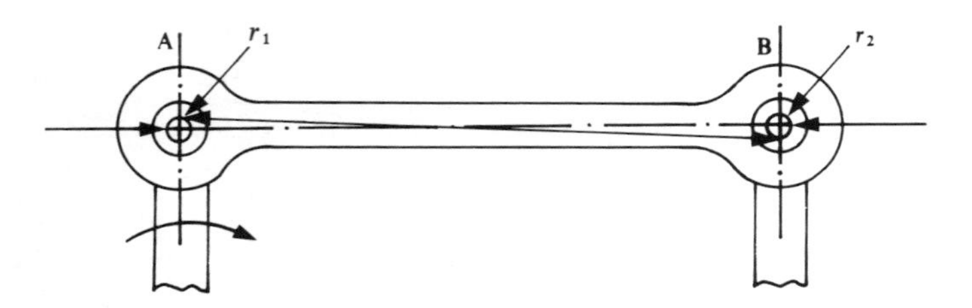

Figure 8.46 Line of action of the forces in a link due to friction.

EXERCISES

8.1 Figure 8.47 shows a dockside crane similar to that in Fig. 1.7. Neglecting the mass of each member, calculate the static torque T required to overcome a load W of 300 kN when $\phi = 60°$ and $135°$, corresponding to the initial and final positions. Dimensions: $a = 14.7$ m; $b = 6.5$ m; $c = 19.3$ m; $d = 6.4$ m; $h = 5.3$ m; $e = 22.3$ m; $f = 16$ m.

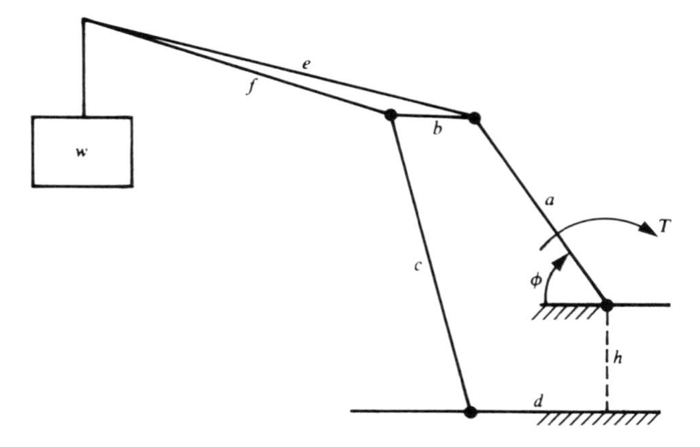

Figure 8.47

8.2 Figure 8.48 shows a four-bar linkage A_0ABB_0C used to operate a pump rod at C. With the data given below calculate the static torque required to overcome the vertically applied load P of 1500 N when $\phi = 30°$. Calculate also all forces and reactions. Dimensions:
Data $A_0A = 150$ mm; $AB = 610$ mm; $BB_0 = 300$ mm; $B_0 = 150$ mm. Coordinates of B_0 are (610 mm, 150 mm). Neglect the weights of the links.

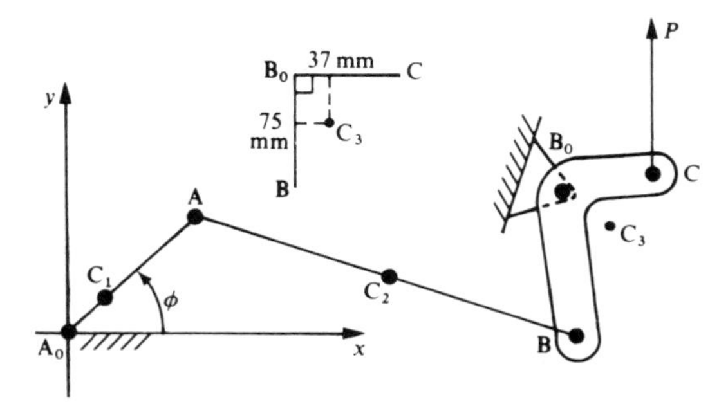

Figure 8.48

8.3 For the double slider–crank mechanism shown in Fig. 8.49 calculate the force F at A to maintain static equilibrium. Dimensions: $AB = 152$ mm; $C_0B = CD = 127$ mm; $BC = 42$ mm.

Calculate also the bending moment in BC_0.

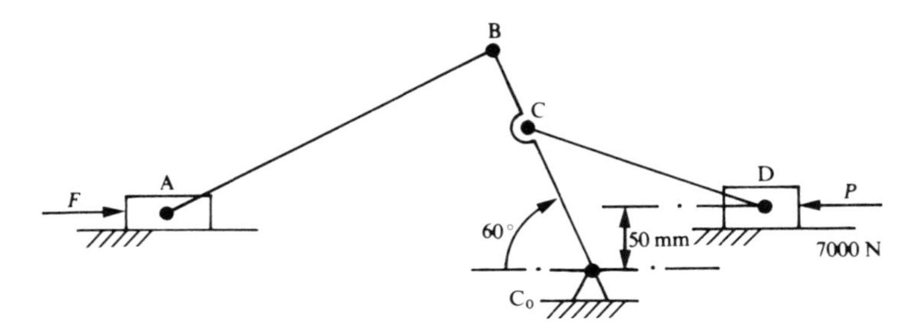

Figure 8.49

8.4 If in Exercise 8.2 the crank A_0A is rotating at 75 rad/s anticlockwise calculate the torque required at A_0 to overcome inertia effects using the following additional data:

Mass of $BB_0C = 20$ kg; C_3 is as shown.
Radius of gyration of BB_0C about $C_3 = 120$ mm.
Mass of $A_0A = 2$ kg, $A_0C_1 = 50$ mm.
Radius of gyration of A_0A about $C_1 = 40$ mm.
Mass of $AB = 8$ kg, $AC_2 = 310$ mm.
Radius of gyration of AB about $C_2 = 175$ mm.
C_1, C_2, and C_3 are the centres of mass.

8.5 Figure 8.50 shows a mechanism which produces an oscillating output with a dwell when the input shaft at A_0 rotates at a constant speed of 1440 rev/min. Dimensions are in millimetres. The links may be assumed to be slender rods of mass 3.25 kg/m, the flat, triangular coupler has a mass of 1.75 kg.

When $\phi = 90°$ calculate the torque T required to overcome inertia effects neglecting friction.

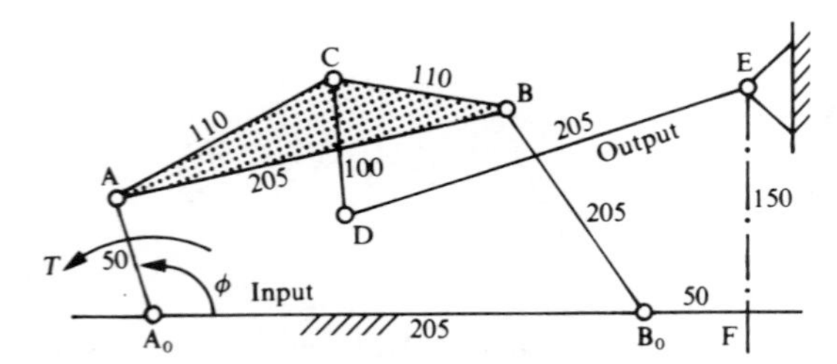

Figure 8.50

8.6 For the quick-return mechanism in the position shown (Fig. 8.51) calculate:
(*a*) the torque T required to overcome a load P of 200 N, neglecting friction at the joints except at the slide D where $\mu = 0.3$, and inertia effects;
(*b*) the torque required to overcome inertia effects given that $m_1 = 3.5$ kg, $I_1 = 0.012$ kg m^2; $m_2 = 8.75$ kg, $I_2 = 0.135$ kg m^2; $m_3 = 1.75$ kg, $I_3 = 1.4 \times 10^{-3}$ kg m^2.

The centres of mass in each case are half-way along the lengths of the links, and the mass of the slider at B has been allowed for.

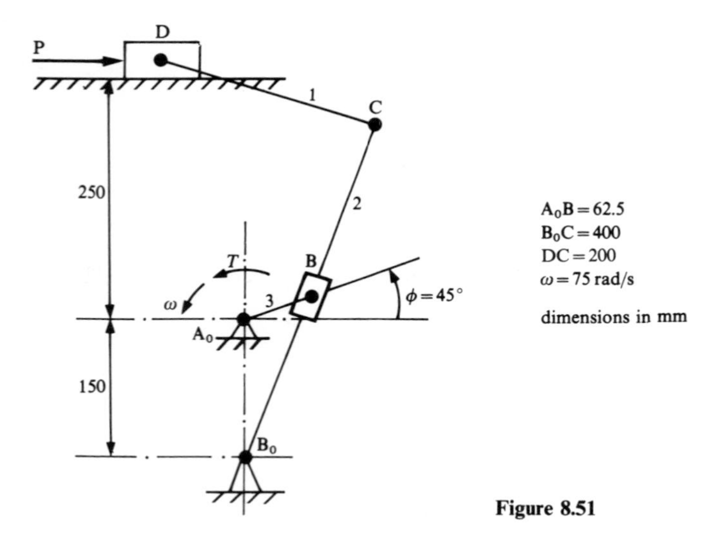

$A_0B = 62.5$
$B_0C = 400$
$DC = 200$
$\omega = 75$ rad/s

dimensions in mm

Figure 8.51

8.7 For the cam follower mechanism shown in Fig. 8.52 calculate the torque at A_0 to overcome inertia effects when the cam is rotating at 3000 rev/min and at the instant when $\phi = 60°$.

Calculate also the bearing force at A_0. The follower together with the rod and the spring have a combined mass 0.5 kg and the cam whose centre of mass is at C_1 may be assumed to be made from a 10-mm-thick flat steel plate. When the angle ϕ is 90° the roller is only just in contact with the cam. Friction may be neglected.

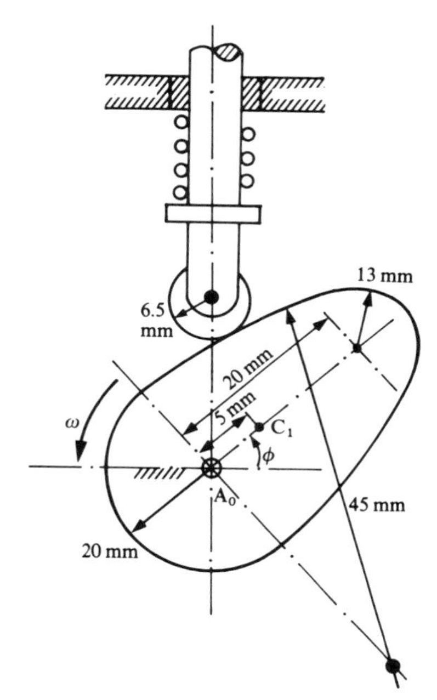

Figure 8.52

8.8 For the four-bar linkage shown in Fig. 8.53 plot the effective inertia and calculate the angular velocity ω_1 when $\phi = 60°$, starting from rest, if a constant torque of 7.5 N m is applied at A_0.

Plot the variation in ω_1 for $\phi = 0°$ to $\phi = 360°$ and hence calculate the acceleration of the linkage at $\phi = 60°$

Data: $m_1 = 1.25$ kg; $I_1 = 1.1 \times 10^{-3}$ kg m^2; $m_2 = 2.50$ kg, $I_2 = 55 \times 10^{-3}$ kg m^2, $AC_2 = 0.3$ m; $m_3 = 1.75$ kg, $I_3 = 25 \times 10^{-3}$ kg m^2.

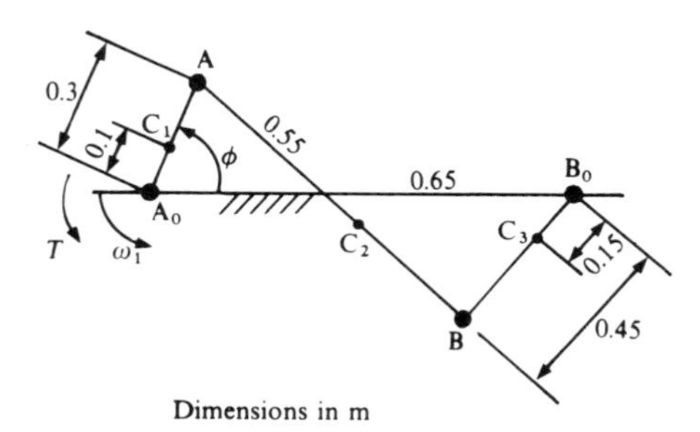

Figure 8.53

8.9 For the four-bar linkage shown in Fig. 8.54 calculate the torque required at A_0 to overcome the inertia loads when the crank A_0A rotates at 120 rad/s and makes an angle of 60° with the x-axis. Data: $I_{A_0} = 0.0056$ kg m^2, $I_2 = 0.25$ kg m^2, $I_{B0} = 0.175$ kg m^2; $m_1 = 0.75$ kg, $m_2 = 5$ kg, $m_3 = 3$ kg; $A_0A = 0.15$ m; $AB = 0.6$ m; $BB_0 = 0.45$ m; $AC_2 = 0.25$ m; $B_0C_3 = 0.2$ m.

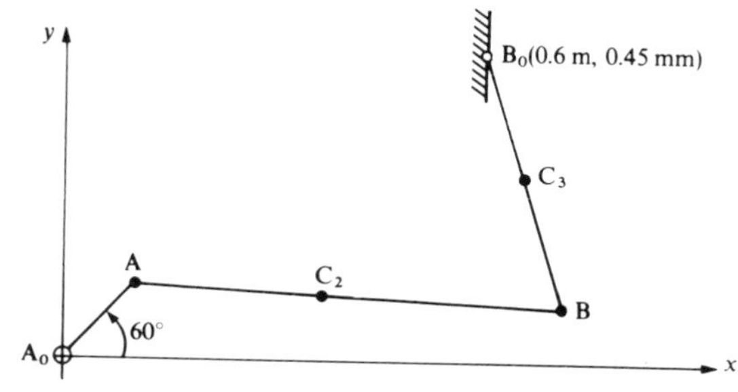

Figure 8.54

CHAPTER

NINE

DYNAMICS OF ROBOTIC MANIPULATORS

9.1 INTRODUCTION: LAGRANGE'S EQUATIONS

In studying the dynamic response of a system having several degrees of freedom, Lagrange's equations are generally employed, and are widely used in engineering and physics. In Lagrange's equations it is convenient to use a set of n independent coordinates, known as generalized coordinates $q_1, q_2, q_3, \ldots, q_n$, such that when q_i is given a small increment δ_{q_i} it will have no effect on the magnitude of the remaining $(n-1)$ coordinates. These coordinates correspond to the number of degrees of freedom of the system. For example, consider the simple linkage shown in Fig. 9.1. Its configuration at any instant is completely defined by the following *independent coordinates*:

$$\alpha \text{ and } \phi \qquad \text{or } \alpha \text{ and } \theta \qquad \text{or } \alpha \text{ and } \psi$$

OA_0B_0 is a rigid link. Thus any one of these pairs can be taken as generalized coordinates for this system.

Lagrange's equations are valid for holonomic systems. If $q_1, q_2, \ldots, q_n$ is the minimum number of independent coordinates, then the system is said to be holonomic. In a non-holonomic system the coordinates cannot all vary independently, i.e., the number of degrees of freedom is less than the minimum number of coordinates needed to specify the configuration of

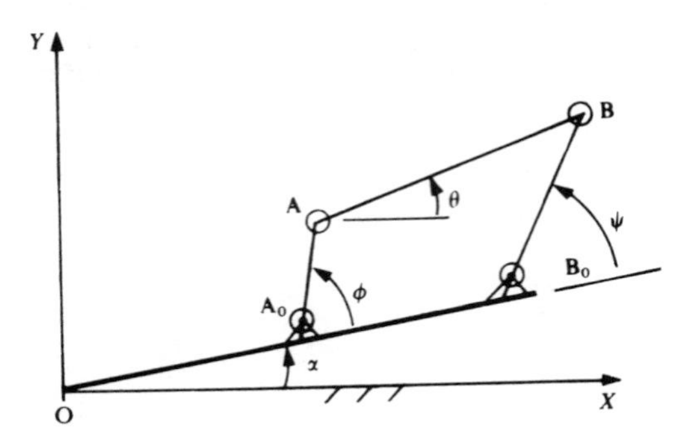

Figure 9.1 Diagram illustrating possible coordinates to define the positions of the members of a linkage.

the system. In these cases Lagrange's equations are modified by the introduction of a number of Lagrange multipliers. Mechanisms are in general holonomic.

Lagrange's equations are based on the kinetic and potential energies of the system as well as with the generalized forces associated with the generalized coordinates q_i.

Let T = total kinetic energy of the system
V = total potential energy of the system
Q_i = generalized force corresponding to the coordinate q_i

We give here without proof Lagrange's equations for a holonomic system; for a proof the reader may consult any text on analytical mechanics.

$$\frac{\mathrm{d}}{\mathrm{d}t}\left(\frac{\partial T}{\partial \dot{q}_i}\right)-\frac{\partial T}{\partial q_i}=Q_i \qquad i=1,2,\ldots,n \tag{9.1}$$

If the system is conservative, then $Q_i = -\partial V/\partial q_i$ and Eq. (9.1) becomes

$$\frac{\mathrm{d}}{\mathrm{d}t}\left(\frac{\partial T}{\partial \dot{q}_i}\right)-\frac{\partial T}{\partial q_i}+\frac{\partial V}{\partial q_i}=0 \tag{9.2}$$

This equation applies to a particle, a system of particles, a rigid body or a system of rigid bodies.

The following simple example illustrates the application of these equations.

Example 9.1 Figure 9.2 shows a device for testing delicate instruments. The instruments are placed in a cage C which is made to rotate at a speed Ω about the vertical axis OZ by means of the variable-speed motor M_1. At the same time the frame F rotates at a speed ω about the axis OX by means of a variable-speed geared motor M_2.

Using the data given below, calculate the maximum value of the torque M to be applied by the motor M_2. The frame F is light compared with the cage and its contents.

Data: Moments of inertia of the cage and its contents about the principal axes 1, 2, and 3 as shown:

$$I_1 = 12\ \mathrm{kg\,m^2} \qquad I_2 = I_3 = 14.5\ \mathrm{kg\,m^2}$$

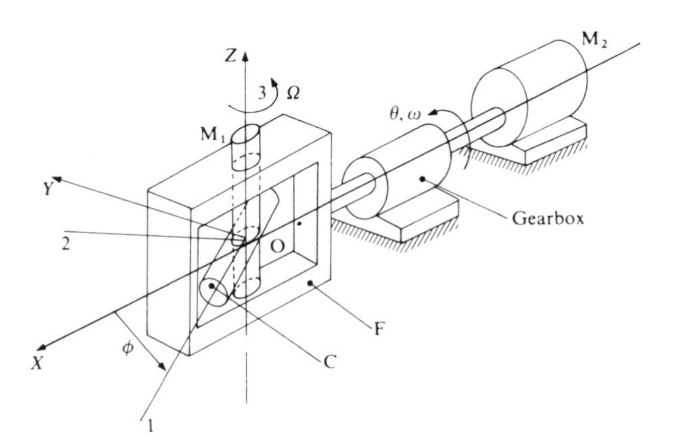

Figure 9.2

Motor speeds at a particular instant:
Motor $M_2 = 720$ rev/min Gear ratio $= 6:1$ Motor $M_1: 1440$ rev/min
SOLUTION (Fig. 9.3) The generalized coordinates, i.e., independent coordinates, are θ for the motion of the frame and ϕ for the position of the cage.

For the cage, the kinetic energy is

$$T = \tfrac{1}{2}(I_1\omega_1^2 + I_2\omega_2^2 + I_3\omega_3^2)$$

where $\omega_1 = \omega \cos \phi$

$\omega_2 = -\omega \sin \phi$

$\omega_3 = \dot{\phi} = \Omega.$

Hence

$$T = \tfrac{1}{2}(I\omega^2 \cos^2 \phi + I_2\omega^2 \sin^2 \phi + I_3\Omega^2)$$

The potential energy, $V = 0$ in this case.

Generalized moment: $Q_i = M$, the moment to be applied by M_2.

Substituting in Eq. (9.1) yields

$$\frac{\partial T}{\partial \theta} = 0 \qquad \frac{\partial T}{\partial \dot{\theta}} = \frac{\partial T}{\partial \omega} = I_1\omega \cos^2 \phi + I_2\omega \sin^2 \phi$$

and

$$\frac{\mathrm{d}}{\mathrm{dt}}\left(\frac{\partial T}{\partial \dot{\theta}}\right) = 2(-I_1 \cos \phi \sin \phi + I_2 \sin \phi \cos \phi)\omega\Omega$$

Hence $(I_2 - I_1)\omega\Omega \sin 2\phi = M$

The maximum value occurs when $\phi = 45°$, giving

$$M_{\max} = (I_2 - I_1)\omega\Omega$$

$$\omega = 2\pi \times 720/60/6 = 12.57 \text{ rad/s}$$

$$\Omega = 2\pi \times 1440/60 = 150.8 \text{ rad/s}$$

Hence

$$M_{\max} = (14.5 - 12) \times 12.57 \times 150.8$$

$$= \underline{4737 \text{ N m}}$$

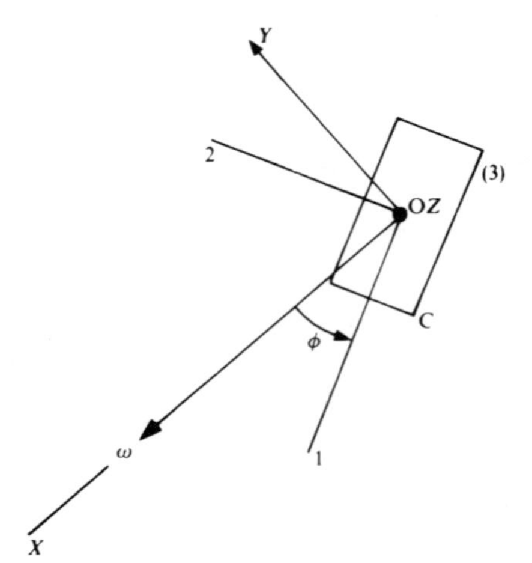

Figure 9.3

Note:
$$\text{Power required} = M_{\max}\omega = 4737 \times 12.57 = \underline{59.5\ \text{kW}}$$

9.2 GIBBS–APPELL'S EQUATION

A method of analysis which applies to holonomic and non-holonomic systems was discovered by Gibbs in 1879 and later studied by Appell in 1899. It is based on the so-called 'energy of acceleration', which we propose to prove for a system of particles.

For a single particle of mass dm whose motion is referred to axes x, y, and z fixed in space and subjected to forces X, Y, and Z, then during a virtual displacement ∂x, ∂y, and ∂z we have

$$(\ddot{x}\,\partial x + \ddot{y}\,\partial y + \ddot{z}\,\partial z)\,\mathrm{d}m = X\,\partial x + Y\,\partial y + Z\,\partial z$$

and for the whole system we have

$$\Sigma(\ddot{x}\,\partial x + \ddot{y}\,\partial y + \ddot{z}\,\partial z)\,\mathrm{d}m = \Sigma(X\,\partial x + Y\,\partial y + Z\,\partial z)$$

If the system can be described by means of independent generalized coordinates $q_1, q_2, \ldots, q_i, \ldots, q_n$ so that $x = x(q_i)$, $y = y(q_i)$, and $z = z\,(q_i)$, then during a virtual displacement ∂q_i we have

$$\partial x = \frac{\partial x}{\partial q_i}\,\partial q_i \qquad \partial y = \frac{\partial y}{\partial q_i}\,\partial q_i \qquad \partial z = \frac{\partial z}{\partial q_i}\,\partial q_i$$

Substituting in the equation above yields

$$\Sigma\left(\ddot{x}\frac{\partial x}{\partial q_i} + \ddot{y}\frac{\partial y}{\partial q_i} + \ddot{z}\frac{\partial z}{\partial q_i}\right)\partial q_i\,\mathrm{d}m = \Sigma\left(X\frac{\partial x}{\partial q_i} + Y\frac{\partial y}{\partial q_i} + Z\frac{\partial z}{\partial q_i}\right)\partial q_i$$

The bracketed term on the right-hand side is the generalized force, or moment, Q_i.

It can be shown that $\dfrac{\partial x}{\partial q_i} = \dfrac{\partial \dot{x}}{\partial \dot{q}_i} = \dfrac{\partial \ddot{x}}{\partial \ddot{q}_i}$

Noting that
$$\ddot{x}\frac{\partial \ddot{x}}{\partial \ddot{q}_i} = \frac{\partial}{\partial \ddot{q}_i}(\tfrac{1}{2}\ddot{x}^2) \quad \text{etc.} \ldots$$

we get
$$\frac{\partial}{\partial \ddot{q}_i}\,\Sigma[\tfrac{1}{2}\,\mathrm{d}m(\ddot{x}^2 + \ddot{y}^2 + \ddot{z}^2)] = Q_i$$

This is the Gibbs–Appell equation, which we may write thus:

$$\frac{\partial G}{\partial \ddot{q}_i} = Q_i \tag{9.3}$$

where G is the 'energy of acceleration' of the system.

If some of the forces (or moments) are derived from a potential V,

then
$$\frac{\partial V}{\partial q_i} = -Q'_i$$

and the Gibbs–Appell equation becomes

$$\frac{\partial G}{\partial \ddot{q}_i} + \frac{\partial V}{\partial q_i} = Q_i \tag{9.4}$$

where the generalized force (or moment) Q_i does not contain potential energy terms. When Gibbs–Appell equations are applied the accelerations will in general take place in different directions, thus if a_x, a_y, and a_z are the accelerations of a particle in mutually perpendicular directions x, y, and z, then

$$a^2 = a_x^2 + a_y^3 + a_z^2$$

and for a whole body
$$\frac{\partial G}{\partial \ddot{q}_i} = \int \left(a_x \frac{\partial a_x}{\partial \ddot{q}_i} + a_y \frac{\partial a_y}{\partial \ddot{q}_i} + a_z \frac{\partial a_z}{\partial \ddot{q}_i} \right) \mathrm{d}m \tag{9.5}$$

It may be convenient to refer accelerations to a set of moving axes. Thus consider a set of fixed axes Oxz and a set of moving axes $Ox'z'$ pitching at an angle ϕ, e.g., a robotic arm. If a_1 and a_3 are the accelerations in the x'- and z'-directions respectively, then

$$\ddot{x} = a_1 \cos\phi - a_3 \sin\phi \qquad \text{and} \qquad \ddot{z} = a_1 \sin\phi + a_3 \cos\phi$$

giving
$$\ddot{x}^2 + \ddot{z}^2 = a_1^2 + a_3^2$$

In the case of a revolute arm (Fig. 9.7), for example, one set of moving axes may be chosen for the inner arm and a different set for the outer arm. If the acceleration in a particular direction, y say, does not contain terms in $\ddot{q}_i$, then a_y will not enter into the analysis. As with Lagrange's equations, Eqs (9.3) and (9.4) apply for a system of particles, a rigid body, or a system of rigid bodies.

The example below illustrates the application of Eq. 9.3.

Example 9.2 Figure 9.4 shows a mass M suspended to another mass m by means of a light rod of length l. Springs of total stiffness K restrain the mass m. Such a system could represent the oscillations of a load suspended from the jib of a crane, the mass m representing the travelling carriage and the spring the stiffness of the cable which moves it along the jib. It could equally well represent a machine on a composite mat of horizontal stiffness K, which is subjected to vibrations due to inherent imbalance, the suspended mass being a dynamic vibration absorber. Derive the equations of motion.

Solution (Fig. 9.5) The generalized coordinates are x and θ, where x is the displacement of the mass m and θ that of the rod. The accelerations are $\ddot{x}$, $l\ddot{\theta}$, and $l\dot{\theta}^2$ for the mass M,

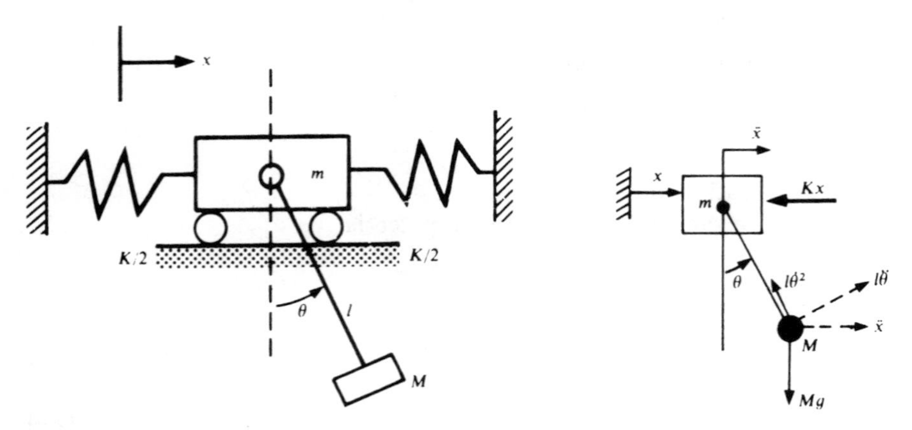

Figure 9.4 **Figure 9.5**

and $\ddot{x}$ for the mass m as shown. Hence the accelerations of the mass M perpendicular to and along the rod are $(l\ddot{\theta}+\ddot{x}\cos\theta)$ and $(l\dot{\theta}-\ddot{x}\sin\theta)$ respectively. Therefore the 'energy of acceleration' of the system is:

$$G=\tfrac{1}{2}M(l\ddot{\theta}+\ddot{x}\cos\theta)^2+\tfrac{1}{2}M(l\dot{\theta}^2-\ddot{x}\sin\theta)^2+\tfrac{1}{2}m\ddot{x}^2$$

Differentiating with respect to $\ddot{\theta}$ and $\ddot{x}$ we get

$$\frac{\partial G}{\partial\ddot{\theta}}=M(l\ddot{\theta}+\ddot{x}\cos\theta)l$$

also $$Q_\theta=-Mgl\sin\theta$$

$$\frac{\partial G}{\partial\ddot{x}}=M(l\ddot{\theta}+\ddot{x}\cos\theta)\cos\theta+M(l\ddot{\theta}^2-\ddot{x}\sin\theta)(-\sin\theta)+m\ddot{x}$$

and $$Q_x=-Kx$$

(In the case of an imbalance producing a periodic force $P\cos t$,

$$Q_x=-Kx+P\cos\omega t$$

Substituting in Eq. (9.3) yields the two equations of motion:

$$(m+M)\ddot{x}+M(l\ddot{\theta}\cos\theta-l\dot{\theta}^2\sin\theta)+Kx=0$$

and $$l\ddot{\theta}+\ddot{x}\cos\theta+g\sin\theta=0$$

These equations apply for large angles but if the oscillations are small we get

$$(m+M)\ddot{x}+Ml\ddot{\theta}+Kx=0 \qquad (\text{or}=P\cos\omega t \text{ due to forced oscillations})$$

and $$\ddot{x}+l\ddot{\theta}+g\theta=0$$

Solutions to these two equations will yield the two natural frequencies of the system.

9.3 THE DYNAMICS OF A ROBOTIC MANIPULATOR

In this introduction to the dynamics of a robotic manipulator we consider two particular types of robots that are widely used in industry. They are:

(a) the polar-coordinate type shown in Fig. 9.6, also referred to as the extending arm; and
(b) the hinged-arm or revolute coordinate type shown in Fig. 9.7.

In the next sections we propose to obtain equations for the torques required to drive the manipulator, leaving out the stiffness and damping in the system. It is important to know the value of the torque requirement for given motions in order to select suitable prime movers, whether hydraulic, pneumatic, or electric.

9.3.1 Analysis of Type (a): Polar Arm

This type of robot is usually operated hydraulically, e.g., the Unimate robot; the arm extends radially, rotates in its plane about a horizontal axis through O, referred to as 'shoulder' freedom, and rotates about the vertical OY axis. At the end of the arm a gripper carries an object to be transferred to some defined position, or accurately positioned as in the case of an automatic

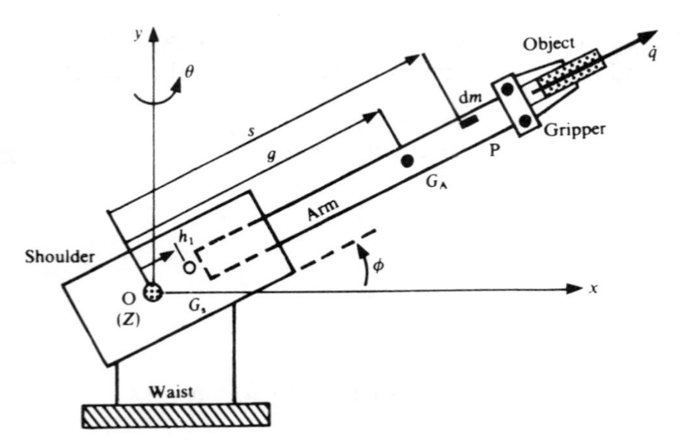

Figure 9.6 Diagram of the polar type of a robotic manipulator.

assembly process. Small robots can position objects with an accuracy of about 0.1 mm while larger ones have an error of around 1 mm.

In some robots the gripper can also rotate about an axis parallel to that of the arm, thereby adding another degree of freedom.

Referring to Fig. 9.6, let

ϕ be the elevation of the arm;
s be the coordinate of a particle of mass dm at P on the arm;
q be the coordinate of the centre of mass G_A of the arm;
θ be the coordinate for the rotation about OY.

The coordinate of P is $x = s \cos\phi$, hence $\dot{x} = \dot{s}\cos\phi - s\dot{\phi}\sin\phi$.

(a) for the arm Accelerations of P along the axes shown in Fig. 9.8 are:

along 1, $\quad a_1 = \ddot{s} - s\dot{\phi}^2 - x\dot{\theta}^2\cos\phi$

along 2, $\quad a_2 = x\ddot{\theta} + 2\dot{x}\dot{\theta}$

along 3, $\quad a_3 = s\ddot{\phi} + 2\dot{s}\dot{\phi} + x\dot{\theta}^2\sin\phi$

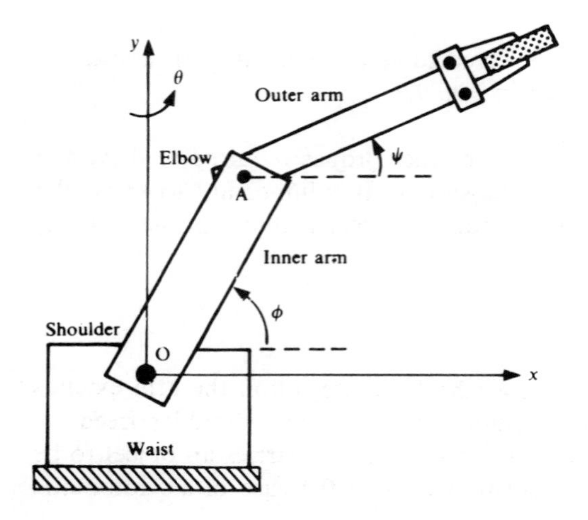

Figure 9.7 Diagram of a revolute or hinged type of a robotic manipulator.

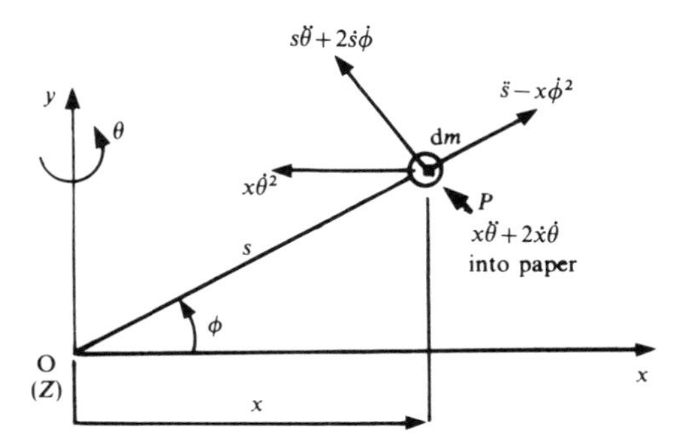

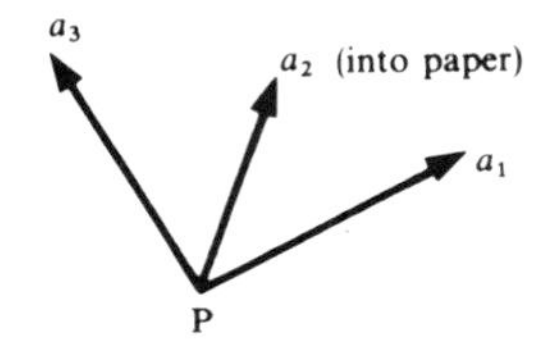

Figure 9.8 Accelerating frame.

Substituting for x and $\dot{x}$ yields

$$a_1 = \ddot{s} - s\dot{\phi}^2 - s\dot{\theta}^2 \cos^2 \phi$$
$$a_2 = s \cos \phi\ddot{\theta} + 2\dot{s}\dot{\theta} \cos \phi - 2s\dot{\phi}\dot{\theta} \sin \phi$$
$$a_3 = s\ddot{\phi} + 2\dot{s}\dot{\phi} + s\dot{\theta}^2 \cos \phi \sin \phi$$

since $\dot{s} = \dot{q}$ and $\ddot{s} = \ddot{q}$ we get

$$a_1 = \ddot{q} - s\dot{\phi}^2 - s\dot{\theta}^2 \cos^2 \phi$$
$$a_2 = s \cos \phi\ddot{\theta} + 2\dot{q}\dot{\theta} \cos \phi - 2s\dot{\phi}\dot{\theta} \sin \phi$$
$$a_3 = s\ddot{\phi} + 2\dot{q}\dot{\phi} + s\dot{\theta}^2 \cos \phi \sin \phi$$

(b) For the shoulder

$$\dot{s} = \ddot{s} = 0$$
$$a_1 = -s\dot{\phi}^2 - s\dot{\theta}^2 \cos^2 \phi$$
$$a_2 = s \cos \phi\ddot{\theta} - 2s\dot{\phi}\dot{\theta} \sin \phi$$
$$a_3 = s\ddot{\phi} + s\dot{\theta}^2 \cos \phi \sin \phi$$

Also, the potential energy V is given by

$$V = m_1 h_1 g \sin \phi + q m_2 g \sin \phi$$

hence

$$\frac{\partial V}{\partial \phi} = m_1 h_1 g \cos \phi + q m_2 g \cos \phi$$

$$\frac{\partial V}{\partial q} = m_2 g \sin \phi$$

$$\frac{\partial V}{\partial \theta} = 0$$

We now require the partial derivatives with respect to each coordinate for the arm and the shoulder and substitute in Eq. (9.5).

(a) ϕ coordinate

(i) For the shoulder, $$\frac{\partial a_1}{\partial \ddot{\phi}}=\frac{\partial a_2}{\partial \ddot{\phi}}=0 \qquad \frac{\partial a_3}{\partial \ddot{\phi}}=s$$

(ii) For the arm, $$\frac{\partial a_1}{\partial \ddot{\phi}}=\frac{\partial a_2}{\partial \ddot{\phi}}=0 \qquad \frac{\partial a_3}{\partial \ddot{\phi}}=s$$

Hence using the suffix 'S' for the shoulder and 'A' for the arm we have:

$$\frac{\partial G}{\partial \ddot{\phi}}=\int_{S} a_3\frac{\partial a_3}{\partial \ddot{\phi}}\,dm+\int_{A} a_3\frac{\partial a_3}{\partial \ddot{\phi}}\,dm$$

$$=\int_{S} s(s\ddot{\phi}+s\dot{\theta}^2\cos\phi\,\sin\phi)\,dm+\int_{A} s(s\ddot{\phi}+2\dot{q}\dot{\phi}+s\dot{\theta}^2\cos\phi\,\sin\phi)\,dm$$

$$=\ddot{\phi}\int_{S} s^2\,dm+\tfrac{1}{2}\dot{\theta}^2\sin 2\phi\int_{S} s^2\,dm+\ddot{\phi}\int_{A} s^2\,dm+2\dot{q}\dot{\phi}\int_{A} s\,dm+\tfrac{1}{2}\dot{\theta}^2\sin 2\phi\int_{A} s^2\,dm$$

$$=\ddot{\phi}I_S+\tfrac{1}{2}\dot{\theta}^2\sin 2\phi I_S+\ddot{\phi}I_A+2\dot{q}\dot{\phi}qm_A+\tfrac{1}{2}\dot{\theta}^2\sin 2\phi I_A$$

$$\Rightarrow \qquad \frac{\partial G}{\partial \ddot{\phi}}=(I_S+I_A)(\ddot{\phi}+\tfrac{1}{2}\dot{\theta}^2\sin 2\phi)+\tfrac{1}{2}\dot{q}\dot{\phi}qm_A$$

Substituting in Eq. (9.4), the torque Q_0 required about the horizontal axis through O is given by:

$$Q_0=\frac{\partial G}{\partial \ddot{\phi}}+\frac{\partial V}{\partial \phi} \quad \text{i.e.} \quad \underline{Q_0=(I_S+I_A)(\ddot{\phi}+\tfrac{1}{2}\dot{\theta}^2\sin 2\phi)+2\dot{q}\dot{\phi}qm_A+(m_Sh_S+qm_A)g\cos\phi} \qquad (9.6)$$

(b) q coordinate:

(i) For the shoulder, $$\frac{\partial a_1}{\partial \ddot{q}}=\frac{\partial a_2}{\partial \ddot{q}}=\frac{\partial a_3}{\partial \ddot{q}}=0$$

(ii) For the arm, $$\frac{\partial a_1}{\partial \ddot{q}}=1 \qquad \frac{\partial a_2}{\partial \ddot{q}}=\frac{\partial a_3}{\partial \ddot{q}}=0$$

Therefore $$\frac{\partial G}{\partial \ddot{q}}=\int_{A} a_1\frac{\partial a_1}{\partial \ddot{q}}\,dm=\int_{A}(\ddot{q}-s\dot{\phi}^2-s\dot{\theta}^2\cos^2\phi)\,dm=\ddot{q}m_A-qm_A(\dot{\phi}^2+\dot{\theta}^2\cos^2\phi)$$

Hence the force F_q required is given by

$$F_q=\frac{\partial G}{\partial \ddot{q}}+\frac{\partial V}{\partial q}$$

Therefore $$\underline{F_q=\ddot{q}m_A-qm_A(\dot{\phi}^2+\dot{\theta}^2\cos^2\phi)+\sin\phi m_Ag} \qquad (9.7)$$

(c) θ equation:

(i) For the shoulder, $\dfrac{\partial a_1}{\partial \ddot{\theta}}=\dfrac{\partial a_3}{\partial \ddot{\theta}}=0 \qquad \dfrac{\partial a_2}{\partial \ddot{\theta}}=s \cos \phi$

(ii) For the arm, $\dfrac{\partial a_1}{\partial \ddot{\theta}}=\dfrac{\partial a_3}{\partial \ddot{\theta}}=0 \qquad \dfrac{\partial a_2}{\partial \ddot{\theta}}=s \cos \phi$

Therefore

$$\frac{\partial G}{\partial \ddot{\theta}}=\int_{\mathrm{S}} a_2 \frac{\partial a_2}{\partial \ddot{\theta}} \, \mathrm{d}m+\int_{\mathrm{A}} a_2 \frac{\partial a_2}{\partial \ddot{\theta}} \, \mathrm{d}m$$

$$=\int_{\mathrm{S}} s \cos \phi(s \cos \phi \ddot{\theta}-2 s \dot{\phi} \dot{\theta} \sin \phi) \, \mathrm{d}m+\int_{\mathrm{A}} s \cos \phi(s \cos \phi \ddot{\theta}+2 \dot{q} \dot{\theta} \cos \phi-2 s \dot{\phi} \dot{\theta} \sin \phi) \, \mathrm{d}m$$

$$=\ddot{\theta} \cos^2 \phi \int_{\mathrm{S}} s^2 \, \mathrm{d}m-\dot{\phi} \dot{\theta} \sin 2\phi \int_{\mathrm{S}} s^2 \, \mathrm{d}m+\ddot{\theta} \cos^2 \phi \int_{\mathrm{A}} s^2 \, \mathrm{d}m+2 \dot{q} \dot{\theta} \cos^2 \phi \int_{\mathrm{A}} s \, \mathrm{d}m$$

$$-\dot{\phi} \dot{\theta} \sin 2\phi \int_{\mathrm{A}} s^2 \, \mathrm{d}m$$

$$=(I_{\mathrm{S}}+I_{\mathrm{A}})(\ddot{\theta} \cos^2 \phi-\dot{\phi} \dot{\theta} \sin 2\phi)+2 \dot{q} \dot{\theta} \cos^2 \phi q m_{\mathrm{A}}$$

Hence the torque T_θ required about the vertical axis is given by:

$$\underline{T_\theta=(I_{\mathrm{S}}+I_{\mathrm{A}})(\ddot{\theta} \cos^2 \phi-\dot{\phi} \dot{\theta} \sin 2\phi)+2 \dot{q} \dot{\theta} \cos^2 \phi q m_{\mathrm{A}}} \qquad (9.8)$$

Example 9.3 Calculate the torque required at the shoulder and the force required on the extending arm for the polar-coordinate robot shown in Fig. 9.6. The shoulder pitches about its centre of mass which coincides with the axis through O; the inertia in pitch is equal to 150 kg m^2. The extending arm has a mass of 90 kg and carries a load of 65 kg, which can be treated as a point mass. The arm may be considered to be a uniform rod of length 2.2 m. Other data: $\phi=20°$; $\dot{\phi}=0.4$ rad/s; $\ddot{\phi}=2$ rad/s^2; $h_1=0$; distance from O to the gripper $=2.5$ m; $\dot{\theta}=0.5$ rad/s; $v=0.4$ m/s; $\dot{v}=1.5$ m/s^2.

SOLUTION (Fig. 9.9) $I_s=150$ kg, $M_{\text{Load}}=65$ kg, $M_{\text{arm}}=90$ kg. Moment of inertia of the arm about O:

$$I_{\mathrm{A}}=2.5\times 65+14^2\times 90+\tfrac{1}{12}\times 90\times 2.2^2$$
$$=618.95 \text{ kg m}^2$$

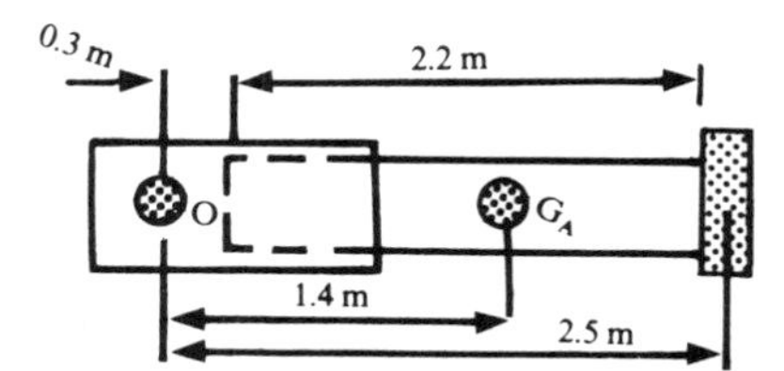

Figure 9.9

Position of the centre of mass of the arm and load from O:

$$q(65+90)=2.5\times 65+1.4\times 90$$
$$\therefore q=1.861\ \text{m}$$

Total mass: $$M_A=M_{\text{Load}}+M_{\text{arm}}=65+90=155\ \text{kg}$$

also from the data: $$\dot{q}=v=0.4\ \text{m/s},\quad \ddot{q}=\dot{v}=1.5\ \text{m/s}^2$$

The torque required at the axis O is given by Eq. (9.6); hence substituting numerical values we get:

$$Q_O=[150+618.95](2+0.5\times 0.5^2\sin 40)+2\times 0.4\times 0.4\times 1.861\times 155$$
$$+1.861\cos 20\times 155\times 9.81$$
$$=4.35\ \text{kN m}$$

The force required to move the arm and its load for the prescribed motion is given by Eq. (9.7); substituting numerical values yields

$$F_q=1.5\times 155-1.861\times 155(0.4^2+0.5^2\cos^2 20)+\sin 20\times 155\times 9.81=0.642\ \text{kN}$$

9.3.2 Analysis of Type (b): Hinged Arm

The coordinates are ϕ, ψ, and θ as shown in Fig. 9.7.

Let L_1 = length of inner arm
m_1 = mass of inner arm and h_1 the CG position from O
L_2 = length of outer arm
m_2 = total mass of outer arm and h_2 the CG position from A
s = coordinate along each arm to points P and Q as shown

Figure 9.10 shows the reference frame for the accelerations.
(i) Inner arm:

$$a_1=-s\dot{\phi}^2-x\dot{\theta}^2\cos\phi$$
$$a_2=x\ddot{\theta}+2\dot{x}\dot{\theta}$$
$$a_3=s\ddot{\phi}+x\dot{\theta}^2\sin\phi$$

where $x=s\cos\phi$ and $\dot{x}=-s\dot{\phi}\sin\phi$.
(ii) Outer arm:

$$a_1=-s\dot{\psi}^2-x\dot{\theta}^2\cos\psi-L_1\dot{\phi}^2\cos(\psi-\phi)+L_1\ddot{\phi}\sin(\psi-\phi)$$
$$a_2=x\ddot{\theta}+2\dot{x}\dot{\theta}$$
$$a_3=s\ddot{\psi}+x\dot{\theta}^2\sin\psi+L_1\dot{\phi}^2\sin(\psi-\phi)+L_1\ddot{\phi}\cos(\psi-\phi)$$

where in this case $$x=L_1\cos\phi+s\cos\psi$$

hence $$\dot{x}=-L_1\dot{\phi}\sin\phi-s\dot{\psi}\sin\psi$$

The total potential energy is:

$$V=(m_1h_1+m_2L_1)g\sin\phi+m_2h_2g\sin\psi$$

The derivatives are as follows:

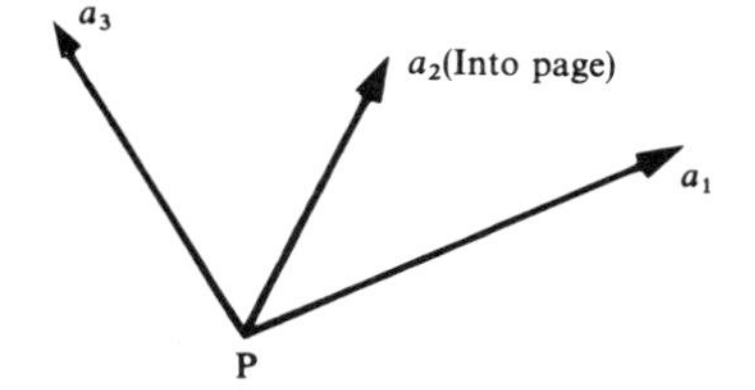

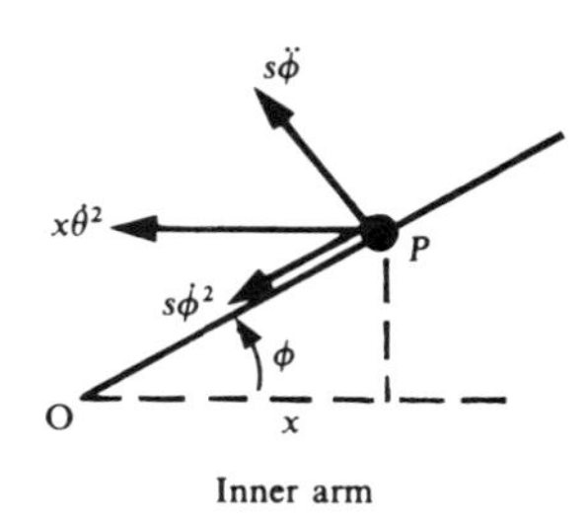

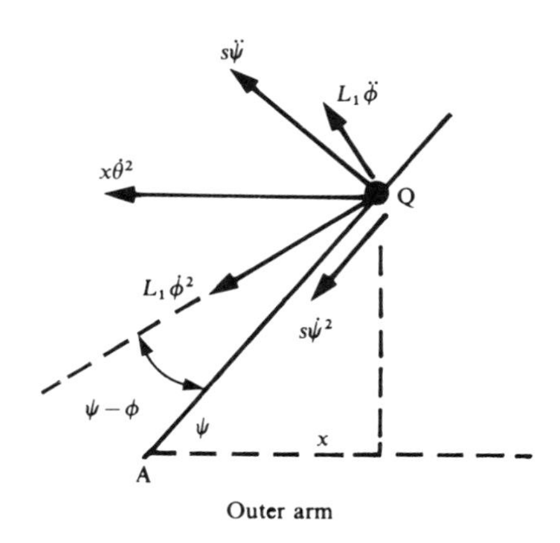

Figure 9.10 Accelerations of the inner and outer arms.

(a) ϕ coordinate

(i) Inner arm: $\dfrac{\partial a_1}{\partial \ddot{\phi}}=\dfrac{\partial a_2}{\partial \ddot{\phi}}=0 \qquad \dfrac{\partial a_3}{\partial \ddot{\phi}}=s$

(ii) Outer arm: $\dfrac{\partial a_1}{\partial \ddot{\phi}}=L_1 \sin(\psi-\phi) \qquad \dfrac{\partial a_2}{\partial \ddot{\phi}}=0 \qquad \dfrac{\partial a_3}{\partial \ddot{\phi}}=L_1 \cos(\psi-\phi)$

Substituting in Eq. (9.5), we find:

$$\frac{\partial G}{\partial \ddot{\phi}}=\int_{\text{inner}}[a_3 s]\,dm+\int_{\text{outer}}[a_1(L_1 \sin\beta)]\,dm+\int_{\text{outer}}[a_3(L_1 \cos\beta)]\,dm$$

also
$$\frac{\partial V}{\partial \phi}=(m_1h_1+m_2L_1)g \cos\phi$$

Hence substitution in the Gibbs–Appell equation gives the torque about the shoulder axis O:

$$(I_1+m_2L_1^2)\ddot{\phi}+I_1 \sin\phi \cos\phi\dot{\theta}^2-m_2h_2L_1 \sin\beta\dot{\psi}^2$$
$$+L_1 \sin\phi(L_1m_2 \cos\phi+m_2h_2 \cos\psi)\dot{\theta}^2+L_1m_2h_2 \cos\beta\ddot{\psi}$$
$$+(m_1h_1+m_2L_1)g \cos\phi=Q_\phi \tag{9.9}$$

where $I_1=\int s^2 \, dm$ = moment of inertia of the inner arm about O and $\beta=\psi-\phi$.

(b) ψ-coordinate

(i) Inner arm: all derivatives are zero since there are no ψ-terms.
(ii) Outer arm: the a_1 and a_2 derivatives are zero, but

$$\frac{\partial a_3}{\partial\ddot{\psi}}=s \qquad \text{also} \qquad \frac{\partial V}{\partial\psi}=m_2h_2g \cos\psi$$

Substituting in the Gibbs–Appell equation yields the torque about the elbow axis A:

$$I_2\ddot{\psi}+(m_2h_2L_1 \cos\phi+I_2 \cos\psi)\sin\psi\dot{\theta}^2+m_2h_2(L_1 \sin\beta\dot{\phi}^2+L_1 \cos\beta\ddot{\phi})$$
$$+m_2h_2g \cos\psi=Q_\psi \tag{9.10}$$

where I_2 = moment of inertia of the outer arm about the elbow axis A, and including the load carried by the gripper.

(c) θ-coordinate

(i) Inner arm: $\frac{\partial a_1}{\partial\ddot{\theta}}=\frac{\partial a_3}{\partial\ddot{\theta}}=0$ and $\frac{\partial a_2}{\partial\ddot{\theta}}=x$

(ii) Outer arm: $\frac{\partial a_1}{\partial\ddot{\theta}}=\frac{\partial a_3}{\partial\ddot{\theta}}=0$ and $\frac{\partial a_2}{\partial\ddot{\theta}}=x$ also $\frac{\partial V}{\partial\theta}=0$

By substituting in the Gibbs–Appell equation we obtain the following expression for the torque Q_θ required about the y-axis:

$$I_y\ddot{\theta}-\dot{\theta}[\dot{\phi}(I_1+m_2L_1^2)\sin 2\phi+2L_1m_2h_2\dot{\psi} \cos\phi \sin\psi+2L_1m_2h_2\dot{\phi} \cos\psi \sin\phi+I_2 \sin 2\psi\dot{\psi}]=Q_\theta \tag{9.11}$$

where I_y = moment of inertia of the inner and outer arms about the vertical y-axis

$$=(I_1+m_2L_1^2)\cos^2\phi+2L_1m_2h_2 \cos\phi \cos\psi+I_2 \cos^2\psi$$

The total torque to be provided by a prime mover at the shoulder about a horizontal axis through O is given by:

$$Q_{\text{shoulder}}=Q_\phi+Q_\psi$$

Example 9.4 A small robot of the hinged-arms type has the following dimensions:

	Inner arm	Outer arm
length, m	0.35	0.40
mass, kg	5	4.5

The gripper carries a load of 1.2 kg at a distance of 0.45 m from the elbow. A constant torque of 15 N m is provided by the prime mover of which 6 N m is available at the elbow.

Calculate the velocity and the acceleration of the inner arm at the instant when it is in a vertical position and the outer arm is horizontal. The robot is slewing at a constant rate of 0.5 rad/s and the outer arm is being elevated at the constant rate of 1.5 rad/s. Calculate also the torque required to provide the slewing motion.

Treat the arms as slender rods.

SOLUTION From the data we have: $\phi=90°$, $\psi=0$, and hence $\beta=0-90=-90°$.

Equation (9.9) becomes $(I_1+m_2L_1^2)\ddot{\phi}+m_2h_2L_1\dot{\psi}^2+m_2h_2L_1\dot{\theta}^2=Q_\phi$

Equation (9.10) becomes $-m_2h_2L_1\dot{\phi}^2+m_2h_2g=Q_\psi$

Equation (9.11) becomes $2L_1m_2h_2\dot{\phi}\dot{\theta}=Q_\theta$, torque to provide the slewing motion

$m_1=5$ kg $L_1=0.35$ m $m_2=m_{\text{arm}}+m_{\text{load}}=4.5+1.2=5.7$ kg $L_2=0.4$ m $L_{\text{load}}=0.45$

Position of the centre of gravity of the outer arm and load from the shoulder:

$$h_2=\frac{4.5\times0.2+1.2\times0.45}{5.7}=0.253\text{ m}$$

$I_1=\frac{1}{3}m_1L_1^2=\frac{1}{3}\times5\times0.35^2=0.204\text{ kg }m^2$ $I_1+m_2L_1^2=0.204+5.7\times0.35^2=0.902\text{ kg m}^2$

$m_2h_2L_1=5.7\times0.253\times0.35=0.505$ $m_2h_2g=5.7\times0.253\times9.81=14.15$

$\ddot{\psi}=\ddot{\theta}=0$ $\dot{\psi}=1.5$ rad/s $\dot{\theta}=0.5$ rad/s $Q_\phi=Q_{\text{shoulder}}$ $-Q_\psi=15-6=9$ N m

Substituting in the above equations we find:

$$9=0.902\ddot{\phi}+0.505\times1.5^2+0.505\times0.5^2 \text{ hence } \ddot{\phi}=8.6\text{ rad/s}^2$$

$$-5.05\dot{\phi}^2+14.15=6 \quad\text{hence}\quad \dot{\phi}=0.678\text{ rad/s}$$

$$Q_\theta=2\times0.505\times0.678\times0.5=0.34\text{ N m}$$

9.3.3 Examples of Practical Robotic Manipulators

1. Figure 9.11 shows an experimental/pedagogical manipulator (with the cover removed) developed at Ecole Centrale, Paris, by Professor J. Hervé and students. It is a hinged-arm or revolute type with six revolute joints (R), hence six degrees of freedom; the gripper has three degrees of freedom: pitch, roll, and yaw. All the movements of the manipulator, including those of the gripper, are derived from toothed belts driven by a number of motors.

   Rotations of these motors are controlled by a computer in order to enable the gripper to reproduce a predetermined path defined by the movement of the arm shown on the left of the picture. The path is thus defined by hundreds of points entered in the memory of the computer which interpolates between points. Other paths can be recorded to cater for various operations. It is in fact what is normally referred to as a 'playback' robot, i.e., a manipulator which is able to perform an operation by reading out stored information for an operating sequence, including positions and the like, which it learned beforehand by being taken manually through the routine.[6]
2. Figure 9.12 shows an industrial robot also of the revolute type with six degrees of freedom by means of six revolute joints (R), three being those of the gripper. This type simulates the human arm and is frequently referred to as an 'anthropomorphic' robot.
3. An industrial version of the polar type is shown in Fig. 9.13; it also has six degrees of freedom, three being those of the gripper. The robot rotates about a vertical axis through

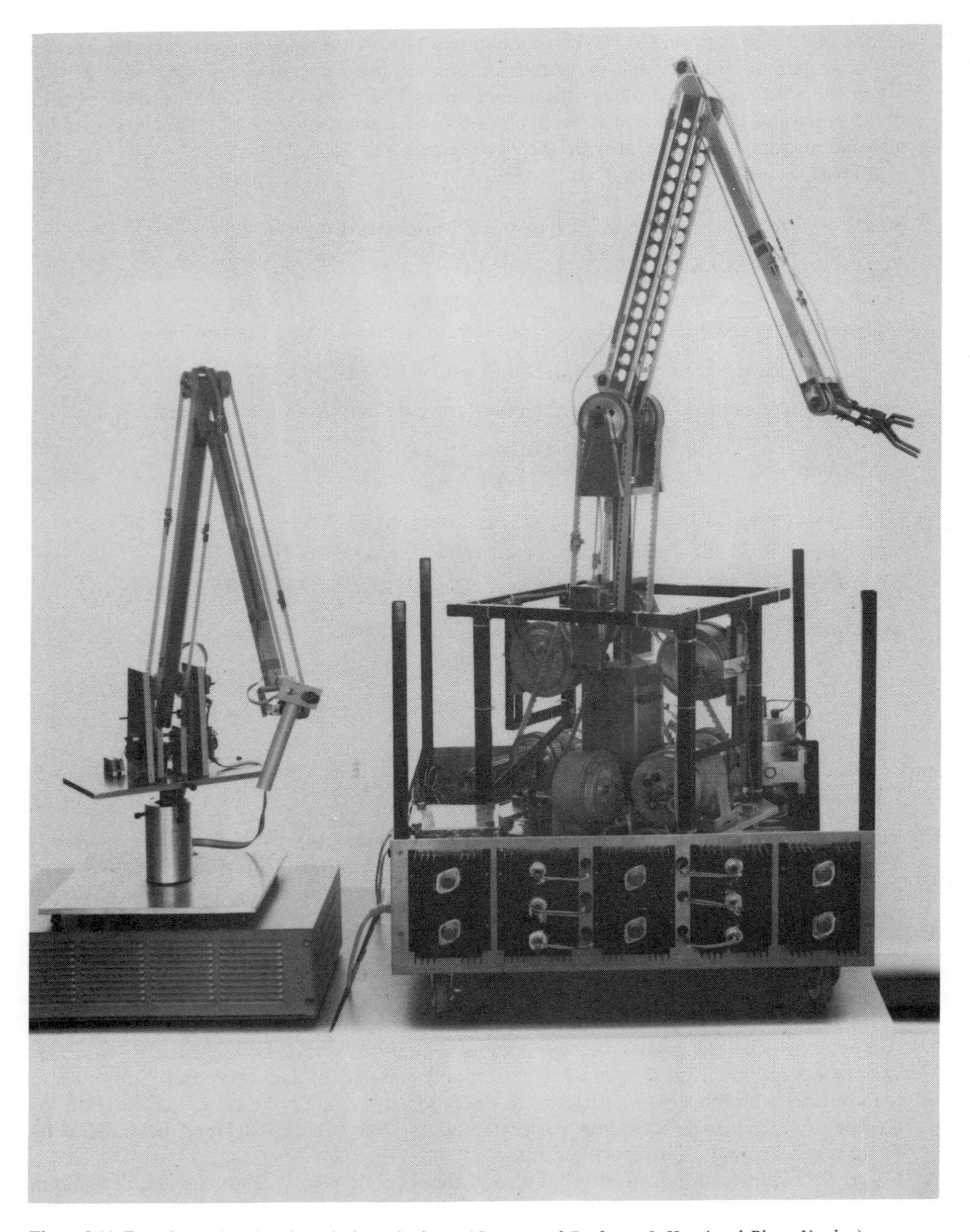

Figure 9.11 Experimental and pedagogical manipulator. (*Courtesy of Professor J. Hervé and Photo Vandor.*)

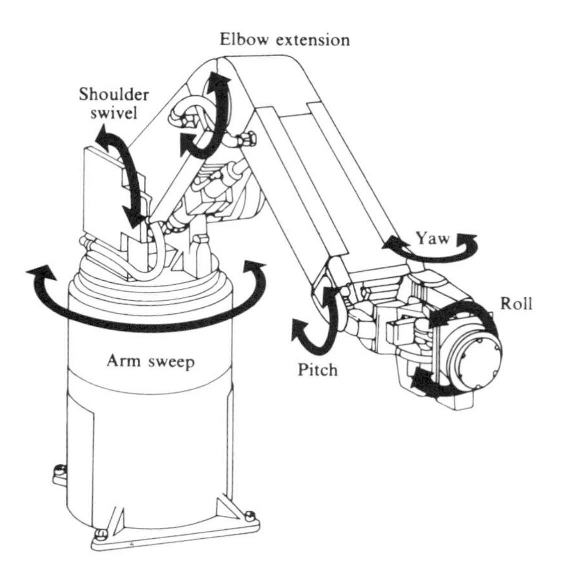

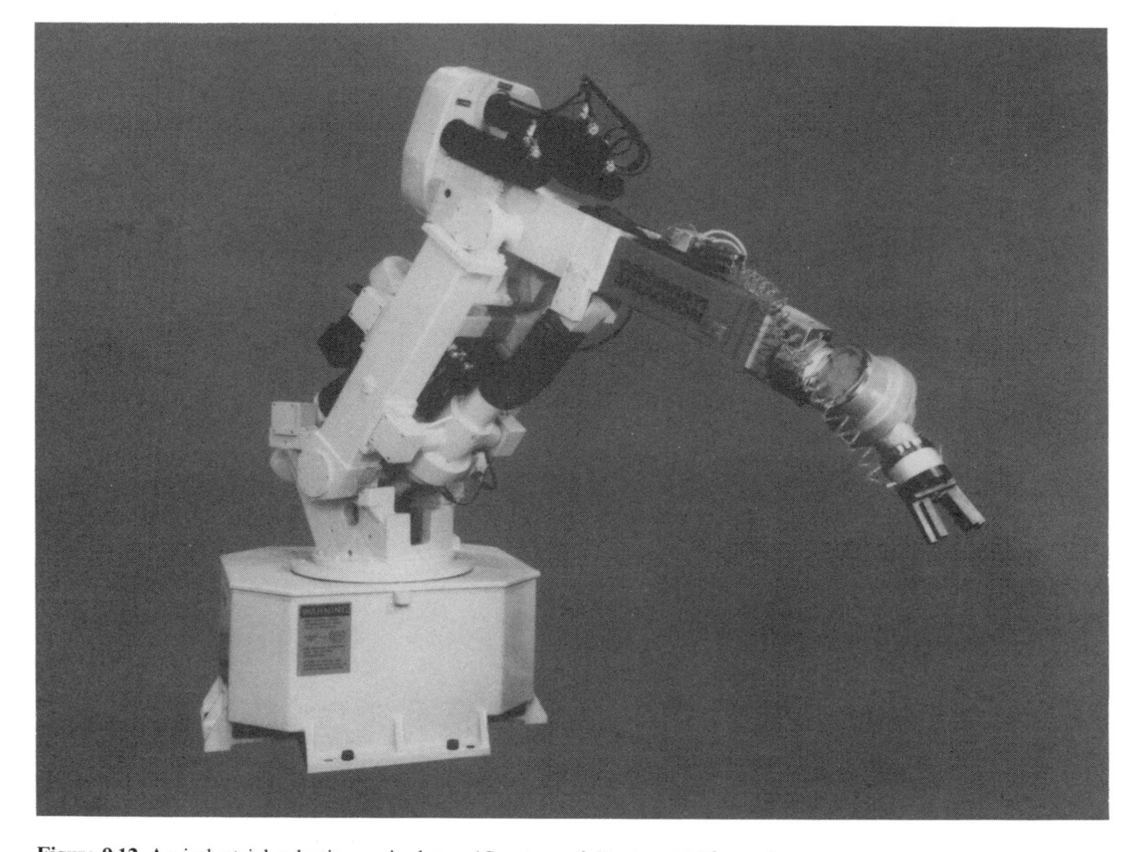

Figure 9.12 An industrial robotic manipulator. (*Courtesy of Cincinatti-Milacron.*)

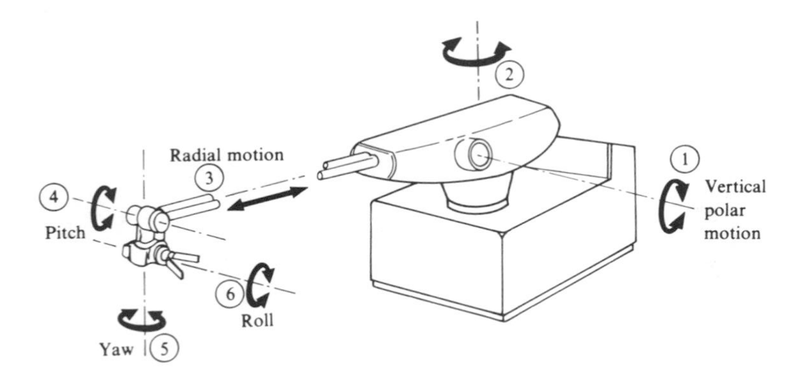

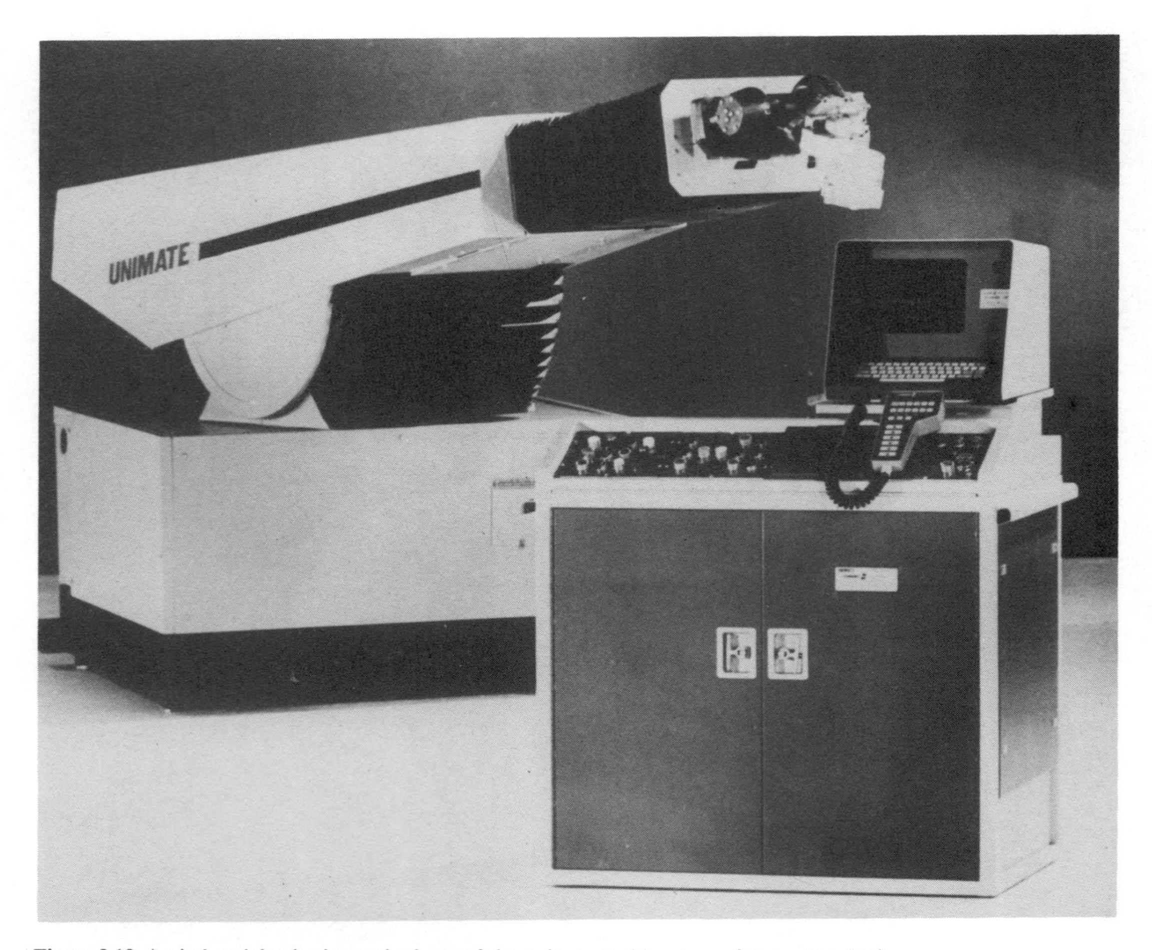

Figure 9.13 An industrial robotic manipulator of the polar type. (*Courtesy of Unimation Inc.*)

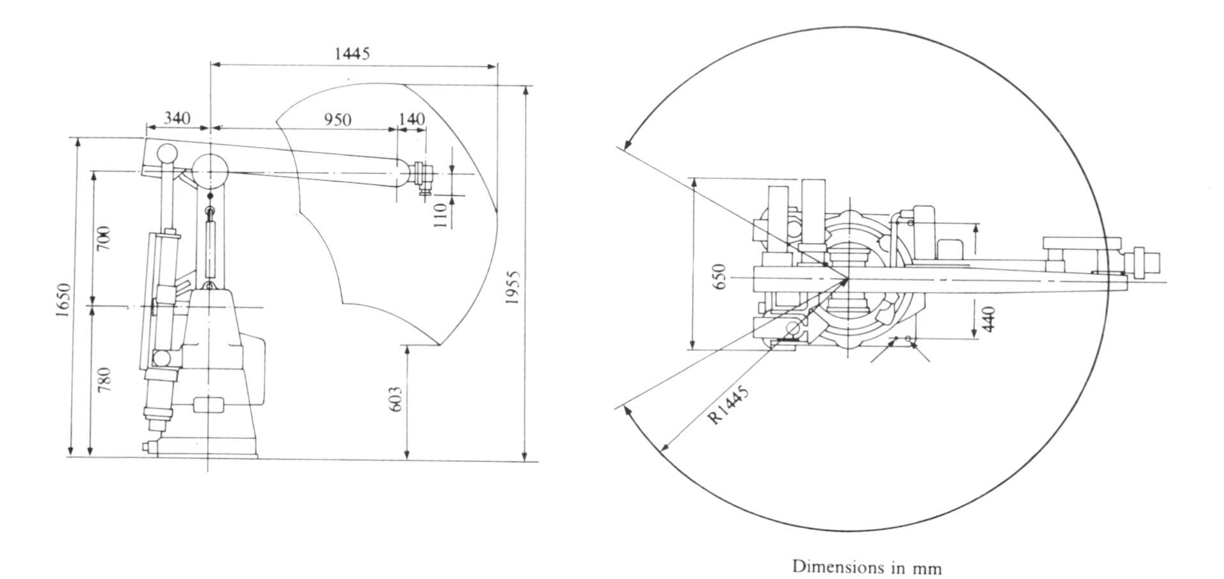

Dimensions in mm

Figure 9.14 Application of the four-joint parallelogram linkage mechanism: (*a*) dimensions and operating space; (*b*) M6060 Mk II robot welding commercial vehicle axles. (*Courtesy of GEC Robots Systems Limited.*)

its waist and the shoulder can rotate in a vertical plane. These two movements are obtained by revolute joints (R), while the arm can move in or out by means of a prismatic joint (P).

The reader should realize that the above manipulators are mechanisms of the 'open-chain' type, unlike the 'closed-chain' type such as the four-bar linkage; in fact other industrial manipulators are based on four-bar linkages as illustrated in Fig. 9.14. In this case the robot is based on the parallelogram linkage (which is a special case of the four-bar linkage) with the coupler link extended, and is used for welding, gluing, component transfer, and other tasks.

EXERCISES

9.1 Referring to Example 9.2 in the text, derive the equations of motion using Lagrange's equations and compare the workings with the Gibbs–Appell method.

9.2 Figure 9.15 is a simplification of a mechanism used for grinding material such as grain. The grinder consists essentially of one or more heavy wheels which run freely on a shaft hinged to a central driving shaft SS'.

Taking body axes Oxyz as shown calculate the moment to be applied by the driving motor M in order to give the grinder an acceleration α assuming no slipping of the wheel. Data: $r=250$ mm: $R=850$ mm; mass of wheel $=45$ kg; $\alpha=1.25$ rad/s^2; $\phi=20°$; gear ratio $=3{:}1$; moment of inertia of the axle about SS' $=2.75$ kg m^2. The wheel may be approximated to a thin disc.

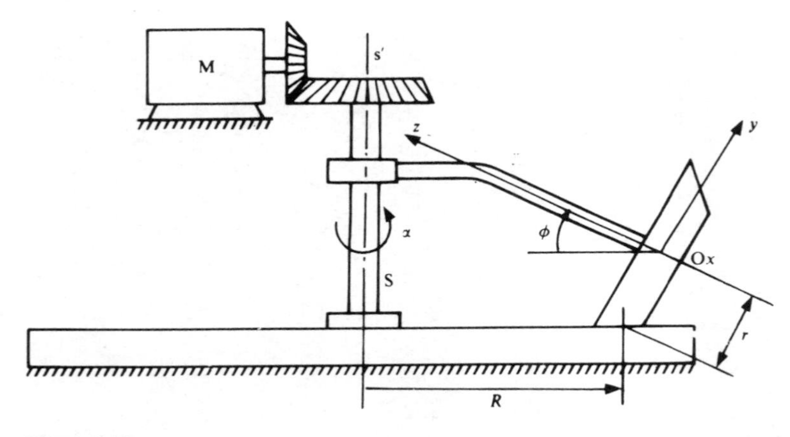

Figure 9.15

9.3 Derive Eqs (9.6) to (9.8) for the polar arm manipulator using Lagrange's equations.

9.4 Derive Eqs (9.9) to (9.11) for the hinged arm manipulator using Lagrange's equations. What are the differences in the amount of work involved compared with the Gibbs–Appell approach?

9.5 Using the Gibbs–Appell method, derive the equations of motion governing small oscillations of a hinged robotic arm (Fig. 9.16). Consider three freedoms only: shoulder motion ϕ, elbow rotation ψ (from horizontal), and wrist yaw α.

Neglect centripetal accelerations, indicating when this assumption is valid.

In the static configuration $\phi=40°$, $\psi=-20°$ and $\alpha=45°$, calculate the moments required to produce $\ddot{\phi}=-2$ rad/s^2, $\ddot{\psi}=1.5$ rad s^2, and $\ddot{\alpha}=-4$ rad/s^2.

Treat the links as uniform plates. The distance l_w is negligible. A point load of 3.25 kg is situated half-way along the gripper.

	Mass, kg	Length, m	Width, m
Inner arm	3	0.5	0.1
Outer arm	2.7	0.45	0.1
Gripper	0.7	0.2	0.05

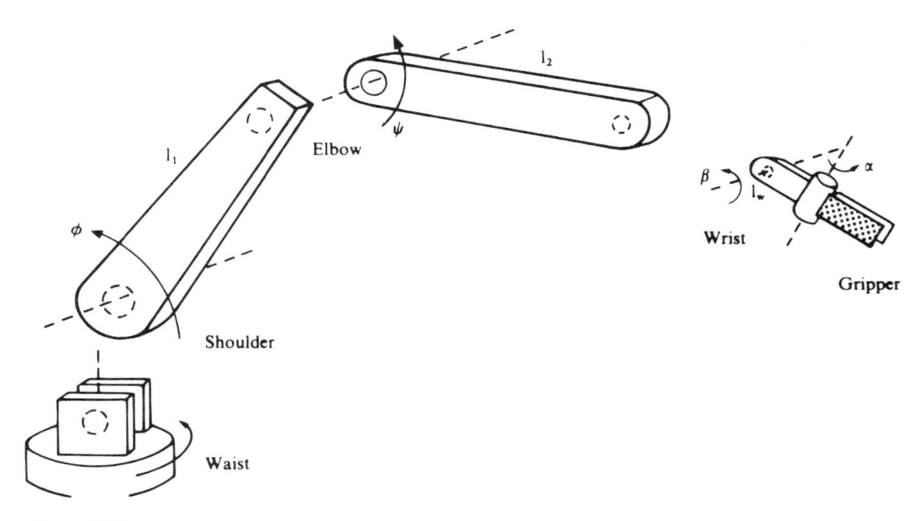

Figure 9.16

9.6 An assembly robot is of revolute arm form, having shoulder freedom ϕ and elbow freedom ψ together with rotation θ about a vertical axis.

Use the Gibbs–Appell method to formulate the equations of motion for the two freedoms ϕ and ψ.

Calculate the torque required at (*a*) the elbow and (*b*) the shoulder when

$$\phi = 35^\circ \qquad \dot{\phi} = -2.5\ \text{rad/s} \qquad \ddot{\phi} = 2\ \text{rad/s}^2$$
$$\psi = -45^\circ \qquad \dot{\psi} = 1.2\ \text{rad/s} \qquad \ddot{\psi} = 1.5\ \text{rad/s}^2$$
$$\dot{\theta} = 1.8\ \text{rad/s} \qquad \ddot{\theta} = 2.2\ \text{rad/s}^2$$

Treat the arms as uniform rods. A load of 0.7 kg, which can be treated as a point mass, is carried by the gripper.

	Length (m)	Mass (kg)
Inner	0.20	5
Outer	0.30	6

9.7 A hydraulically powered robotic arm (Fig. 9.17) has three degrees of freedom: rotation θ about a vertical axis Oz, rotation ϕ about a horizontal axis Ox at right angles to the arm, and a radial movement q of the arm A. A point load mass M is carried at the end of the arm.

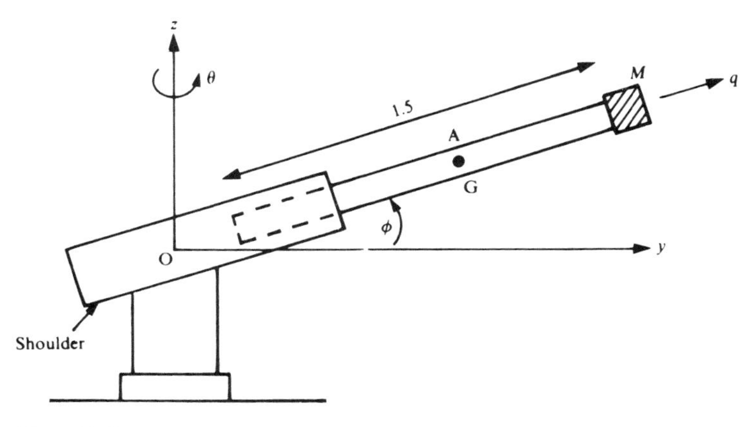

Figure 9.17

State the Gibbs–Appell equations and use to derive the non-linear differential equations governing motion in the three degrees of freedom, including a linear damping term in each freedom, but ignoring feedback control.

Hence derive the torque required about the vertical axis to produce $\ddot{\theta}=3\ \mathrm{rad/s^2}$, when $\dot{\theta}=1.5\ \mathrm{rad/s}$, $\dot{\phi}=-0.8\ \mathrm{rad/s}$, $\dot{q}=0$, $\phi=20°$, and $\theta=0°$.
Data: shoulder: inertia about body axes $=40\ \mathrm{kg\,m^2}$;
arm: length 1.5 m; mass (excluding load) 52 kg;
centre of mass G (without load) 1.75 m from 0. Inertias about body axes through G both $10.5\ \mathrm{kg\,m^2}$. Damping constant for motion about the waist (θ freedom) $=33.7\ \mathrm{kN\,m\,s/rad}$. Load mass $=20$ kg.

CHAPTER

TEN

MECHANICAL ERRORS: TOLERANCES

10.1 INTRODUCTION: LINEAR AND NON-LINEAR FUNCTIONS

The performance of any mechanism will be affected by the deformation of its elements due to the loads involved, temperature changes, humidity, tolerances on the dimensions, clearances in joints, and out-of-squareness of such things as pins.

Since no mechanism can be manufactured to absolute perfection, the influence of tolerances and clearances will be important, particularly in precision mechanisms; it is this aspect we wish to consider now. For the time being however we restrict our discussion to tolerances only, i.e., tolerances on lengths between pivots. In certain practical cases we can allow for clearances by introducing an equivalent tolerance on the distance between holes or pin centres.[6,7]

In a mechanism the position of any point on a link, a pin or a slot is a function of its geometry. This position will depend on a number of lengths and angles and may be quite complicated; the functions involved will be linear, non-linear, or a combinaton of both. This is illustrated by three examples as shown in Fig. 10.1(*a*), (*b*), and (*c*).

Case (a) Three components are fitted end to end as shown in Fig. 10.1(*a*). The δl's are the tolerances imposed for manufacture. The length L of the assembly is a linear function of the individual length so that $L = l_1 + l_2 + l_3$ without tolerances, and $L \pm \Delta l = l_1 + l_2 + l_3 \pm \delta l_1 \pm \delta l_2 \pm \delta l_3$ owing to the tolerances.

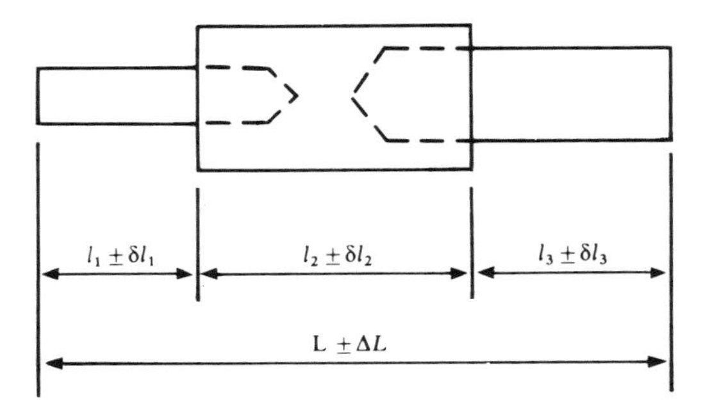

Figure 10.1(a) Illustration of tolerances in the case of three elements in series.

Hence the 'total' tolerance on the length L is $\Delta l = \Sigma(\pm\delta l)$.

Case (b) A lever AOB is pivoted at O and actuated by a cam as shown in Fig. 10.1(*b*). The position of the pin relative to the frame or casting is given by

$$X = a \cos\left[\alpha - \arcsin\left(\frac{y-R}{b}\right)\right]$$

and

$$Y = a \sin\left[\alpha - \arcsin\left(\frac{y-R}{b}\right)\right]$$

These are two non-linear functions; both lengths and the angle are subject to the manufacturing tolerances shown. The dimensions a, b, R, y, and α are referred to as 'primary dimensions'; they are the manufacturing dimensions. The angle β js not a primary dimension, since it depends on y, R, and b.

Case (c) Consider the four-bar linkage shown in Fig. 10.1(*c*) where ϕ is the input position. The output position ψ is given by

$$\psi = 2 \arctan\left(\frac{A \pm \sqrt{A^2 + B^2 - C^2}}{B + C}\right)$$

where

$$A = \sin\phi$$

$$B = \cos\phi - \frac{d}{a}$$

$$C = \frac{d}{c}\cos\phi - \left(\frac{a^2 - b^2 + c^2 + d^2}{2ac}\right)$$

This shows quite clearly that ψ is another non-linear function. The 'total' tolerance on ψ, i.e., $\Delta\psi$ will be a function of the individual tolerances shown. The input ϕ is not subject to a tolerance, unless it is derived from another mechanism, e.g., a six-bar linkage like the shoe-stitching machine where the output of the first four-bar linkage becomes the input of the following linkage; see Sec. 1.2(*g*).

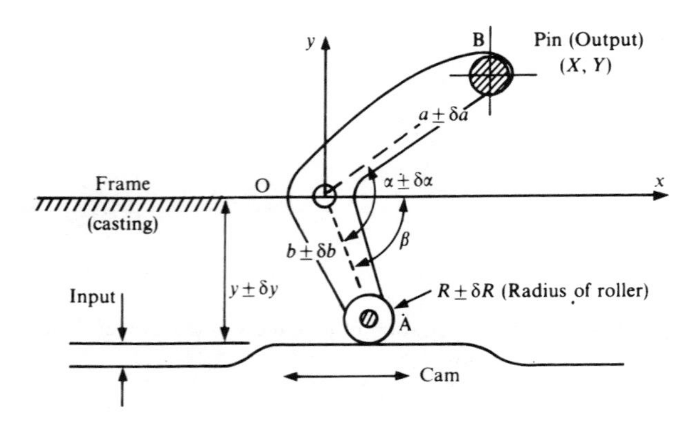

Figure 10.1(b) Illustration of tolerances in the case of a cranked lever operated by a cam.

and

$$e = 2L_T \sin\left(\frac{\theta_T}{2}\right)$$

θ_T is a very small angle so that

$$e \doteq L_T\theta_T/57.3 \tag{10.4}$$

Since θ_T is measured in degrees we need the conversion factor of 57.3.

Also $\quad L_T = \sqrt{a^2 + b^2} \quad$ from Fig. 10.3 $\quad$ (10.5)

$\quad\quad Z = \sqrt{X_s^2 + Y_s^2} \quad$ from Fig. 10.4 $\quad$ (10.6)

Hence

$$\psi = \arccos((R_s + a)/Z) \tag{10.7}$$

$$\theta = \arccos((L_T^2 + Z^2 - R_s^2)/2L_T Z) \tag{10.8}$$

$$\gamma = \arctan(b/a) \tag{10.9}$$

1. *Design values.* (All dimensions are in mm and angles in degrees.)

$$\left.\begin{aligned} a &= 13.50 \\ b &= 16.90 \end{aligned}\right\} \gamma = \arctan(16.9/13.50) = 51.30^\circ$$

$L_T = \sqrt{13.50^2 + 16.90^2} = 21.63$ from Eq. (10.5)

$Z = \sqrt{13.35^2 + 21.05^2} = 24.93$ from Eq. (10.6)

$\psi = \arccos((6.00 + 13.50)/24.93) = 38.54$ from Eq. (10.7)

$\theta = \arccos((21.63^2 + 24.93^2 - 6^2)/(2 \times 21.63 \times 24.93)) = 12.39$ from Eq. (10.8)

$\theta_T = 51.38 - 12.39 - 38.54 = 0.45$ from Eq. (10.3)

Hence $\quad e = 21.63 \times 0.45/57.3 = \underline{0.17}$ from Eq. (10.4)

2. *Tolerances on the position of S.* Since the value of e is critical we will assume that the worse combination of tolerances prevails and use the differentials of the various functions to calculate the tolerance δ to be allocated to the position of S, i.e., $X_s \pm \delta$ and $Y_s \pm \delta$.

From eq. (10.5) we have

$$L_T^2 = a^2 + b^2$$

Hence

$$L_T\delta L_T = |a\delta a| + |b\delta b|$$

Solving for δL_T and substituting numerical values yields

$$\delta L_T = (a\delta a + b\delta b)/L_T = (13.50 \times 0.05 + 16.9 \times 0.05)/21.63 = \pm 0.07$$

Also, from Eq. (10.6),

$$Z^2 = X_s^2 + Y_s^2$$

Hence

$$Z\delta Z = |X\delta X_s| + |Y\delta Y_s| = (X_s + Y_s)\delta$$

Solving for δ yields

$$\delta = \frac{Z\delta Z}{X_s + Y_s} \tag{10.10}$$

We now need a value for δZ. From the diagram (Fig. 10.4) we see that the limiting value for e would be $e=0$, which corresponds to $\theta=0$, in which case

$$Z=R_s+L_T$$

Hence
$$\delta Z=|\delta R_s|+|\delta L_T|$$

Substituting numerical values, we find

$$\delta Z=0.05+0.07=\pm 0.12$$

By substituting in Eq. (10.10) we obtain

$$\delta=\frac{24.93\times 0.12}{21.05+13.35}=\pm 0.09$$

It therefore follows that to ensure $e>\theta$ we must have

$$\underline{\delta<0.09\ \text{mm}}$$

In practice we would most probably specify

$$\underline{\delta=\pm 0.05\ \text{mm}}$$

In illustrating the steps required in the evaluation of δ the above example has been simplified, but in the actual case with which the author was involved other key dimensions had to be taken into account which considerably increased the amount of work.

Example 10.2 Figure 10.5 shows two elements of the turntable mechanism: (1) a programme lever which can be positioned in any one of three positions in order to play a record automatically, or to reject a record being played, or to repeat a record, and (2) a small pin T fixed to the underside of the cam (part of the large gear wheel) whose function is to activate the programme lever at the right time during a cycle.

In the particular position shown in the figure we must ensure that under adverse tolerance combinations the pin T will, as the cam rotates, make contact with the face of the

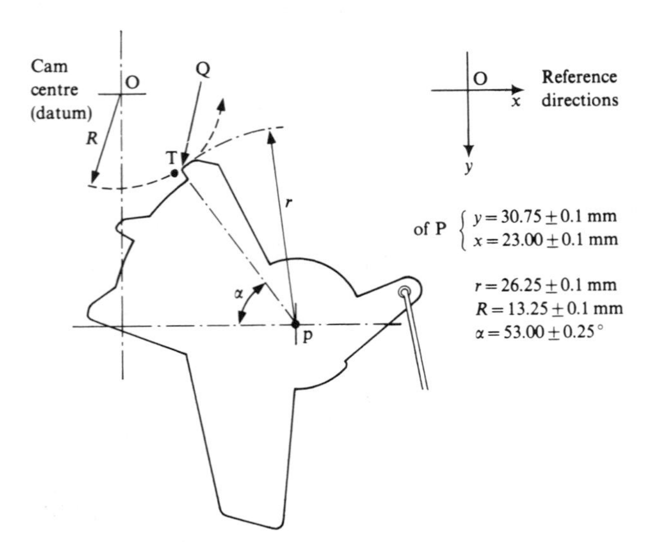

Figure 10.5 Diagram of the programme lever of the turntable mechanism.

tooth Q and hence place the programme lever in a 'waiting position', i.e., until the end of play of a particular record.

The dimensions shown are those specified on the manufacturing drawings.

SOLUTION This example illustrates a relatively simple situation where we can easily deduce the worse combination of tolerances. In this case we need to be certain that the distance $l = OQ$ (Fig. 10.6) will always be less than the radius R of the pin T if it is to make contact with Q on the programme lever, i.e., we must ensure that $l_{max} < R_{min}$.

It is easy to see that the worse combination of tolerances will occur with $y + \delta y$, $x + \delta x$, $R - \delta R$, $\alpha \pm \delta\alpha$, $r - \delta r$.

1. *Design values*

$$L = \sqrt{x^2 + y^2} = \sqrt{23.00^2 + 30.75^2} = 38.40 \text{ mm}$$

and

$$\beta = \arctan(y/x) = \arctan(30.75/23.00) = 53.20°$$

$$\theta = \beta - \alpha = 53.20 - 53.00 = 0.20°$$

also $l^2 = r^2 + L^2 - 2rL \cos\theta$ by the cosine rule

hence

$$l = \sqrt{26.25^2 + 38.40^2 - 2 \times 26.25 \times 38.40 \cos 0.2} = \underline{12.15 \text{ mm}}$$

Since $R = \underline{13.25 \text{ mm}}$, it follows that $l < R$ by $\underline{1.1 \text{ mm}}$

2. *Adverse tolerance combination*

$$L' = \sqrt{23.10^2 + 30.85^2} = 38.54 \text{ mm}$$

$$\beta' = \arctan(30.85/23.10) = 53.17°$$

$$\theta' = \beta' - \alpha' = 53.17 - 53.25 = 0.08°$$

Hence

$$l' = \sqrt{26.15^2 + 38.54^2 - 2 \times 26.15 \times 38.54 \cos 0.08} = \underline{12.39 \text{ mm}}$$

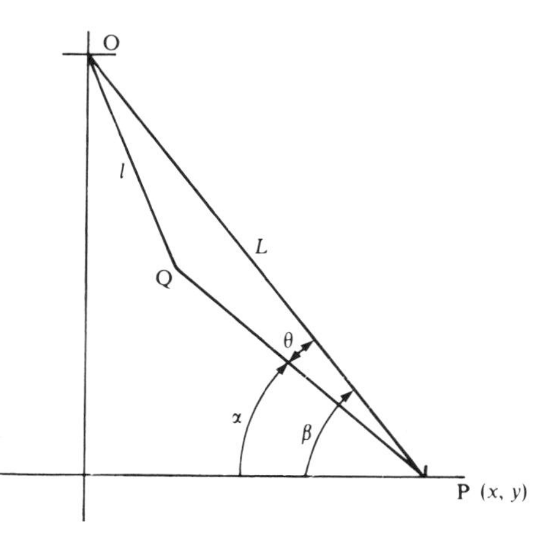

Figure 10.6 Diagram showing the required parameters for the analysis of the programme lever shown in Fig. 10.5.

Since $\qquad R' = 13.15$, it follows that

$$l' < R' \qquad \text{i.e.} \qquad l_{max} < R_{min} \text{ by } \underline{0.76 \text{ mm}}$$

Hence the required condition is satisfied and the device will operate satisfactorily provided that none of the tolerances imposed by the designer are exceeded during production of the parts; quality control in mass production usually ensures that this situation does not occur.

3. *Use of the total derivative.* In more complex cases it is not so easy to deduce the worse combination of tolerances. We therefore propose to illustrate the use of the total derivative in the above example to calculate the effect of adverse tolerance combination on the performance of the device.

Referring to the equations in (1) above and using the symbol Δ for the total derivative we have:

$$\Delta L = \frac{1}{L}(x\delta x + y\delta y) = \frac{\delta}{L}(x + y)$$

where $\delta = \delta x = \delta y$.
Substituting numerical values yields

$$\Delta L = \pm\frac{0.1}{38.40}(23.00 + 30.75) = \underline{\pm 0.14 \text{ mm}}$$

Also

$$\Delta\beta = \pm\frac{(x - y)\delta}{x^2}\cos^2\beta$$

$$= \pm\left(\frac{23.00 - 30.75}{23.00}\right) \times 0.1 \times \cos^2 53.20 \times 57.3 = \underline{\pm 0.03}^\circ$$

$$\Delta\theta = |\delta\beta| + |\delta\alpha| = 0.03 + 0.25 = \pm\underline{0.28}^\circ$$

$$\Delta l = \frac{1}{l}[(r - l\cos\theta)\delta r + (l - r\cos\theta)\Delta l + rl\sin\theta\Delta\theta]$$

Hence substituting numerical values we obtain

$$\Delta l = \frac{1}{12.15}[(26.25 - 38.4\cos 0.2) \times 0.1 + (38.4 - 26.25\cos 0.2) \times 0.14 + 26.25 \times 38.4 \times \sin 0.2 \times 0.28/57.3]$$

$$\doteq (1.21 + 1.70 + 0.02)/12.15 = \underline{\pm 0.24 \text{ mm}}$$

Hence $l_{max} = 12.15 + 0.24 = 12.39$ mm

$R_{min} = 13.25 - 0.10 = 13.15$ mm

It therefore follows that $l_{max} < R_{min}$ by $\underline{0.76 \text{ mm}}$ as in (2).

10.3 TOLERANCES IN THE FOUR-BAR LINKAGE

We now consider the influence of tolerances on the output position of the four-bar linkage (Fig. 10.7) using the total derivative. We recall Freudenstein's equation, namely

$$K_1 \cos\phi + K_2 \cos\psi - K_3 = \cos(\phi - \psi)$$

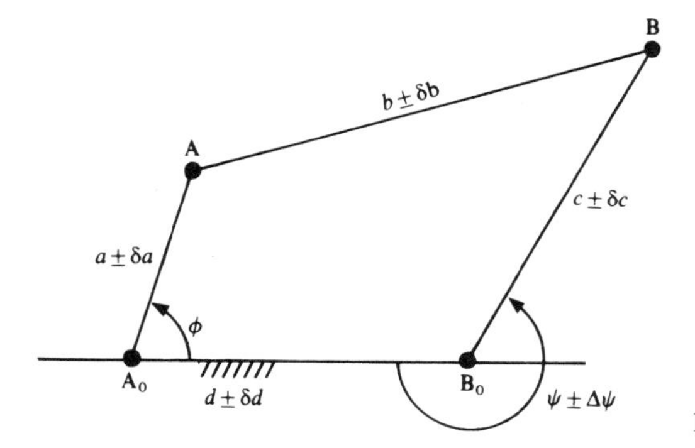

Figure 10.7 Tolerances on the links of a four-bar linkage.

where $K_1 = \frac{d}{c}$

$K_2 = \frac{d}{a}$

$K_3 = \frac{a^2 - b^2 + c^2 + d^2}{2ac}$

Expanding the right-hand term, we have

$$K_1 \cos \phi + K_2 \cos \psi - K_3 = \cos \phi \cos \psi + \sin \phi \sin \psi$$

Rearranging yields

$$\sin \phi \sin \psi + \left(\cos \phi - \frac{d}{a}\right) \cos \psi = \frac{d}{c} \cos \phi - \frac{a^2 - b^2 + c^2 + d^2}{2ac}$$

$$\Leftrightarrow 2ac \sin \phi \sin \psi + (2ac \cos \phi - 2cd) \cos \psi = 2ad \cos \phi - (a^2 - b^2 + c^2 + d^2)$$

This may be more conveniently expressed as follows:

$$D \sin \psi + E \cos \psi = F$$

where $D = 2ac \sin \phi$

$E = 2ac \cos \phi - 2cd$

$F = 2ad \cos \phi - (a^2 - b^2 + c^2 + d^2)$

To calculate the influence of the tolerances on the output position ψ we differentiate this equation implicitly getting

$$D \cos \psi \, \Delta\psi + \Delta D \sin \psi + \Delta E \cos \psi - E \sin \psi \Delta\psi = \Delta F$$

Solving for $\Delta\psi$ yields

$$\Delta\psi = \frac{\Delta F - \Delta D \sin \psi - \Delta E \cos \psi}{D \cos \psi - E \sin \psi} \tag{10.11}$$

Since D, E, and F are functions of a, b, c, d, and ϕ we require the values of ΔD, ΔE, and ΔF, remembering that, in general, since ϕ is an *input* it is not subject to a tolerance unless it is derived from the output of a previous mechanism or linkage or other device.

1. Consider the influence on ψ due to δa, i.e., $\Delta\psi_a$:

$$\Delta D = 2c \sin\phi\, \delta a \qquad \Delta E = 2c \cos\phi\, \delta a \qquad \Delta F = (2d \cos\phi - 2a)\, \delta a$$

Hence substitution in Eq. (10.3) gives

$$\Delta\psi_a = 2\frac{(d \cos\phi - a) - c \cos(\phi - \psi)}{G}\delta a$$

where $G = D \cos\psi - E \sin\psi$.

2. Similarly for δb,

$$\Delta\psi_b = \frac{2b}{G}\delta b$$

3. For δc,

$$\Delta\psi_c = -2\frac{c + a \cos(\phi - \psi) - d \cos\psi}{G}\delta c$$

4. Finally for δd,

$$\Delta\psi_d = 2\frac{a \cos\phi - d + c \cos\psi}{G}\delta d$$

In the case of the worse combination of tolerance we have

$$\Delta\psi_{max} = \pm(|\Delta\psi_a| + |\Delta\psi_b| + |\Delta\psi_c| + |\Delta\psi_d|)$$

If statistical conditions prevail we can use the following value:

$$\Delta\psi_s = \sqrt{\Delta\psi_a^2 + \Delta\psi_b^2 + \Delta\psi_c^2 + \Delta\psi_d^2}$$

which is often referred to as the RMS value (root mean square value).

If a tolerance on the input is imposed (or as a result of a previous device), then

$$\Delta\psi_\phi = \frac{2a(c \sin(\phi - \psi) - d \sin\phi)\, \delta\phi}{G}$$

10.4 INFLUENCE OF TOLERANCES ON THE OUTPUT OF A MECHANISM TO GENERATE THE FUNCTION DISCUSSED IN SEC. 5.7.2

The function was $\psi = 240 + 0.095\phi^{1.5}$ degrees

Using the relationships just derived, the mechanical error (resulting tolerance) on the output ψ of the above mechanism has been calculated for typical values of tolerances which would be allocated for manufacturing purposes. The dimensions quoted on the manufacturing drawings would read:

$$a = 99.00 \pm 0.10 \text{ mm}$$
$$b = 218.25 \pm 0.15 \text{ mm}$$

$$c = 75.50 \pm 0.10 \text{ mm}$$
$$d = 150.00 \pm 0.15 \text{ mm}$$

The results obtained from a computer printout are shown in Table 10.1 as a function of the input position $20 < \phi < 65$ degrees in 5° intervals. The corresponding output angles ψ are also given.

It can be seen that the maximum mechanical error is 0.314 degree. Recalling from Table 5.4 the structural error, i.e., 0.063 degree, we see that the mechanical error is $0.314/0.063 \simeq 5$; five times the structural error.

Tables 10.1 to 10.6 show the influence of the individual tolerances, and we observe that the tolerances on b and d have the greatest effect on the output position. If, in practice, it is necessary to reduce the overall error on the output then we could tighten the tolerances on b and d and relax them on a and c. This is illustrated in Table 10.6, where in order to reduce the overall mechanical error to 0.25° the tolerances on a and c have been increased and those on b and d reduced.

It must be remembered that the allocation of tolerances has an influence on the cost of production and in practice a balance between performance and cost must be established.

Table 10.7 shows the influence of a tolerance of $\pm 0.25°$ (a typical practical value) on the input angle ϕ. We see that in such a situation the resulting error on the output position of the mechanism defined by the angle ψ has practically doubled in value.

Table 10.1

```
**************************************
 MECHANICAL ERROR IN THE 4-BAR LINKAGE
DUE TO IMPOSED MANUFACTURING TOLERANCES
**************************************

THE LINK LENGTHS ARE:
CRANK= 99 COUPLER= 218.25 FOLLOWER= 75.5 FIXED LINK= 150 mm
THE TOLERANCES ARE:
DELTA PHI= 0 DELTA A= .1 DELTA B= .15 DELTA C= .1 DELTA D= .15 mm
```

INPUT ANGLE PHI,DEG.	OUTPUT ANGLE PSI,DEG.	MECH. ERROR DEG.	STAT. ERROR DEG.
20	111.62	.314	.177
25	108.16	.308	.175
30	104.39	.302	.173
35	100.33	.294	.172
40	95.99	.287	.17
45	91.38	.279	.169
50	86.49	.281	.168
55	81.33	.284	.168
60	75.9	.288	.169
65	70.17	.294	.172

OVERALL STAT. MECH. ERROR,DEG.= .293

Table 10.2

THE TOLERANCES ARE:
DELTA PHI= 0 DELTA A= .1 DELTA B= 0 DELTA C= 0 DELTA D= 0 mm

INPUT ANGLE PHI,DEG.	OUTPUT ANGLE PSI,DEG.	MECH. ERROR DEG.	STAT. ERROR DEG.
20	111.62	.067	.067
25	108.16	.064	.064
30	104.39	.061	.061
35	100.33	.058	.058
40	95.99	.055	.055
45	91.38	.052	.052
50	86.49	.048	.048
55	81.33	.045	.045
60	75.9	.042	.042
65	70.17	.039	.039

OVERALL STAT. MECH. ERROR,DEG.= .054

Table 10.3

THE TOLERANCES ARE:
DELTA PHI= 0 DELTA A= 0 DELTA B= .15 DELTA C= 0 DELTA D= 0 mm

INPUT ANGLE PHI,DEG.	OUTPUT ANGLE PSI,DEG.	MECH. ERROR DEG.	STAT. ERROR DEG.
20	111.62	.116	.116
25	108.16	.115	.115
30	104.39	.115	.115
35	100.33	.114	.114
40	95.99	.113	.113
45	91.38	.113	.113
50	86.49	.114	.114
55	81.33	.114	.114
60	75.9	.115	.115
65	70.17	.117	.117

OVERALL STAT. MECH. ERROR,DEG.= .115

Table 10.4

THE TOLERANCES ARE:
DELTA PHI= 0 DELTA A= 0 DELTA B= 0 DELTA C= .1 DELTA D= 0 mm

INPUT ANGLE PHI,DEG.	OUTPUT ANGLE PSI,DEG.	MECH. ERROR DEG.	STAT. ERROR DEG.
20	111.62	.016	.016
25	108.16	.013	.013
30	104.39	.01	.01
35	100.33	7E-03	7E-03
40	95.99	3E-03	3E-03
45	91.38	0	0
50	86.49	4E-03	4E-03
55	81.33	9E-03	9E-03
60	75.9	.014	.014
65	70.17	.02	.02

OVERALL STAT. MECH. ERROR,DEG.= .011

Table 10.5

THE TOLERANCES ARE:
DELTA PHI= 0 DELTA A= 0 DELTA B= 0 DELTA C= 0 DELTA D= .15 mm

INPUT ANGLE PHI,DEG.	OUTPUT ANGLE PSI,DEG.	MECH. ERROR DEG.	STAT. ERROR DEG.
20	111.62	.114	.114
25	108.16	.114	.114
30	104.39	.114	.114
35	100.33	.114	.114
40	95.99	.113	.113
45	91.38	.113	.113
50	86.49	.114	.114
55	81.33	.114	.114
60	75.9	.115	.115
65	70.17	.117	.117

OVERALL STAT. MECH. ERROR,DEG.= .114

Table 10.6

THE TOLERANCES ARE:
DELTA PHI= 0 DELTA A= .15 DELTA B= .1 DELTA C= .15 DELTA D= .1 mm

INPUT ANGLE PHI,DEG.	OUTPUT ANGLE PSI,DEG.	MECH. ERROR DEG.	STAT. ERROR DEG.
20	111.62	.279	.15
25	108.16	.271	.147
30	104.39	.262	.143
35	100.33	.251	.139
40	95.99	.241	.136
45	91.38	.23	.132
50	86.49	.232	.13
55	81.33	.235	.128
60	75.9	.239	.128
65	70.17	.245	.129

OVERALL STAT. MECH. ERROR,DEG.= .249

Table 10.7

****
MECHANICAL ERROR IN THE 4-BAR LINKAGE
DUE TO IMPOSED MANUFACTURING TOLERANCES
****

THE LINK LENGTHS ARE:
CRANK= 99 COUPLER= 218.25 FOLLOWER= 75.5 FIXED LINK= 150 mm

EFFECT OF A TOLERANCE ON THE INPUT PHI

THE TOLERANCES ARE:
DELTA PHI= .25° DELTA A= .15 DELTA B= .1 DELTA C= .15 DELTA D= .1 mm

INPUT ANGLE PHI,DEG.	OUTPUT ANGLE PSI,DEG.	MECH. ERROR DEG.	STAT. ERROR DEG.
20	111.62	.445	.223
25	108.16	.452	.233
30	104.39	.457	.242
35	100.33	.461	.252
40	95.99	.465	.262
45	91.38	.467	.272
50	86.49	.483	.282
55	81.33	.5	.294
60	75.9	.518	.307
65	70.17	.539	.321

OVERALL STAT. MECH. ERROR,DEG.= .48

EXERCISES

10.1 Figure 10.8 shows two of the many elements that make up a precision timing mechanism. The cam is pivoted at A and the lever at B. In the position shown the pin C is located in the recess E, and the position of the pin D relative to the fixed pivot F is critical in the operation of the mechanism. If all dimensions, except for pivots B and F, have a tolerance of ± 0.1 mm, calculate the most probable deviation of the pin D from its nominal position relative to F.

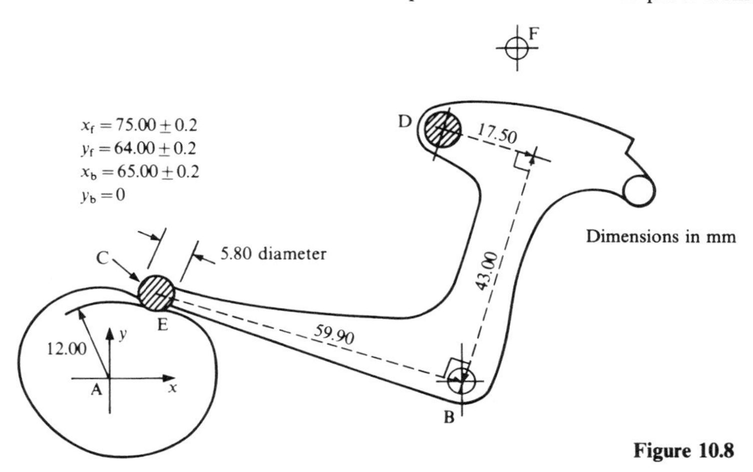

Figure 10.8

10.2 A cam segment AB is used to position the lever L, which slides in the groove G as shown in Fig. 10.9. For the position shown sketch an equivalent slider–crank mechanism and calculate the tolerances you should allocate on its three dimensions if the tolerance on the positioning of the lever as measured by d is to remain within ± 0.250 mm.

If most of the friction is between the lever and the groove calculate the power loss when the cam has an angular velocity of 5 rad/s and the friction force is 1.5 N.

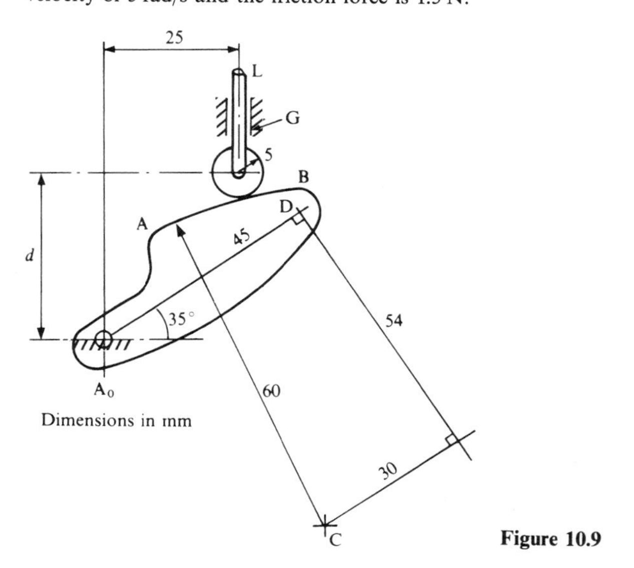

Figure 10.9

10.3 Figure 10.10 shows a four-bar linkage whose link proportions are such that the path of point C on the coupler ABC is a straight line parallel to A_0B_0 within the range of operation defined by $90° \leqslant \phi \leqslant 270°$. Dimensions: $a = 12.5$ mm; $b = c = l = 50$ mm; $d = 37.5$ mm.

Show that the angular position θ of the coupler ABC can be expressed in the form

$$D \sin\theta + E \cos\theta = F$$

where D is a function of ϕ and E and F are functions of the link lengths and of the angle ϕ.

Calculate the maximum error in the coordinate y_C of the point C when $\phi = 150°$ owing to tolerances of ± 0.1 mm in the lengths c and d. Would you expect this error to remain constant within the range of operation of the linkage? Give your reasons.

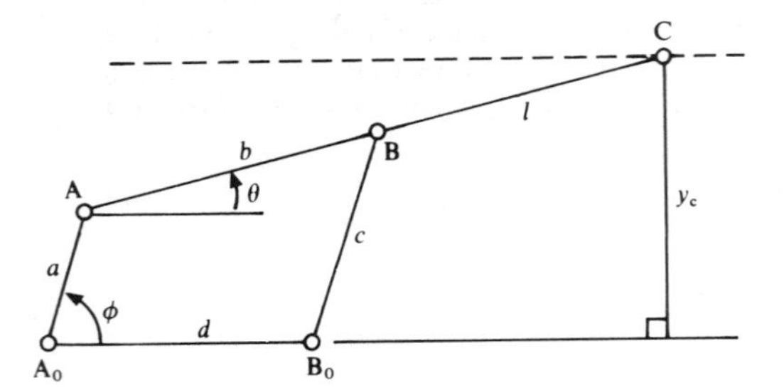

Figure 10.10

10.4 The angular position of the cam OA in a precision mechanism is derived from the rectilinear movement of a slider B constrained to move in the groove G as shown in Fig. 10.11.

Derive all algebraic expressions which would be necessary to calculate the maximum possible positional error of the output angle ϕ of lever OA due to manufacturing tolerances $\pm\Delta a$, $\pm\Delta b$, $\pm\Delta c$ and $\pm\Delta d$ when the slider has been moved into position C.

The displacement BC of the slider is not subject to a tolerance. In a particular case the design dimensions to obtain the desired movement of the cam were: $a = 24$ mm, $b = 49$ mm, $c = 7.5$ mm, and $d = 70$ mm, with specified tolerances of ± 0.1, ± 0.15, ± 0.075, and ± 0.15 mm respectively. Calculate the maximum error on a statistical basis in the position of the cam when the slider is at B and at C given that $l = 15$ mm.

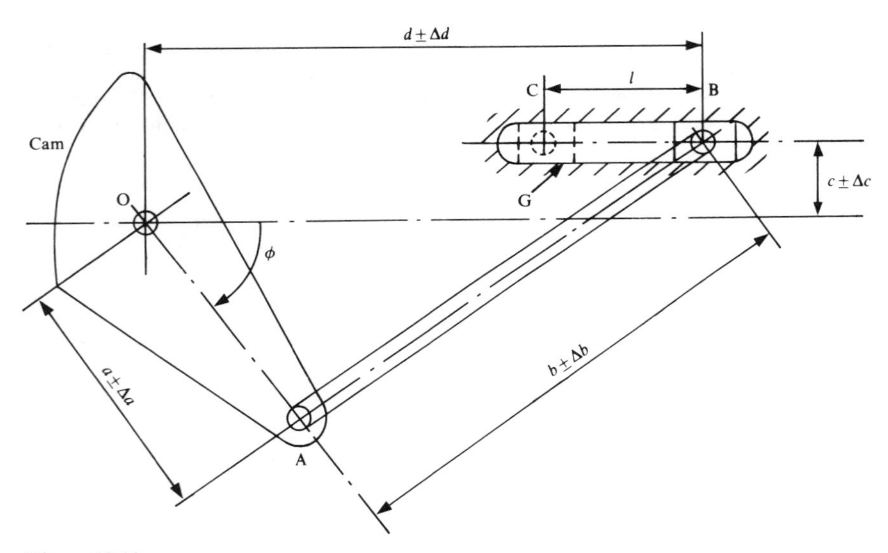

Figure 10.11

10.5 Referring to Exercise 3.10 on page 72 calculate the maximum error in the angular position of the output link D_0D when $\phi = 90°$. The tolerances are ± 0.15 mm on all the dimensions.

10.6 In the case of the turntable mechanism illustrated in Fig. 10.12, one important aspect is to ensure that the stylus will always be positioned within the starting band of either a 7-inch or a 12-inch record under adverse tolerance combinations.

From the manufacturing drawings the dimensions are shown in the figure. Calculate the tolerances to be imposed on θ_s in order to ensure that the stylus will fall inside the starting bands.

Use the x–y coordinate system indicated with the origin at A_0, the coordinates of B_0 are 147.25 mm and 100.00 mm, and those of D_0 are 145.40 mm and 55.15 mm with tolerances of ± 0.15 mm.

Note: Once the stylus has been brought into position the record size selector is moved out of the way to ensure free movement of the pick-up arm.

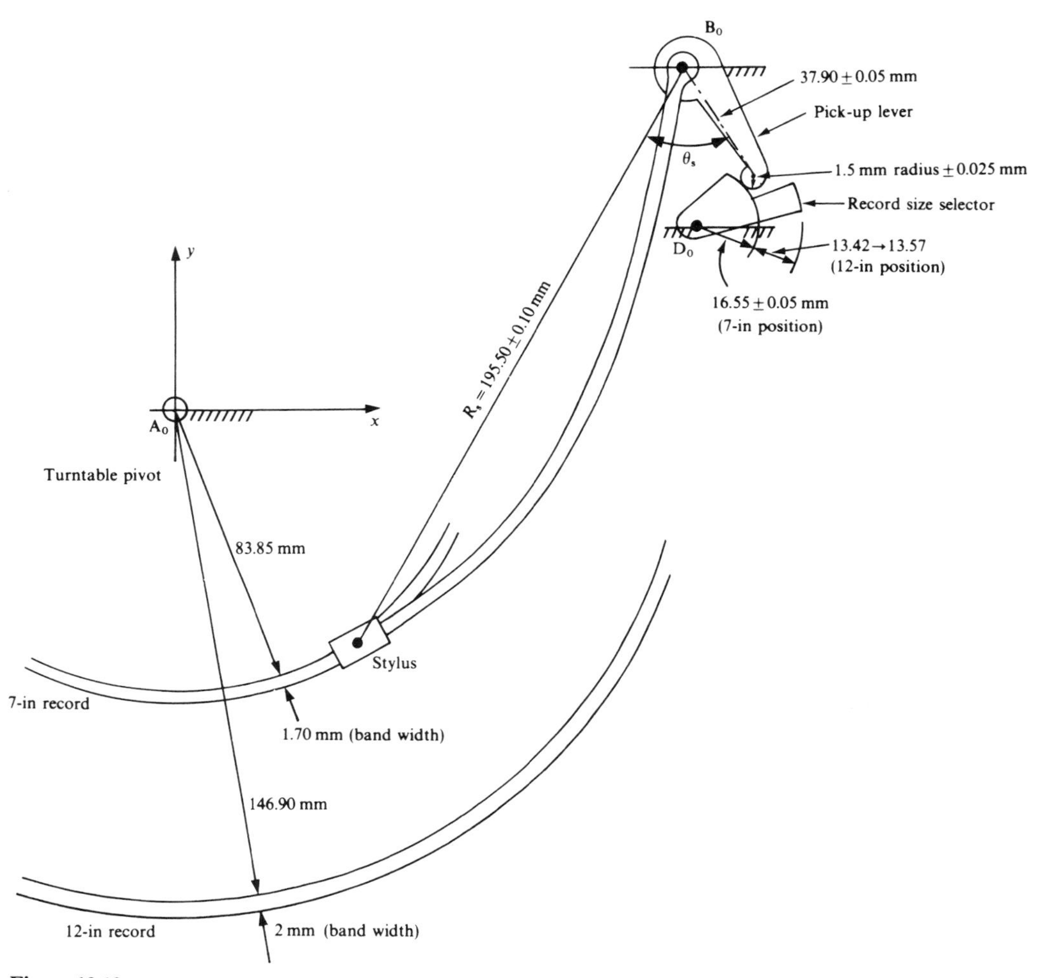

Figure 10.12

10.7 The precision mechanism shown in Fig. 10.13 has the following dimensions and proposed manufacturing tolerances in mm: $a = 25 \pm 0.1$; $b = 53.5 \pm 0.15$; $c = 45 \pm 0.1$; $d = 62.5 \pm 0.15$; $L = 48.5 \pm 0.15$; $\alpha = 27.5° \pm 0.25°$; $e = 75 \pm 0.2$; $f = 80 \pm 0.2$; $A_0E = 115 \pm 0.25$.

The output is taken from F on the crank DEF.

(*a*) Calculate the design value of the total angular movement γ of the output crank.

(*b*) Plot the variation in the maximum and statistical error on the value of the angle γ due to the imposed manufacturing tolerances for one complete revolution of the input crank A_0A.

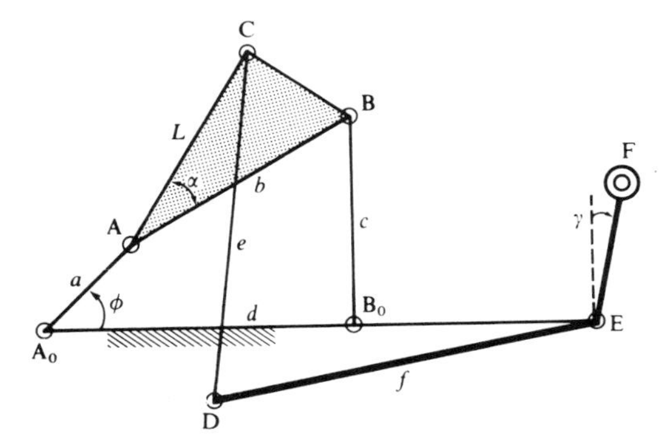

Figure 10.13

APPENDIX

A

COMPUTER PROGRAMS

A.1 ANALYSIS OF FOUR-BAR AND SIX-BAR LINKAGES (FORTRAN)

```
***********************************************************************
*                           PROGRAM LINKS
* Given the link lengths of a 4 bar linkage and the input angle
* PHI(0) and its derivatives PHI(1) to PHI(3), the program calculates
* the output angle and derivatives PSI(0) to PSI(3) and the coupler angle
* and derivatives THETA(0) to THETA(2).Optionally it also gives the
* motion of a point on the coupler or follower , and the motion of a
* further pair of links between this point and a fixed point.
***********************************************************************
      PROGRAM LINKS
      LOGICAL OPEN4,COUPL,FOLLOW,SIXBAR,OPEN6,BETCAL,REPLY,NEWPHI,NEWPHD
     +,AOK,NEWALL
      REAL A,B,C,D,E,F,AL,ALPHA,PHI(0:3),THETA(0:3),PSI(0:3),BETA(0:2),
     +GAMMA(0:2),XE,YE,PI,CONV,XC(0:2),YC(0:2)
      CHARACTER*60 QU1,QU2,QU3,QU4,QU5,QU6,QU7,QU8,QU9,QUA*42,QUB*30,
     +FAULT
*======================================================================
      PI=3.14139
      CONV=PI/180.0
      QU1='Is the linkage open rather than crossed? [y or n] '
      QUA='Do you require motion of a point C on the '
      QU2=QUA//'coupler? '
      QU3=QUA//'follower? '
      QU4='Is your linkage a 6 bar one? '
      QU5='Are links CD and DE open rather than crossed? '
      QU6='Do you require beta and derivatives? '
      QUB='Do you wish to enter new value'
      QU7=QUB//' for phi? '
      QU8=QUB//'s for derivatives of phi? '
      QU9=QUB//'s for everything? '
*======================================================================
```

```
      WRITE(*,600)
  600 FORMAT('1*********************************************************
     +***************'//'                 FOUR BAR LINKAGE CALCULATIONS'
     +//' **************************************************************
     +**********'//' Entered lengths are assumed to be in metres,angles
     +in degrees'/' angular velocities in rad/s, ang accel in rad/s2'//)
*=======================================================================
  100 CONTINUE
*=======================================================================
      WRITE(*,*) ' Enter link lengths as indicated, in metres'
      WRITE(*,*) ' Crank Ao A = '
      READ(*,*) A
      WRITE(*,*) ' Coupler A B = '
      READ(*,*) B
      WRITE(*,*) ' Follower B Bo = '
      READ(*,*) C
      WRITE(*,*) ' Fixed link Ao Bo = '
      READ(*,*) D
*=======================================================================
* Determine users requirements
*-----------------------------------------------------------------------
      OPEN4=REPLY(QU1)
      COUPL=REPLY(QU2)
      IF(COUPL) THEN
         FOLLOW=.FALSE.
      ELSE
         FOLLOW=REPLY(QU3)
      ENDIF
      IF(COUPL.OR.FOLLOW) THEN
         WRITE(*,*) ' Enter length AC in metres '
         READ(*,*) AL
         WRITE(*,*) ' Enter angle alpha in degrees '
         READ(*,*) ALPHA
         ALPHA=ALPHA*CONV
         SIXBAR=REPLY(QU4)
         IF(SIXBAR) THEN
            OPEN6=REPLY(QU5)
            WRITE(*,*) ' Enter coords XE,YE of fixed point E,and ',
     +                 'link lengths CD and DE in metres'
            WRITE(*,*) ' XE= '
            READ(*,*) XE
            WRITE(*,*) ' YE= '
            READ(*,*) YE
            WRITE(*,*) ' CD= '
            READ(*,*) E
            WRITE(*,*) ' DE= '
            READ(*,*) F
            BETCAL=REPLY(QU6)
         ENDIF
      ELSE
         SIXBAR=.FALSE.
      ENDIF
```

```
      NEWPHI=.TRUE.
      NEWPHD=.TRUE.
*=====================================================================
 101   CONTINUE
*=====================================================================
         IF(NEWPHI) THEN
            WRITE(*,*) ' Enter input angle PHI in degrees '
            READ(*,*) PHI(0)
            PHI(0)=PHI(0)*CONV
         ENDIF
         IF(NEWPHD) THEN
            WRITE(*,*) ' Enter derivatives of PHI as follows'
            WRITE(*,*) ' Velocity (rad/s)= '
            READ(*,*) PHI(1)
            WRITE(*,*) ' Acceleration(rad/s2)= '
            READ(*,*) PHI(2)
            WRITE(*,*) ' Jerk (rad/s3)= '
            READ(*,*) PHI(3)
         ENDIF
*=====================================================================
         AOK=.TRUE.
         FAULT=' '
         CALL BAR4(A,B,C,D,OPEN4,PHI,THETA,PSI,AOK,FAULT)
         IF(AOK) THEN
            IF(COUPL) CALL POINTC(A,0.0,AL,ALPHA,PHI,THETA,XC,YC)
            IF(FOLLOW) CALL POINTC(-C,D,AL,ALPHA,PSI,PSI,XC,YC)
            IF(SIXBAR) THEN
             CALL BAR6(XC,YC,XE,YE,E,F,OPEN6,.FALSE.,FAULT)
             IF(BETCAL) CALL BAR6(XC,YC,XE,YE,-F,-E,.NOT.OPEN6,
     +              .TRUE.,FAULT)
            ENDIF
         ENDIF
         WRITE(*,*) FAULT
*=====================================================================
         NEWALL=REPLY(QU9)
         IF(.NOT.NEWALL) THEN
            NEWPHI=REPLY(QU7)
            NEWPHD=REPLY(QU8)
         ENDIF
*=====================================================================
      IF((NEWPHI.OR.NEWPHD).AND.(.NOT.NEWALL)) GO TO 101
*=====================================================================
      IF(NEWALL) GO TO 100
*=====================================================================
      END
**********************************************************************
**********************************************************************
      LOGICAL FUNCTION REPLY(QU)
      CHARACTER QU*60,ANS*1
*=====================================================================
      WRITE(*,'(1X,A60)') QU
      READ(*,'(A1)') ANS
```

```
      REPLY=ANS.EQ.'Y'.OR.ANS.EQ.'y'
*=======================================================================
      RETURN
      END
************************************************************************
************************************************************************
      REAL FUNCTION TANFUN(FA,FB,FC,FD,OPENX)
      REAL FA,FB,FC,FD
      LOGICAL OPENX
*=======================================================================
      FD=SQRT(FD)
      IF(.NOT.OPENX) FD=-FD
      TANFUN=2.0*ATAN2(FA+FD,FB+FC)
*=======================================================================
      RETURN
      END
************************************************************************
************************************************************************
      SUBROUTINE BAR4(A,B,C,D,OPEN4,PHI,THETA,PSI,AOK,FAULT)
      REAL A,B,C,D,PHI(0:3),THETA(0:3),PSI(0:3),FK1,FK2,FK3,SINPHI,
     +COSPHI,FB,FC,FD2,COSPSI,SINPSI,BCOSTH,BSINTH,PI,CONV,TANFUN,G,GXX
      LOGICAL OPEN4,AOK
      CHARACTER*60 FAULT
      INTEGER I
      PI=3.14159
      CONV=PI/180.0
*=======================================================================
* Calculate psi
*-----------------------------------------------------------------------
      FK1=D/C
      FK2=D/A
      FK3=(A*A-B*B+C*C+D*D)/(2.0*A*C)
      SINPHI=SIN(PHI(0))
      COSPHI=COS(PHI(0))
      FB=COSPHI-FK2
      FC=FK1*COSPHI-FK3
      FD2=SINPHI*SINPHI+FB*FB-FC*FC
      IF(FD2.LT.0.0) THEN
       AOK=.FALSE.
       FAULT=' Impossible input'
      ELSE
       PSI(0)=TANFUN(SINPHI,FB,FC,FD2,OPEN4)
*=======================================================================
* Calculate theta from horizontal and vertical projections of link AB
*-----------------------------------------------------------------------
       COSPSI=COS(PSI(0))
       SINPSI=SIN(PSI(0))
       BCOSTH=D-A*COSPHI-C*COSPSI
       BSINTH=-C*SINPSI-A*SINPHI
       THETA(0)=ATAN2(BSINTH,BCOSTH)
*=======================================================================
* Calculate derivatives of psi and theta
```

```
*-----------------------------------------------------------------------
       CALL DERIV(FK1,FK2,SINPHI,COSPHI,PHI,PSI,.TRUE.,G)
       CALL DERIV(D/B,FK2,SINPHI,COSPHI,PHI,THETA,.FALSE.,GXX)
*=======================================================================
* Print the transmission ratio and velocity ratio
*-----------------------------------------------------------------------
       WRITE(*,600)  (PSI(0)-THETA(0)+PI)/CONV,G
*=======================================================================
* Print other results
*-----------------------------------------------------------------------
       WRITE(*,601) PHI(0)/CONV,(PHI(I),I=1,3),THETA(0)/CONV,(THETA(I)
     +    ,I=1,2),PSI(0)/CONV,(PSI(I),I=1,3)
      ENDIF
      RETURN
*=======================================================================
  600 FORMAT(' Transmission ratio=',F7.1,3X,'Velocity ratio=',G14.4/)
  601 FORMAT(6X,'Angle(deg)  Vel(rad/s)  Accel(rad/s2)  Jerk(rad/s3)'/
     +' Phi   ',F9.1,3E13.4/' Theta',F9.1,2E13.4/' Psi   ',F9.1,3E13.4/)
      END
************************************************************************
************************************************************************
      SUBROUTINE DERIV(FKX,FK2,SINPHI,COSPHI,PHI,ANGLE,JERK,G)
      REAL FKX,FK2,SINPHI,COSPHI,PHI(0:3),ANGLE(0:3),SINDIF,COSDIF,
     +SINANG,COSANG,Z1,Z2,G,GD,Z1D,Z2D,G1,ZSUB,Z1DD,Z2DD,G2
      LOGICAL JERK
*=======================================================================
* Calculate velocity
*-----------------------------------------------------------------------
      SINDIF=SIN(PHI(0)-ANGLE(0))
      COSDIF=COS(PHI(0)-ANGLE(0))
      SINANG=SIN(ANGLE(0))
      COSANG=COS(ANGLE(0))
      Z1=SINDIF-FKX*SINPHI
      Z2=SINDIF+FK2*SINANG
      G=Z1/Z2
      ANGLE(1)=G*PHI(1)
*=======================================================================
* Calculate acceleration
*-----------------------------------------------------------------------
      GD=1.0-G
      Z1D=GD*COSDIF-FKX*COSPHI
      Z2D=GD*COSDIF+G*FK2*COSANG
      G1=(Z2*Z1D-Z1*Z2D)/(Z2*Z2)
      ANGLE(2)=G1*PHI(1)*PHI(1)+G*PHI(2)
*=======================================================================
* Calculate jerk of psi only
*-----------------------------------------------------------------------
      IF(JERK) THEN
       ZSUB=-GD*GD*SINDIF-G1*COSDIF
       Z1DD=ZSUB+FKX*SINPHI
       Z2DD=ZSUB-(G*G*SINANG-G1*COSANG)*FK2
       G2=(Z2*Z1DD-Z1*Z2DD)/(Z2*Z2)-2.0*G1*Z2D/Z2
```

```
        ANGLE(3)=G2*PHI(1)**3+3.0*G1*PHI(1)*PHI(2)+G*PHI(3)
       ENDIF
*=====================================================================
       RETURN
       END
**********************************************************************
**********************************************************************
       SUBROUTINE POINTC(AC,D,AL,ALPHA,ANG1,ANG2,XC,YC)
       REAL AC,D,ALPHA,ANG1(0:3),ANG2(0:3),XC(0:2),YC(0:2),ASP,ACP,SUMANG
      +,ASSUM,ACSUM
*=====================================================================
       ASP=AC*SIN(ANG1(0))
       ACP=AC*COS(ANG1(0))
       SUMANG=ANG2(0)+ALPHA
       ASSUM=SIN(SUMANG)*AL
       ACSUM=COS(SUMANG)*AL
       XC(0)=ACP+ACSUM
       YC(0)=ASP+ASSUM
       XC(1)=-ASP*ANG1(1)-ASSUM*ANG2(1)
       YC(1)=ACP*ANG1(1)+ACSUM*ANG2(1)
       XC(2)=-ACP*ANG1(1)*ANG1(1)-ACSUM*ANG2(1)*ANG2(1)-ASP*ANG1(2)-ASSUM
      +*ANG2(2)
       YC(2)=-ASP*ANG1(1)*ANG1(1)-ASSUM*ANG2(1)*ANG2(1)+ACP*ANG1(2)+ACSUM
      +*ANG2(2)
*=====================================================================
       WRITE(*,600) XC,YC
   600 FORMAT(5X,'Coordinate(m)  Velocity(m/s)  Acceleration(m/s2)'/
      +' Xc',3E15.4/' Yc',3E15.4/)
*=====================================================================
       RETURN
       END
**********************************************************************
**********************************************************************
       SUBROUTINE BAR6(XC,YC,XE,YE,E,F,OPEN6,BETCAL,FAULT)
       REAL XC(0:2),YC(0:2),XE,YE,E,F,GAMMA(0:2),XCE,YCE,FA,FB,
      +FC,FD,TANFUN,DA,DB,DC,DD,DE,SINGAM,COSGAM,SA,SB,SC,SD,SE
       LOGICAL OPEN6,BETCAL
       CHARACTER*60 FAULT
*=====================================================================
* Calculate gamma. If BETCAL is true,the same calculations are used for
* beta and its derivatives by putting E=-F,F=-E when the subroutine is
* called.Opposite sign in TAN formula is needed so put OPEN6=.NOT.OPEN6
*---------------------------------------------------------------------
       XCE=XC(0)-XE
       YCE=YC(0)-YE
       FA=2.0*F*YCE
       FB=2.0*F*XCE
       FC=E*E-F*F-YCE*YCE-XCE*XCE
       FD=FA*FA+FB*FB-FC*FC
       IF(FD.LT.0.0) THEN
        FAULT=' Links CD and DE have impossible specification'
       ELSE
```

```
      GAMMA(0)=TANFUN(FA,FB,FC,FD,OPEN6)
*=========================================================================
* Calculate derivatives of gamma or beta
*-------------------------------------------------------------------------
      DA=2.0*F*YC(1)
      DB=2.0*F*XC(1)
      DC=-2.0*(YCE*YC(1)+XCE*XC(1))
      SINGAM=SIN(GAMMA(0))
      COSGAM=COS(GAMMA(0))
      DD=DC-DB*COSGAM-DA*SINGAM
      DE=FA*COSGAM-FB*SINGAM
      GAMMA(1)=DD/DE
      SA=2.0*F*YC(2)
      SB=2.0*F*XC(2)
      SC=-2.0*(XCE*XC(2)+XC(1)*XC(1)+YCE*YC(2)+YC(1)*YC(1))
      SD=GAMMA(1)*(DB*SINGAM-DA*COSGAM)+SC-SB*COSGAM-SA*SINGAM
      SE=DA*COSGAM-DB*SINGAM-GAMMA(1)*(FA*SINGAM+FB*COSGAM)
      GAMMA(2)=(DE*SD-DD*SE)/(DE*DE)
*=========================================================================
* Print results
*-------------------------------------------------------------------------
      GAMMA(0)=GAMMA(0)*57.296
      IF(BETCAL) THEN
         WRITE(*,600) GAMMA
      ELSE
         WRITE(*,601) GAMMA
      ENDIF
     ENDIF
     RETURN
*=========================================================================
  600 FORMAT(' Beta ',F9.1,2E13.4/)
  601 FORMAT(6X,'Angle(deg)  Vel(rad/s)  Accel(rad/s2)'/' Gamma',F9.1,
     +2E13.4)
*=========================================================================
     END
*************************************************************************
```

A.2 ANALYSIS OF FOUR-BAR AND SIX-BAR LINKAGES (BASIC)

```
10 PRINT"ANALYSIS OF 4-BAR"
20 PRINT"AND 6-BAR LINKAGES"
30 PRINT"********************"
50 REM   THE PROGRAMME CALCULATES:
60 REM       POSITION,VELOCITY,
70 REM       ACCELERATION,JERK AND
80 REM       TRANSMISSION ANGLE FOR
90 REM   A GIVEN INPUT POSITION,VELOCITY,
100 REM       ACCELERATION AND JERK
110 REM    REFER TO 5.11.5 IN THE TEXT
120 PRINT
130 PRINT "ENTER LINK LENGTHS IN UNITS OF YOUR CHOICE"
140 PRINT
150 INPUT "CRANK A0A=";A
```

```
160 INPUT "COUPLER AB=";B
170 INPUT "FOLLOWER BB0=";C
180 INPUT "FIXED LINK A0B0=";D
190 K1=D/C:K2=D/A:K3=(A*A-B*B+C*C+D*D)/2/A/C
200 PRINT"ENTER PHI DOT,PHI DOUBLE DOT,JERK"
210 PRINT
220 INPUT"PHI DOT IN RAD/SEC";W1
230 INPUT"PHI DOUBLE DOT IN RAD/SEC↑2";AC
240 INPUT"JERK IN RAD/SEC↑3";J1
250 PRINT
260 INPUT"INPUT CRANK ANGLE PHI IN DEGREES";P1
270 P1=P1*π/180
280 PRINT"IS LINKAGE OPEN (O) OR CROSSED (C) ?"
290 INPUT Z$
300 IF Z$="0" GOTO 320
310 IF Z$="C" GOTO 370
320 GOSUB 1680
330 P3=2*ATN((A1+D1)/(B1+C1))
340 GOSUB 1730
350 P2=2*ATN((H1-H4)/(H2+H3))
360 GOTO 410
370 GOSUB 1680
380 P3=2*ATN((A1-D1)/(B1+C1))
390 GOSUB 1730
400 P2=2*ATN((H1+H4)/(H2+H3))
410 PRINT "PHI="P1*180/π;"PSI="P3*180/π;"THETA="P2*180/π
420 L1=SIN(P1-P3)-K1*SIN(P1)
430 L2=SIN(P1-P3)+K2*SIN(P3)
440 VR=INT(L1*1000/L2)/1000
450 PRINT"VELOCITY RATIO="VR
460 W3=VR*W1
470 PRINT"OUTPUT VELOCITY,RAD/S="W3
480 L3=(1-VR)*COS(P1-P3)-K1*COS(P1)
490 L4=(1-VR)*COS(P1-P3)+VR*K2*COS(P3)
500 A3=(L2*L3-L1*L4)/L2/L2*W1*W1+VR*AC
510 PRINT"OUTPUT ACCELERATION,RAD/S↑2="INT(A3*1000)/1000
520 Z1=(L2*L3-L1*L4)/L2/L2
530 Y1=-(1-VR)*(1-VR)*SIN(P1-P3)-Z1*COS(P1-P3)
540 Z3=Y1+K1*SIN(P1)
550 Z4=Y1-VR*VR*K2*SIN(P3)+Z1*K2*COS(P3)
560 Z2=(L2*Z3-L1*Z4)/L2/L2
570 Z5=2*Z1*L4/L2
580 J3=(Z2-Z5)*W1*W1*W1+3*Z1*AC*W1+VR*J1
590 PRINT"OUTPUT JERK,RAD/S↑3="INT(J3*1000)/1000
600 O=(B*B+C*C-A*A-D*D+2*A*D*COS(P1))/2/B/C
610 U1=-ATN(O/SQR(-O*O+1))+π/2
620 U=INT(U1*180/π*100)/100
630 PRINT"TRANSMISSION ANGLE MU,DEG.="U
640 L5=SIN(P1-P2)-D*SIN(P1)/B
650 L6=SIN(P1-P2)+K2*SIN(P2)
660 W2=L5*W1/L6
670 PRINT"COUPLER VELOCITY THETA DOT,RAD/S="INT(W2*1000)/1000
680 L7=(1-L5/L6)*COS(P1-P2)-D*COS(P1)/B
690 L8=(1-L5/L6)*COS(P1-P2)+L5*K2*COS(P2)/L6
700 A2=(L6*L7-L5*L8)/L6/L6*W1*W1+L5/L6*AC
710 PRINT"COUPLER ACCELERATION THETA DOUBLE DOT,RAD/S↑2="INT(A2*1000)/1000
720 PRINT"DO YOU REQUIRE THE MOTION OF A POINT  C  ON:"
730 PRINT"COUPLER (CO), FOLLOWER(FO) OR NEITHER (N) ?"
740 INPUT Z$
750 IF Z$="CO" GOTO 780
760 IF Z$="FO" GOTO 930
770 IF Z$="N"  GOTO 1590
780 INPUT"LENGTH AC=";L
790 INPUT"ANGLE ALPHA IN DEGREES=";AL
```

```
800 AL =AL*π/180
810 XC=A*COS(P1)+L*COS(P2+AL)
820 YC=A*SIN(P1)+L*SIN(P2+AL)
830 UC=-A*SIN(P1)*W1-L*SIN(P2+AL)*W2
840 VC= A*COS(P1)*W1+L*COS(P2+AL)*W2
850 X1=-A*(COS(P1)*W1*W1+SIN(P1)*AC)
860 X2=-L*(COS(P2+AL)*W2*W2+SIN(P2+AL)*A2)
870 AX=X1+X2
880 Y1=A*(-SIN(P1)*W1*W1+COS(P1)*AC)
890 Y2=L*(-SIN(P2+AL)*W2*W2+COS(P2+AL)*A2)
900 AY=Y1+Y2
910 GOSUB 1780
920 GOTO 1070
930 INPUT"LENGTH BC=";L
940 INPUT"ANGLE ALPHA=";AL
950 AL=AL*π/180
960 XC=D-C*COS(P3)+L*COS(P3+AL)
970 YC=-C*SIN(P3)+L*SIN(P3+AL)
980 UC=(C*SIN(P3)-L*SIN(P3+AL))*W3
990 VC=(-C*COS(P3)+L*COS(P3+AL))*W3
1000 X1=(C*COS(P3)-L*COS(P3+AL))*W3*W3
1010 X2=(C*SIN(P3)-L*SIN(P3+AL))*A3
1020 AX=X1+X2
1030 Y1=(C*SIN(P3)-L*SIN(P3+AL))*W3*W3
1040 Y2=(-C*COS(P3)+L*COS(P3+AL))*A3
1050 AY=Y1+Y2
1060 GOSUB 1780
1070 PRINT"IS YOUR LINKAGE A 6-BAR:YES (Y) OR NO (N) ?"
1080 INPUT E$
1090 IF E$="Y" GOTO 1110
1100 IF E$="N" GOTO 1590
1110 PRINT"ENTER COORDINATES XE AND YE OF FIXED POINT E,"
1120 PRINT"AND LINK LENGTHS CD AND DE"
1130 PRINT
1140 INPUT "XE=";XE
1150 INPUT "YE=";YE
1160 INPUT "LENGTH OF LINK CD";E
1170 INPUT "LENGTH OF LINK DE";F
1180 AE=2*F*(YC-YE)
1190 BE=-2*F*(XE-XC)
1200 CE=-(XE-XC)*(XE-XC)-(YC-YE)*(YC-YE)-F*F+E*E
1210 DE=AE*AE+BE*BE-CE*CE
1220 PRINT"WHAT IS THE CONFIGURATION OF LINK CDE:OPEN (O) OR CROSSED (C)?"
1230 INPUT Z$
1240 IF Z$="O" GOTO 1350
1260 GA=2*ATN((AE-SQR(DE))/(BE+CE))
1270 PRINT"DO YOU REQUIRE BETA,BETA DOT,BETA DOUBLE DOT:YES (Y) OR NO (N)?"
1280 INPUT Z$
1290 IF Z$="Y" GOTO 1310
1300 IF Z$="N" GOTO 1440
1310 GOSUB 1920
1320 BT=2*ATN((A0+D0)/(B0+C0))
1330 GOSUB 1970
1340 GOTO 1440
1350 GA=2*ATN((AE+SQR(DE))/(BE+CE))
1360 PRINT"DO YOU REQUIRE BETA,BETA DOT,BETA DOUBLE DOT:YES (Y) OR NO (N)?"
1370 INPUT Z$
1380 IF Z$="Y" GOTO 1400
1390 IF Z$="N" GOTO 1440
1400 GOSUB 1920
1410 BT=2*ATN((A0-D0)/(B0+C0))
1420 GOSUB 1970
1430 GOTO 1440
1440 AD=2*F*VC
1450 BD=2*F*UC
```

```
 1460 CD=2*(XE-XC)*UC-2*(YC-YE)*VC
 1470 N=CD-BD*COS(GA)-AD*SIN(GA)
 1480 G=AE*COS(GA)-BE*SIN(GA)
 1490 GD=N/G
 1500 SA=2*F*AY
 1510 SB=2*F*AX
 1520 SC=2*(XE-XC)*AX-2*UC*UC-2*(YC-YE)*AY-2*VC*VC
 1530 L1=N/G*(BD*SIN(GA)-AD*COS(GA))+SC-SB*COS(GA)-SA*SIN(GA)
 1540 L2=AD*COS(GA)-BD*SIN(GA)-N/G*(AE*SIN(GA)+BE*COS(GA))
 1550 SG=(G*L1-N*L2)/G/G
 1560 PRINT"OUTPUT ANGLE GAMMA,DEG.="INT(GA*180/π*100)/100
 1570 PRINT"GAMMA DOT, RAD/S="INT(GD*1000)/1000
 1580 PRINT"GAMMA DOUBLE DOT, RAD/S↑2="INT(SG*1000)/1000
 1590 PRINT"DO YOU WISH TO ENTER A NEW VALUE FOR PHI: YES (Y) OR NO (N) ?"
 1600 INPUT Z$
 1610 IF Z$="Y" GOTO 260
 1630 PRINT"DO YOU WISH TO ENTER NEW VALUES ALTOGETHER:YES (Y) OR NO (N) ?"
 1640 INPUT Z$
 1650 IF Z$="Y" GOTO 150
 1670 END
 1680 A1=SIN(P1)
 1690 B1=COS(P1)-K2
 1700 C1=K1*COS(P1)-K3
 1710 D1=SQR(A1*A1+B1*B1-C1*C1)
 1720 RETURN
 1730 H1=SIN(P1)
 1740 H2=COS(P1)-K2
 1750 H3=(C*C-A*A-B*B-D*D)/2/A/B+D*COS(P1)/B
 1760 H4=SQR(H1*H1+H2*H2-H3*H3)
 1770 RETURN
 1780 PRINT"PHI="P1*180/π
 1790 PRINT"COORDINATES OF POINT C:"
 1800 PRINT
 1810 PRINT"X-COORD. XC="XC
 1820 PRINT"Y-COORD. YC="YC
 1830 PRINT"VELOCITIES OF POINT C:"
 1840 PRINT
 1850 PRINT"XC DOT="INT(UC*1000)/1000
 1860 PRINT"YC DOT="INT(VC*1000)/1000
 1870 PRINT
 1880 PRINT"ACCELERATIONS OF POINT C:"
 1890 PRINT"XC DOUBLE DOT="INT(AX*1000)/1000
 1900 PRINT"YC DOUBLE DOT=" INT(AY*1000)/1000
 1910 RETURN
 1920 A0=-2*E*(YC-YE)
 1930 B0=2*E*(XE-XC)
 1940 C0=F*F-E*E-(YC-YE)*(YC-YE)-(XE-XC)*(XE-XC)
 1950 D0=SQR(A0*A0+B0*B0-C0*C0)
 1960 RETURN
 1970 DA=-2*E*VC
 1980 DB=-2*E*UC
 1990 DC=-2*(YC-YE)*VC+2*(XE-XC)*UC
 2000 N1=DC-DB*COS(BT)-DA*SIN(BT)
 2010 G1=A0*COS(BT)-B0*SIN(BT)
 2020 BX=N1/G1
 2030 S3=-2*E*AY
 2040 S4=-2*E*AX
 2050 S5=-2*(YC-YE)*AY-2*VC*VC+2*(XE-XC)*AX-2*UC*UC
 2060 S1=BX*(DB*SIN(BT)-DA*COS(BT))+S5-S4*COS(BT)-S3*SIN(BT)
 2070 S2=-BX*(A0*SIN(BT)+B0*COS(BT))+DA*COS(BT)-DB*SIN(BT)
 2080 BY=(G1*S1-N1*S2)/G1/G1
 2090 PRINT"BETA,DEG.="INT(BT*100*180/π)/100
 2100 PRINT"BETA DOT,RAD/S="INT(BX*1000)/1000
 2110 PRINT"BETA DOUBLE DOT,RAD/S↑2-"INT(BY*1000)/1000
 2120 RETURN
READY.
```

A.3 DESIGN OF FOUR-BAR LINKAGES USING THE METHOD OF LEAST SQUARES (BASIC)

```
10 PRINT"****************************"
20 PRINT" DESIGN OF 4-BAR LINKAGES"
30 PRINT"        USING           "
40 PRINT"THE METHOD OF LEAST SQUARES"
50 PRINT"  FOR THE COORDINATION OF"
60 PRINT"         N POINTS          "
70 PRINT"****************************"
80 PRINT
90 PRINT"N MUST BE EQUAL TO OR GRATER THAN 3"
100 PRINT"___________________________________________"
110 PRINT
120 INPUT"NUMBER OF POINTS N";N
130 PRINT"ENTER:INPUT ANGLES PHI,OUTPUT ANGLES PSI IN DEGREES"
140 PRINT
150 FOR I=1 TO N
160 INPUT W(I),Q(I)
170 NEXT I
180 A1=0
190 A2=0
200 A3=0
210 B2=0
220 B3=0
230 D1=0
240 D2=0
250 D3=0
260 C3=N
270 FOR I=1 TO N
280 W(I)=W(I)*π/180
290 Q(I)=Q(I)*π/180
300 A1=A1+COS(W(I))↑2
310 A2=A2+COS(W(I))*COS(Q(I))
320 A3=A3+COS(W(I))
330 B2=B2+COS(Q(I))↑2
340 B3=B3+COS(Q(I))
350 B1=A2
360 C1=A3
370 C2=B3
380 D1=D1+COS(W(I))*COS(W(I)-Q(I))
390 D2=D2+COS(Q(I))*COS(W(I)-Q(I))
400 D3=D3+COS(W(I)-Q(I))
410 NEXT I
420 E1=B2-A2*B1/A1
430 E2=B3-A3*B1/A1
440 E3=D2-D1*B1/A1
450 F1=C2-A2*C1/A1
460 F2=C3-A3*C1/A1
470 F3=D3-D1*C1/A1
480 G1=F2-E2*F1/E1
490 G2=F3-E3*F1/E1
500 K3=-G2/G1
510 K2=(E3+E2*K3)/E1
520 K1=(D1+A3*K3-A2*K2)/A1
530 PRINT
540 PRINT"THE LINK RATIOS ARE:"
550 PRINT"____________________"
560 PRINT"K1="K1
570 PRINT"K2="K2
580 PRINT"K3="K3
590 PRINT
600 PRINT"CHOOSE A LENGTH FOR THE FIXED LINK A0B0 IN UNITS OF YOUR CHOICE"
```

```
 610 PRINT
 620 INPUT "LENGTH OF FIXED LINK A0B0=";D
 630 PRINT
 640 C=D/K1
 650 A=D/K2
 660 B=SQR(A*A+C*C+D*D-2*A*C*K3)
 670 PRINT"THE LINK LENGTHS ARE:"
 680 PRINT"--------------------"
 690 PRINT"INPUT CRANK A0A="A
 700 PRINT"    COUPLER  AB ="B
 710 PRINT"  FOLLOWER  BB0="C
 720 PRINT"FIXED LINK A0B0="D
 730 PRINT
 740 PRINT"DO YOU WISH TO FIX THE LENGTH OF THE CRANK A0B"
 750 PRINT"INSTEAD OF THE FIXED LINK A0B0 ?: ENTER Y FOR YES, N FOR NO"
 760 PRINT
 770 INPUT Z$
 780 IF Z$="Y" GOTO 800
 790 IF Z$="N" GOTO 910
 800 INPUT "LENGTH OF THE CRANK A0A=";A
 810 PRINT
 820 D0=K2*A
 830 C0=K2*A/K1
 840 B0=A*SQR(1+K2*K2*(1+1/(K1*K1))-2*K2*K3/K1)
 850 PRINT"THE LENGTHS OF THE LINKS ARE:"
 860 PRINT"-------------------------"
 870 PRINT"CRANK A0A="A
 880 PRINT"COUPLER AB="INT(B0*100)/100
 890 PRINT"FOLLOWER BB0="INT(C0*100)/100
 900 PRINT"FIXED LINK A0B0="INT(D0*100)/100
 910 PRINT
 920 PRINT"TRANSMISSION ANGLE MU DEGREES"
 930 PRINT"============================="
 940 FOR I=1 TO N
 950 GOSUB 1060
 960 PRINT"PHI="W(I)*180/π,"MU="U
 970 NEXT I
 980 PRINT
 990 PRINT"ARE THESE VALUES ACCEPTABLE ?"
 1000 PRINT"PRESS Y FOR YES, N FOR NO"
 1010 PRINT
 1020 INPUT Z$
 1030 IF Z$="Y" GOTO 1050
 1040 IF Z$="N" GOTO 130
 1050 END
 1060 X=B*B+C*C
 1070 Y=A*A+D*D
 1080 Z=2*A*D*COS(W(I))
 1090 U1=(X-Y+Z)/(2*B*ABS(C))
 1100 U2=-ATN(U1/SQR(-U1*U1+1))+π/2
 1110 U=INT(U2*100*180/π)/100
 1120 RETURN
READY.
```

A.4 MECHANICAL ERRORS IN THE FOUR-BAR LINKAGE DUE TO IMPOSED MANUFACTURING TOLERANCES (BASIC)

```
10 PRINT"*****************************************"
20 PRINT" MECHANICAL ERROR IN THE 4-BAR LINKAGE"
30 PRINT"DUE TO IMPOSED MANUFACTURING TOLERANCES"
40 PRINT"****************************************"
50 PRINT
```

```
60 PRINT"ENTER LINK LENGTHS IN UNITS OF YOUR CHOICE"
70 PRINT
80 INPUT"CRANK A0A";A
90 INPUT"COUPLER AB";B
100 INPUT"FOLLOWER BB0";C
110 INPUT"FIXED LINK A0B0";D
120 PRINT"THE LINK LENGTHS ARE:"
130 PRINT
140 PRINT"CRANK="A;"COUPLER="B;"FOLLOWER="C;"FIXED LINK="D
150 PRINT
160 PRINT"ENTER THE VALUES OF THE TOLERANCES"
170 PRINT
180 INPUT"DELTA PHI";DP
190 INPUT"DELTA A";DA
200 INPUT"DELTA B";DB
210 INPUT"DELTA C";DC
220 INPUT"DELTA D";DD
230 PRINT"THE TOLERANCES ARE:"
240 PRINT
250 PRINT"DELTA PHI="DP;"DELTA A="DA;"DELTA B="DB;"DELTA C="DC;"DELTA D="DD
260 PRINT
270 PRINT"ENTER INITIAL INPUT CRANK ANGLE PHI IN DEGREES"
280 PRINT
290 INPUT"INITIAL CRANK ANGLE PHI";WS
300 PRINT"ENTER TOTAL ANGULAR MOVEMENT OF CRANK IN DEGREES"
310 PRINT
320 INPUT"TOTAL ANGULAR MOVEMENT OF CRANK";DW
330 PRINT
340 PRINT"ENTER STEP SIZE IN DEGREES"
350 PRINT
360 INPUT"STEP SIZE";DI
370 PRINT
380 M1=0
390 K1=D/C
400 K2=D/A
410 K3=(A*A-B*B+C*C+D*D)/2/A/C
420 PRINT"IS THE LINKAGE OPEN (O) OR CROSSED (C) ?"
430 INPUT Z$
440 IF Z$="O" GOTO 460
450 IF Z$="C" GOTO 580
460 PRINT"-----------------------------------------------------------"
470 PRINT"INPUT       OUTPUT        MECH.       STAT."
480 PRINT"ANGLE        ANGLE        ERROR       ERROR"
490 PRINT"PHI,DEG.    PSI,DEG.       DEG.        DEG.
500 PRINT"-----------------------------------------------------------"
510 FOR W =WS TO WS+DW STEP DI
520 W1=W*π/180
530 GOSUB 800
540 Q1=2*ATN((L1+L4)/(L2+L3))
550 GOSUB 850
560 NEXT W
570 GOTO 690
580 PRINT"-----------------------------------------------------------"
590 PRINT"INPUT       OUTPUT        MECH.       STAT."
600 PRINT"ANGLE        ANGLE        ERROR       ERROR"
610 PRINT"PHI,DEG.    PSI,DEG.       DEG.        DEG.
620 PRINT"-----------------------------------------------------------"
630 FOR W=WS TO WS+DW STEP DI
640 W1=W*π/180
650 GOSUB 800
660 Q1=2*ATN((L1-L4)/(L2+L3))
670 GOSUB 850
680 NEXT W
690 PRINT"-----------------------------------------------------------"
```

```
 700 PRINT
 710 IF DI=0 GOTO 740
 720 PRINT"OVERALL STAT. MECH. ERROR,DEG.="OS
 730 PRINT"------------------------------------------------------------------------"
 740 PRINT"DO YOU WISH TO CHANGE THE VALUES OF THE TOLERANCES:"
 750 PRINT"          YES (Y)   OR   NO   (N) ?"
 760 INPUT Z$
 770 IF Z$="Y" GOTO 180
 780 IF Z$="N" GOTO 790
 790 END
 800 L1=SIN(W1)
 810 L2=COS(W1)-K2
 820 L3=K1*COS(W1)-K3
 830 L4=SQR(L1*L1+L2*L2-L3*L3)
 840 RETURN
 850 G1=2*A*C*SIN(W1)*COS(Q1)
 860 G2=(2*A*C*COS(W1)-2*C*D)*SIN(Q1)
 870 G=G1-G2
 880 QA=2*(D*COS(W1)-A-C*COS(W1-Q1))*DA/G
 890 QP=2*A*(C*SIN(W1-Q1)-J*SIN(W1))*DP*π/180/G
 900 QB=2*B*DB/G
 910 QC=-2*(C+A*COS(W1-Q1)-D*COS(Q1))*DC/G
 920 QD=2*(A*COS(W1)-D+C*COS(Q1))*DD/G
 930 E1=(ABS(QP)+ABS(QA)+ABS(QB)+ABS(QC)+ABS(QD))*180/π
 940 E2=SQR(QP*QP+QA*QA+QB*QB+QC*QC+QD*QD)*180/π
 950 ME=INT(E1*1000)/1000
 960 SE=INT(E2*1000)/1000
 970 02=INT(Q1*180*100/π)/100
 980 M1=M1+E1*E1
 990 IF DI=0 GOTO1030
 1000 N1=DW/DI+1
 1010 X1=SQR(M1/N1)
 1020 OS=INT(X1*1000)/1000
 1030 PRINT W,02,ME,SE
 1040 RETURN
READY.
```

REFERENCES AND BIBLIOGRAPHY

1. J. A. Hrones and G. L. Nelson, Analysis of the Four-Bar Linkage, MIT–Wiley, 1951.
2. M. K. Kloomok and R. V. Muffley, 'Plate cam design', *Prod. Eng.*, **26**, 1955.
3. S. Beck and M. Chapman, Microcomputer program, unpublished undergraduate thesis, University of Bath, 1985.
4. L. A. Jones and R. J. Ansell, unpublished undergraduate thesis, University of Bath, 1986.
5. F. Freudenstein, 'Structural error analysis in plane kinematic synthesis', *ASME J. Eng. Ind.*, Ser. E, **81**, 1, 15–22, 1959.
6. R. E. Garrett and A. S. Hall Jr, 'Effect of Tolerance and clearance in linkage design', *Trans ASME J. Eng. Ind.*, 91B, 198–202, 1969.
7. G. M. Sutherland and B. Roth, 'Mechanism design: accounting for manufacturing tolerances and costs in function generating problems', *ASME J. Eng. Ind.*, **97**, Ser. B, 303–307, 1975.

BIBLIOGRAPHY

1. A. G. Erdman and G. N. Sandor, *Mechanism design*, vols. 1 and 2, Prentice-Hall International, 1984.
2. F. Freudenstein, 'Approximate synthesis of four-bar linkages', *Trans. ASME*, **77**, 853–861, 1955.
3. R. S. Hartenberg and J. Denavit. *Kinematic Synthesis of Mechanisms*, McGraw-Hill, 1964.
4. K. H. Hunt. *Kinematic Geometry of Mechanisms*, Oxford University Press, 1978.
5. H. H. Mabie and C.F. Reinholtz, *Mechanisms and Dynamics of Machinery*, Wiley, 1987.
6. G. H. Martin, *Kinematics and Dynamics of Machines*, McGraw-Hill, 1969.
7. F. N. Nagy an A. Siegler, *Engineering Foundations of Robotics*, Prentice-Hall International, 1987.
8. J. M. Prentis, *Dynamics of Mechanical Systems*, Ellis Horwood, 2nd edn, 1980.
9. J. E. Shigley, and J. J. Uicker Jr, *Theory of Machines and Mechanisms*, McGraw-Hill, 1980.
10. C. W. Stammers, 'An alternative to Lagrange'; *3rd British Conference on the Teaching of Vibration and Noise, Sheffield, England*, 1979.
11. C. W. Stammers, 'The dynamics of a robotic arm: Gibbs–Appell and Lagrange formulation compared', *Conference on Mechanisms, Cranfield, England, 1985*.

ANSWERS TO SELECTED EXERCISES

2.1 122 rad/s,
2.2 72 m/min, 87 m/min
2.3 0.47 m/s, 0.47 rad/s, 0.47 rad/s
2.5 1.45 m/s, 3.35 m/s, 2.2 m/s
2.6 0.23 m/s
2.7 0.18 m/s

3.1 930 rad/s^2, 480 rad/s^2, 961 rad/s^2, 152 m/s^2
3.4 232 mm, 1.02 m/s, 1.77 m/s^2
3.5 2.28 rad/s clockwise, 43 rad/s^2 clockwise
3.7 32.25 rad/s, 515 rad/s^2, 1.5×10^5 rad/s^3
3.10 62.8 rad/s clockwise, 42 rad/s^2 clockwise
3.11 1.01 rad/s clockwise, 0.286 rad/s^2 clockwise

4.5 6 mm, 1.4 m/s, 227 m/s^2
4.6 10.1 m/s, 8400 m/s^2

5.1 15.9 rad/s anticlockwise, 3.5 m/s, 10.9 rad/s^2 anticlockwise, 16.7 rad/s^2 anticlockwise
5.2 118 rad/s^2, 102 rad/s^2, 181 rad/s^2, 62 m/s^2
5.3 23.4 m/s^2
5.5 Possible values: CD = 379 mm, AD = 310 mm, AB = 292 mm
5.6 With $d = 390$ mm, $a = -40.3$ mm, $b = 407$ mm and $c = -102.6$ mm: transmission angle = 56°
5.8 $a = 609$ mm, $b = 232$ mm, $c = 363$ mm
5.10 $a = 48.2$ mm, $b = 75.4$ mm, $c = -22.3$ mm; transmission angles: 32, 40, 48.6 degrees
5.11 angular velocity of DE = 0.212 rad/s, angular acceleration of DE = 0.068 rad/s^2
5.14 209 rad/s, 39 255 rad/s^2

6.1 first gear: 1745 rev/min, 350 N m, 228 N mm
second gear: 3059 rev/min, 200 N m, 77.5 N m
6.2 11.11, 26.66 kN m, 24.26 kN m, 43.3 kN
6.3 45.4, 15.7

6.4 5160 N m
6.5 -555 rev/min to $+835$ rev/min; 3.81 N m
6.7 822 rev/min
6.8 0.6 m/s^2, 687 N m

7.1 -1.25 m/s, -8.9 m/s^2, $2.38(\mathbf{i}-\mathbf{j}-\mathbf{k})$ rad/s, $-8.5\,(\mathbf{j}+\mathbf{k})$ rad/s^2
7.6 80 rad/s, 1920 m/s^2
7.8 1.083 m/s, 0.862 m/s^2

8.1 180 kN m, 420 kN m
8.2 1.1 kN m
8.3 2.3 kN m
8.7 112 N m
8.9 434 N m

9.2 26 N m
9.5 -3.75 N m, 0.574 N m, -0.41 N m
9.7 51.7 kN m

10.1 $x=18.15\pm0.32$ mm, $y=18.19\pm0.24$ mm
10.2 ±0.10 mm maximum, 0.25 W
10.3 ±0.15 mm
10.7 (*a*) 17.5°
(*b*) See Fig. 10.14

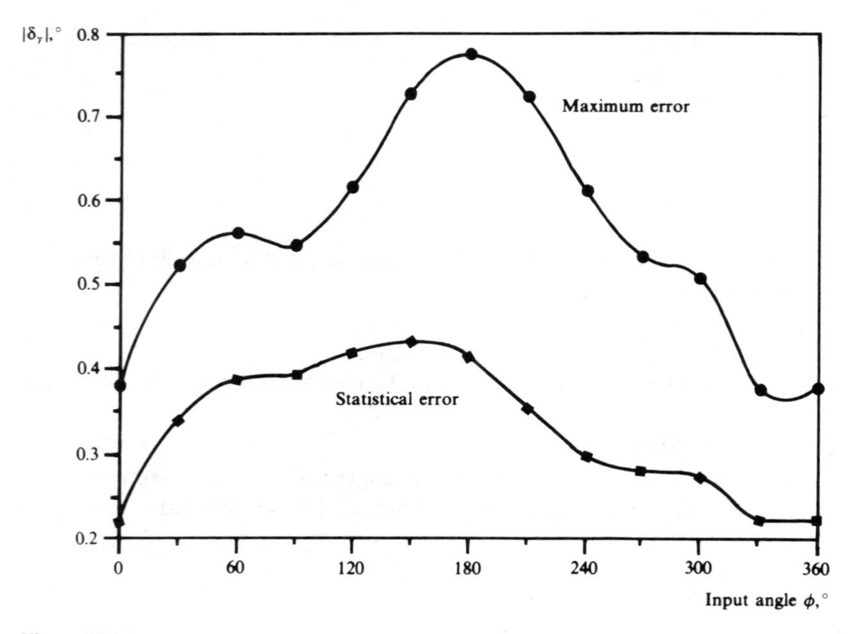

Figure 10.14

INDEX